Nutritional Neurosciences

Series Editor

Mohamed Essa, Sultan Qaboos University, Qaboos, Oman

This book series aims to publish volumes focusing on both basic and clinical research in the field of nutritional neuroscience with a focus on delineating the effect of nutrition on brain function and behavior. The books will examine the role of different nutrients, food agents and supplements (both macro and micro) on brain health, neurodevelopment, neurochemistry, and behaviour. The books will examine the influence of diet, including phytochemicals, antioxidants, dietary supplements, food additives, and other nutrients on the physiology and metabolism of neurons, neurotransmitters and their receptors, cognition, behavior, and hormonal regulations.

The books will also covers the influence of nutrients and dietary supplements on the management of neurological disorders. It details the mechanism of action of phytonutrients on signaling pathways linked with protein folding, aggregation, and neuroinflammation. The books published in the series will be useful for neuroscientists, nutritionists, neurologists, psychiatrists, and those interested in preventive medicine.

More information about this series at https://link.springer.com/bookseries/16639

M. Walid Qoronfleh •
Musthafa Mohamed Essa •
Chidambaram Saravana Babu
Editors

Proteins Associated with Neurodevelopmental Disorders

Springer

Editors
M. Walid Qoronfleh
Research & Policy Division
Q3CG Research Institute (QRI)
Ann Arbor, MI, USA

Musthafa Mohamed Essa
College of Agricultural and Marine
Sultan Qaboos University
Al Khoudh, Oman

Chidambaram Saravana Babu
Department of Pharmacology
JSS Academy of Higher Education and Research
Mysore, Karnataka, India

ISSN 2730-6712 ISSN 2730-6720 (electronic)
Nutritional Neurosciences
ISBN 978-981-15-9780-0 ISBN 978-981-15-9781-7 (eBook)
https://doi.org/10.1007/978-981-15-9781-7

This Springer imprint is published by the registered company Springer Nature Singapore Pte Ltd.
The registered company address is: 152 Beach Road, #21-01/04 Gateway East, Singapore 189721, Singapore

Preface

There is a constant increase in the prevalence of neurodegenerative diseases (such as Alzheimer's disease, Parkinson's disease, and Huntington's disease) and neurodevelopmental disorders [autism spectrum disorder (ASD), attention deficit hyperactivity disorder (ADHD), and intellectual/learning disability (ID)] due to various reasons. ASD and ADHD are two of the most widely studied neurodevelopmental disorders. Neurodevelopmental disorders in general, autism and ADHD in particular, are considered multifactorial disorders in which genetic and environmental factors interact, triggering its development. As neurodevelopmental disorders, both autism and ADHD share some phenotypic similarities, yet are characterized by distinct diagnostic criteria, imposing a major impediment to childhood development and a significant burden on society. Every country is spending huge amounts for medical treatments each year. In the USA for example, over an affected individual's lifetime, costs of care can reach about $3.2 million, while the annual cost to society is an estimated $35 billion. There is an urgent and desperate need to understand the biology and biochemistry of both ASD and ADHD. Understanding the biological bases of autism and ADHD, especially the protein function, may facilitate the design of future experimental treatments and diagnosis tools for these conditions.

This body of work consists of a collection of the most important proteins that have been linked or associated with neurodevelopmental disorders, in particular autism and ADHD. Also, it is a compilation and an update on the proteins and their interactions at the subcellular and molecular levels. There are many proteins that have been suggested to be involved in the pathogenesis of ASD and ADHD, including kinases, bromodomain-containing proteins (BCPs), synaptic proteins, scaffolding proteins, cell adhesion molecules, DNA-binding proteins, and transcription factors to name a few. To our knowledge, there are no books covering the links of proteins associated with ASD and ADHD. To fill the information gap, this book focuses on select proteins (or protein families/pathways) that are linked or associated with ASD and ADHD, which is unique and not covered well as a topic in the literature. Furthermore, the book explores epigenetic factors and emphasizes the implication of the studied proteins as druggable targets and biomarkers or on ADHD and ASD intervention and management via diet/nutrition influence plus the future potential of stem cell therapy.

This book has a comprehensive collection of research studies, which will benefit students at various levels, postdoctoral trainees, researchers in several disciplines (such as genetics, neuroscience, developmental biology, medicine, and nutrition/food science), and many others who are interested in this discipline. The book can also be used as a required or recommended text for related courses taught at universities globally.

Lastly, we hope that this book could be a window for scientists and researchers alike into protein function and neurodevelopmental disorders to further advance this field.

Ann Arbor, MI, USA — M. Walid Qoronfleh
Al Khoudh, Oman — Musthafa Mohamed Essa
Mysore, India — Chidambaram Saravana Babu

Acknowledgments

We are grateful to the book chapters' authors for sharing their expertise to provide evidence-based information on the role of proteins in neurodevelopmental disorders.

We are indebted to our families for their understanding and unconditional support during this endeavor, allowing us to spend extra time on completing the book. Many thanks are due to our teachers who have inspired us to investigate proteins/molecular pathways in health and disease and to those colleagues for many stimulating discussions.

Special thanks to the Editing Refinery, USA, for providing wonderful proofreading and editing services.

Thanks to the staff of Springer Nature for their patience and assistance at various stages of the book publication.

We, as editors, gratefully want to acknowledge our respective institutions for their continued support that allowed us to finish this book successfully.

We declare that there is no potential conflict of interest.

M. Walid Qoronfleh
Musthafa Mohamed Essa
Chidambaram Saravana Babu

Contents

Editors and Contributors

About the Editors

M. Walid Qoronfleh, PhD, MBA is currently the Executive Director of the US-based Q3CG Research Institute and a Senior Director at 21HealthStreet, London, UK. Prior, he was the Director of the Healthcare Research and Policy at Qatar Foundation. Dr. Qoronfleh is engaged in project areas that are of immediate global and provincial healthcare relevance. His research interests are focused on neurodevelopmental disorders, mental health, patient safety, noncommunicable diseases (cancer, diabetes, and obesity), precision medicine, and bioethics. Dr. Qoronfleh has over 20 years of scientific, technology, business, and commercial experience. He has held several senior management and executive positions in the USA, with increasing responsibilities at GlaxoSmithKline, Sanofi, NIH-NCI, Thermo Fisher, NextGen Sciences, SDIX, and University of Michigan-Ann Arbor, with a particular focus on precision medicine. Dr. Qoronfleh is the founder of three biotechnology companies, and he is a co-founder and the managing director of the boutique consulting company Q3CG. Dr. Qoronfleh has published over 100 peer-reviewed research articles in journals of high repute. He is also an editor and an ad hoc reviewer for various scientific journals and a frequent speaker at national and international conferences. He holds graduate degrees from the University of Louisville School of Medicine, Penn State University, and the University of Wisconsin–Madison.

Musthafa Mohamed Essa, PhD is a Professor of Nutrition at Sultan Qaboos University, Oman. He is an Editor in Chief for the International Journal of Nutrition, Pharmacology, Neurological Diseases. He is also acting as an Associate Editor for BMC complementary and alternative medicine and is an Editor for Frontiers in Neuroscience and Frontiers in Biosciences. Prof. Essa is an expert in the field of nutritional neuroscience/neuropharmacology and has published more than 170 papers, 60 book chapters, and 14 books. Prof. Essa is the recipient of several awards, including the best book in the world, National Research Award, Outstanding Neuroscientist, and Distinguished Researcher. He has over 15 years of research and teaching experience in the field of nutritional neuroscience, and under his guidance, five Ph.D. and ten master students have completed their research. He is the Editor for a unique and first-in-kind book series titled "Nutritional Neurosciences" published by Springer Nature.

Chidambaram Saravana Babu, PhD is a Professor in the Dept. of Pharmacology, JSS College of Pharmacy, and Coordinator, Centre for Experimental Pharmacology and Toxicology, JSS Academy of Higher Education & Research, Mysuru, India. He has 12 years of research and teaching experience in the field of pharmacology and toxicology. He has published more than 80 research articles in the field of neuropharmacology and toxicology. He is the recipient of "Outstanding Researcher in Pharmacology," VIRA (September 2018), and Dr. APJ Abdul Kalam Award for Scientific Excellence, Marina Labs, Chennai (September 2017). Recently, he has established a GLP standard preclinical facility in the JSS Academy of Higher Education & Research, Mysuru.

Contributors

Palli Valapila Abdulrouf Department of Research, Women's Wellness and Research Center (WWRC), Hamad Medical Corporation (HMC), Doha, Qatar

Nahla Al Anqodi Dietary Department, Directorate General of Medical Services, Royal Oman Police, Muscat, Oman

Ruqaiya Moosa Al Balushi Department of Food Science and Nutrition, College of Agricultural and Marine Sciences, Sultan Qaboos University, Muscat, Oman

Nader Al-Dewik Department of Research, Women's Wellness and Research Center (WWRC), Hamad Medical Corporation (HMC), Doha, Qatar
Interim Translational Research Institute (iTRI), Hamad Medical Corporation (HMC), Doha, Qatar
Faculty of Health and Social Care Sciences, Kingston University, St. George's University of London, London, UK
Clinical and Metabolic Genetics, Department of Pediatrics, Hamad General Hospital (HGH), Hamad Medical Corporation (HMC), Doha, Qatar
Genomics and Precision Medicine (GPM), College of Health and Life Science (CHLS), Hamad Bin Khalifa University (HBKU), Doha, Qatar

Athanasios Alexiou Novel Global Community Educational Foundation, Hebersham, NSW, Australia
AFNP Med Austria, Wien, Austria

Amal Al-Jurf Department of Special Education and Gifted Students, Ministry of Education, Doha, Qatar

Rana Al-Jurf Department of Research, Women's Wellness and Research Center (WWRC), Hamad Medical Corporation (HMC), Doha, Qatar
Department of Biomedical Sciences, College of Health Sciences, Member of QU Health, Qatar University, Doha, Qatar

Hilal Al-Rifai Department of Pediatrics and Neonatology, Neonatal Intensive Care Unit, Newborn Screening Unit, Women's Wellness and Research Center (WWRC), Hamad Medical Corporation, Doha, Qatar

Gaber El-Saber Batiha Department of Pharmacology and Toxicology, Faculty of Veterinary Medicine, Damanhour University, Damanhour City, Egypt

Pathan Shajahan Begum Department of Zoology, Silver Jubilee Government College, Kurnool, India

Anupom Borah Cellular and Molecular Neurobiology Laboratory, Department of Life Sciences and Bioinformatics, Assam University, Silchar, Assam, India

Chinnasamy Dhanalakshmi Department of Pathology, University of Arizona College of Medicine and College of Pharmacy, Tucson, AZ, USA

Khalid A. Fakhro Department of Human Genetics, Sidra Medicine, Doha, Qatar
Department of Genetic Medicine, Weill Cornell Medical College, Doha, Qatar

Tom Farrell Department of Research, Women's Wellness and Research Center (WWRC), Hamad Medical Corporation (HMC), Doha, Qatar

Sara Hammuda Department of Psychology, Kingston University London, Kingston upon Thames, London, UK

Mohammad Haris Laboratory of Molecular and Metabolic Imaging, Sidra Medicine, Doha, Qatar

Mohamed A. Ismail Department of Research, Women's Wellness and Research Center (WWRC), Hamad Medical Corporation (HMC), Doha, Qatar

Andleeb Khan Department of Pharmacology and Toxicology, College of Pharmacy, Jazan University, Jazan, Saudi Arabia

Anna Maria Louka Faculty of Medicine, University of Thessaly, Larissa, Greece

Thamilarasan Manivasagam Department of Biochemistry and Biotechnology, Faculty of Science, Annamalai University, Chidambaram, Tamil Nadu, India

Sawsan Mohammed Clinical Academic Education Department, Qatar University, College of Medicine, Doha, Qatar

Dimitrios V. Moysidis Faculty of Medicine, Aristotle University of Thessaloniki, Thessaloniki, Greece

Ghada Mubarak Sidra Medicine, Doha, Qatar

Vidya Murugesan Department of Chemistry and Biochemistry, Ramaiah College of Arts, Science and Commerce, Bengaluru, Karnataka, India

Gheyath K. Nasrallah Department of Biomedical Sciences, College of Health Sciences, Member of QU Health, Qatar University, Doha, Qatar

Sabah Nisar Laboratory of Molecular and Metabolic Imaging, Sidra Medicine, Doha, Qatar

Stavros Papadakos Department of Pathology, University of Athens, Athens, Greece

Andreas S. Papazoglou Faculty of Medicine, Aristotle University of Thessaloniki, Thessaloniki, Greece

Asokan Prema Department of Chemistry, Government College of Arts and Science, Tittagudi, Tamil Nadu, India

M. Walid Qoronfleh Q3CG Research Institute (QRI), Research and Policy Division, Ypsilanti, MI, USA
21HealthStreet, Consulting Services, London, UK
Applied Biomedicine, Inc., Doha, Qatar

Senthilkumar Rajagopal Department of Biotechnology, School of Applied Sciences, REVA University, Bengaluru, Karnataka, India
Department of Biotechnology, REVA University, Bengaluru, Karnataka, India

Meerza Abdul Razak Department of Life Sciences, Dr. Buddollas Institute of Life Sciences, Tirupathi, India

Muthanna Samara Department of Psychology, Kingston University London, Kingston upon Thames, London, UK

Dorothy Martha Scordilis National Institutes of Health, Bethesda, MD, USA

Rwik Sen Active Motif Inc., Carlsbad, CA, USA

Anastasiia Shkodina Ukrainian Medical and Stomatological Academy, Poltava, Ukraine

Gadde Venkata Swarnalatha Department of Biochemistry, Rayalaseema University, Kurnool, Andhra Pradesh, India

Arokiasamy Justin Thenmozhi Department of Biochemistry, School of Biological Sciences, Madurai Kamaraj University, Madurai, Tamil Nadu, India
Department of Biochemistry and Biotechnology, Faculty of Science, Annamalai University, Chidambaram, Tamil Nadu, India

Christos Tsagkaris Novel Global Community Educational Foundation, Hebersham, NSW, Australia

Kyriakoula Varmpompiti Department of Neurosciences, Kings College Hospital, London, UK

Salma N. Younes Department of Research, Women's Wellness and Research Center (WWRC), Hamad Medical Corporation (HMC), Doha, Qatar

Farah R. Zahir Department of Medical Genetics, BC Children's and Women's Hospital, University of British Columbia, Vancouver, BC, Canada

Hatem Zayed Department of Biomedical Sciences, College of Health Sciences, Member of QU Health, Qatar University, Doha, Qatar

Ayah Ziyada Genomics and Precision Medicine (GPM), College of Health and Life Science (CHLS), Hamad Bin Khalifa University (HBKU), Doha, Qatar

Principal Molecular Pathways Affected in Autism Spectrum Disorder

1

Salma N. Younes, Rana Al-Jurf, Sara Hammuda, Gheyath K. Nasrallah, Hatem Zayed, M. Walid Qoronfleh, Mohamed A. Ismail, Tom Farrell, Hilal Al-Rifai, Muthanna Samara, and Nader Al-Dewik

S. N. Younes (✉) · M. A. Ismail · T. Farrell
Department of Research, Women's Wellness and Research Center (WWRC), Hamad Medical Corporation (HMC), Doha, Qatar
e-mail: V-SYounes@hamad.qa; MISMAIL4@hamad.qa; TFarrell@hamad.qa

R. Al-Jurf
Department of Research, Women's Wellness and Research Center (WWRC), Hamad Medical Corporation (HMC), Doha, Qatar

Department of Biomedical Sciences, College of Health Sciences, Member of QU Health, Qatar University, Doha, Qatar
e-mail: 199660690@student.qu.edu.qa

S. Hammuda · M. Samara
Department of Psychology, Kingston University London, Kingston upon Thames, London, UK
e-mail: M.Samara@kingston.ac.uk

G. K. Nasrallah · H. Zayed
Department of Biomedical Sciences, College of Health Sciences, Member of QU Health, Qatar University, Doha, Qatar
e-mail: gheyath.nasrallah@qu.edu.qa; hatem.zayed@qu.edu.qa

M. W. Qoronfleh
Q3CG Research Institute (QRI), Research and Policy Division, Ypsilanti, MI, USA

21HealthStreet, Consulting Services, 1 Christian Fields, London, UK

Applied Biomedicine, Inc., Doha, Qatar

H. Al-Rifai
Department of Pediatrics and Neonatology, Neonatal Intensive Care Unit, Newborn Screening Unit, Women's Wellness and Research Center (WWRC), Hamad Medical Corporation, Doha, Qatar
e-mail: Halrifai@hamad.qa

N. Al-Dewik (✉)
Department of Research, Women's Wellness and Research Center (WWRC), Hamad Medical Corporation (HMC), Doha, Qatar

M. W. Qoronfleh et al. (eds.), *Proteins Associated with Neurodevelopmental Disorders*, Nutritional Neurosciences, https://doi.org/10.1007/978-981-15-9781-7_1

Abstract

Autism spectrum disorder (ASD) development is a highly multifaceted process as evidenced by the complexity of the factors involved in the etiology of ASD, including genetic and nongenetic factors. Several forms of ASD result from genetic alterations in genes that regulate the process of protein synthesis. A growing body of evidence suggests that abnormal synaptic protein synthesis might contribute to ASD and ASD-like clinical features. Several reports of different mutated genes responsible for ASD cases and genetic models have emerged, revealing dysregulation of many crucial signaling pathways. In this chapter, the authors summarize the various factors described to contribute to ASD, both genetic and nongenetic, and their association with WNT, SHH, RA, FGF, and BMP/TGF-β signaling pathways. In addition, the authors discuss the scope for additional research for a better understanding of the pathophysiology of ASD in the context of disrupted signaling pathways, which could help open the doors to identify possible gene targets and novel therapeutic strategies.

Keywords

ASD · Autism · Signaling pathways · Genes · Therapeutic targets · Developmental mechanisms

1.1 Introduction

Autism spectrum disorder (ASD) encompasses a heterogenous set of multifactorial challenges in neurodevelopment, classified according to three fundamental features; compromised social correspondence skills, delayed dialect development, and raised stereotyped alternately tedium practices (Mohn et al. 2014; Golden et al. 2017; Abrahams and Geschwind 2008; Geschwind 2008; Sudhof 2008; Zoghbi 2003). ASD is believed to occur due to the complicated processes that involve numerous gene-environment interactions, as evidenced by the association of multiple elements (genetic and nongenetic). People with autism are recognized to have several traits, such as hampered social interactions, communication impairment, and increased tedium behaviors, implying that aberrant signaling pathways during brain development resulted in the disturbance of specific neural circuits in ASD (Styles et al. 2020; Al-Dewik et al. 2020). The disruption of multiple critical signaling pathways,

Interim Translational Research Institute (iTRI), Hamad Medical Corporation (HMC), Doha, Qatar

Faculty of Health and Social Care Sciences, Kingston University, St. George's University of London, London, UK

Clinical and Metabolic Genetics, Department of Pediatrics, Hamad General Hospital (HGH), Hamad Medical Corporation (HMC), Doha, Qatar

Genomics and Precision Medicine (GPM), College of Health and Life Science (CHLS), Hamad Bin Khalifa University (HBKU), Doha, Qatar
e-mail: Naldewik@hamad.qa; Nader.Al-Dewik@kingston.ac.uk

including WNT (Bae and Hong 2018; Kalkman 2012; Mulligan and Cheyette 2017), BMP (Zhang et al. 2017; Kashima et al. 2016; Li et al. 2016; Sajan et al. 2011), SHH (Halepoto et al. 2015; Patel et al. 2017), and retinoic acid (RA) (Niculae and Pavăl 2016), has been discovered in research using genetic models in ASD individuals. While direct evidence has not been found, indirect evidence of abnormal FGF or TGF-β signaling in ASD exists (Ansari et al. 2017; Chen et al. 2017; Iwata and Hevner 2009). The success of therapy treatments is severely constrained due to a lack of data on the etiology of ASD and the mechanisms involved. In this chapter the authors aim to summarize current knowledge on the various ASD-associated factors (genetic and nongenetic), and their interactions with signaling pathways that are frequently altered in ASD. This chapter also discusses what additional research could be conducted to gain better insights into altered pathways in ASD.

1.2 Principal Signaling Pathways

1.2.1 Altered WNT Signaling in ASD

WNT signaling involves the secretion of cysteine-rich glycolipoproteins that are WNT receptors themselves. The WNT signaling pathway controls important developmental and regulatory processes, including embryonic development and tissue homeostasis, through regulating receptors, such as frizzled (FZD), when the receptors interact with the WNT protein. WNT signaling is required for several developmental and post-developmental neuroscientific processes, including synaptogenesis and CNS regionalization (Tang 2014; Wada and Okamoto 2009; Wodarz and Nusse 1998; Rosso and Inestrosa 2013; Bielen and Houart 2014; Abu-Khalil et al. 2004; Bengoa-Vergniory and Kypta 2015; Burden 2000; Inestrosa and Varela-Nallar 2015; Onishi et al. 2014). Thus, any disturbance in WNT signaling has the potential to cause the development of CNS-related diseases (Mulligan and Cheyette 2017; Okerlund and Cheyette 2011).

Investigations of both genetically engineered animal models and Human Induced Pluripotent Stem Cell (hiPSC) models have shown the critical role of spatiotemporal WNT signaling throughout animal development (Mulligan and Cheyette 2017). It has also been shown that irregularities in WNT may cause several types of mental illnesses, such as autism, schizophrenia, bipolar disorder, and developmental problems (Kalkman 2012; Mulligan and Cheyette 2017; Okerlund and Cheyette 2011; Oron and Elliott 2017; Kwan et al. 2016a; Martin et al. 2013). Many of the genes and epigenetic factors involving ASD have been identified as affecting common biological processes such as epigenetic modification, WNT, and synaptic transmission (Oron and Elliott 2017; Hormozdiari et al. 2015; Krumm et al. 2014).

The WNT pathway is classified into two major pathways: (1) β-catenin-dependent (canonical pathways) and (2) β-catenin-independent (noncanonical pathways), both of which are key players in neural development and related neurodevelopmental disorders (Grainger and Willert 2018; Komiya and Habas 2008). Several genetic variations associated with and/or documented in ASD cases include either fundamental elements of the β-catenin-dependent pathway, such as *CTNNB1* (β-catenin) (Kalkman 2012; Mulligan and Cheyette 2017; Krumm et al. 2014; O'Roak et al. 2012a) and adenomatous polyposis coli (*APC*) (Mohn et al. 2014), or β-catenin-independent pathway, such as *PRICKLE2* (Sowers et al. 2013), implying that these two pathways are key players in ASD etiology. A list of selected ASD-associated genes that the authors have found to influence ASD-related pathways is presented in Table 1.1. Possible interactions between the ASD-associated genes and WNT pathway are illustrated in Fig. 1.1.

1.3 Genetic Etiologies

Core elements of the canonical WNT signaling In humans, the components of the β-catenin-dependent route includes many WNT ligands, frizzled receptors, LDL receptor family co-receptors, and intracellular and extracellular modulators (Klaus and Birchmeier 2008). WNT ligands are modified cysteine-rich proteins, and to function, they need to be glycosylated and palmitoylated (Komiya and Habas 2008). WNT1, WNT2, WNT3, and WNT9B are known to be involved with ASD. ASD individuals usually harbor a unique missense genetic variant in *WNT1* (S88R, rs61758378) that has demonstrated higher activation of the WNT/β-catenin pathway compared to WNT1 wild type i.e. without this variant (Martin et al. 2013). Beyond that, ASD Individuals have been found to have uncommon pathogenic variants in *WNT2*, *WNT3*, and *WNT9B* (Wassink et al. 2001; Marui et al. 2010; Lin et al. 2012; Levy et al. 2011). Therefore, it is surprising that in people with ASD, WNT3 expression in the prefrontal cortex is more pronounced (Chow et al. 2012). The above finding suggests that WNT signaling is overactive and that overactivation could play a key role in ASD etiology.

Several studies have revealed that WNT1 is crucial for cerebellar and midbrain development in animal models (Thomas and Capecchi 1990; McMahon and Bradley 1990; McMahon et al. 1992). Furthermore, it has been shown that WNT2 is necessary for the development of cortical dendrites as well as the production of dendritic spines. Moreover, it was shown that *WNT2* expression could be regulated by a protein referred to as brain-derived neurotrophic factor (BDNF), which is a key player in neuron survival and growth (Hiester et al. 2013). It has also been shown that altered dendritic spines were associated with neurodegenerative and neurodevelopmental problems (Hiester et al. 2013). In addition, WNT3 has been shown to be required for gastrulation and for the regulation of hippocampus neurogenesis (Lie et al. 2005; Liu et al. 1999). Increased inhibitory synaptic density and decreased excitatory synapse counts have been seen in cortical layer 6 neurons during late mouse gestation when the function of T-brain-1, a T-box transcription factor and one of the high-confidence ASD-associated genes, is lost (Fazel Darbandi et al. 2018). WNT9b is known to promote lip/palate formation and fusion (Jin et al. 2012; Juriloff et al. 2006), although it is not known what function it plays in neurodevelopment.

Table 1.1 ASD-related genes influencing WNT, SHH, RA, FGF, and BMP/TGF-β signaling pathways

ASD causal genes	Species	Chromosome	Genetic score[a]	Gene category[a]	Molecular function	Associated signaling pathway	Relevance to ASD	Ref
ALDH1A3	Human	15q26.3	3	Rare single-gene pathogenic variant, syndromic	This gene encodes an enzyme that uses retina as a substrate	RA	Loss of function leads to autistic traits	Moreno-Ramos et al. (2015)
ALDH5A1	–	6p22.3	1	Rare single-gene pathogenic variant, syndromic	Oxidoreductase	–	Gene has been linked to syndromic autism	De Rubeis et al. (2014); Iossifov et al. (2014)
ANK3	Mouse	10q21.2	1	Rare single-gene pathogenic variant, genetic association	Cadherin binding; cytoskeletal protein-binding source; ion channel-binding source; spectrin binding source	WNT (canonical)	Increases proliferation of neural progenitor cells and nuclear β-catenin	Durak et al. (2015)
APC	Mouse	5q22.2	–	–	Beta-catenin binding	WNT (canonical)	Increased repetitive behaviors, reduction in social interest, and other autistic like behaviors; memory and learning difficulties as well	Mohn et al. (2014)
CD38	Human and mouse	4p15.32	3	Rare single-gene pathogenic variant, genetic	Cell adhesion, signal transduction,	RA	Upregulation of CD38	Riebold et al. (2011); Kim et al. (2016)

(continued)

Table 1.1 (continued)

ASD causal genes	Species	Chromosome	Genetic score[a]	Gene category[a]	Molecular function	Associated signaling pathway	Relevance to ASD	Ref
				association, functional	and calcium signaling			
CHD1	–	5q15-q21.1	3	Rare single-gene pathogenic variant	Chromatin regulator, DNA binding, helicase, hydrolase	–	Missense and frameshift variants in the *CHD1* gene have been identified in ASD probands (Neale et al. 2012; Iossifov et al. 2014). Heterozygous *CHD1* missense variants were identified by the authors in patients with autism	Iossifov et al. (2014); Neale et al. (2012); Pilarowski et al. (2018)
CHD2	–	15q26.1	1	Rare single-gene pathogenic variant, syndromic	Chromatin regulator, DNA binding, helicase, hydrolase	–		
CHD3	–	17p13.1	1	Rare single-gene pathogenic variant, syndromic	Chromatin regulator, DNA binding, helicase, hydrolase	–	Two missense and one in-frame deletion variant were identified in the *CHD3* gene in ASD probands	Iossifov et al. (2014); Yuen et al. (2016, 2017)

CHD7	–	8q12.2	1	Rare single-gene pathogenic variant, syndromic	Chromatin regulator, DNA binding, helicase, hydrolase	–	Gene has been associated with syndromic autism. A rare pathogenic variant in the *CHD7* gene has been identified in an individual with ASD	O'Roak et al. (2012a)
CHD8	Mouse	14q11.2	1	Rare single-gene pathogenic variant, syndromic, functional	Chromatin remodeling	WNT (canonical)	Lethal embryonically	Nishiyama et al. (2009)
CHD8	Mouse	14q11.2	1	Rare single-gene pathogenic variant, syndromic, functional	Chromatin remodeling	WNT (canonical)	Craniofacial abnormalities, macrocephaly, and behavioral issues; WNT signaling is upregulated in the NAc region of the brain	Platt et al. (2017)
CTNNB1	Mouse	3p22.1	1	Rare single-gene pathogenic variant, syndromic	Activator; cell-cell adhesion and gene transcription	WNT (canonical)	Object recognition and social interactions are impaired; repetitive behaviours are increased, and spatial memory is enhanced	Dong et al. (2016)
CTNNB1	Mouse	3p22.1	1	Rare single-gene pathogenic	Activator; cell-cell adhesion	WNT (canonical)	Caudal axis bending, spina bifida	Zhao et al. (2014)

(continued)

Table 1.1 (continued)

ASD causal genes	Species	Chromosome	Genetic score[a]	Gene category[a]	Molecular function	Associated signaling pathway	Relevance to ASD	Ref
				variant, syndromic	and gene transcription		aperta, and tail truncation	
CTNNB1	Human and mouse	3p22.1	1	Rare single-gene pathogenic variant, syndromic	Activator; cell-cell adhesion and gene transcription	WNT (canonical)	Craniofacial abnormalities, neuronal loss and hair follicle defects	Dubruc et al. (2014)
DHCR7	Mouse	11q13.4	1	Rare single-gene pathogenic variant, syndromic	Oxidoreductase	SHH	Impairment of SMO and a decrease in SHH signaling	Blassberg et al. (2016)
DIXDC1	Human and mouse	11q23.1	3	Rare single-gene pathogenic variant, functional	This gene regulates positively WNT signaling pathway and is present in the centrosome, located at the base of the spindle, together with gamma tubulin	WNT (canonical)	Dendrite impairment and spine growth, positive modulator of WNT signaling	Kwan et al. (2016b)
DLX2	–	2q31.1	3	Genetic association	Encodes for a homeobox transcription factor, activator,	TGF-β/BMP signaling	The genetic association has been reported in ASD CARC cohorts	Liu et al. (2009)

					developmental protein, DNA binding			
DLX3	–	17q21.33	2	Rare single-gene pathogenic variant	Developmental protein, DNA binding; likely to play a regulatory role in the development of the ventral forebrain and might play a role in craniofacial patterning and morphogenesis	TGF-β/BMP signaling	De novo missense variants in the *DLX3* gene have been identified in two ASD probands (Iossifov et al. 2014)	Iossifov et al. (2014)
DLX5	Mouse	7q21.3	–	–	Activator, developmental protein, DNA binding	TGF-β/BMP signaling	Bmper upregulation	Sajan et al. (2011)
DLX6	–	7q21.3	3	Rare single-gene pathogenic variant	Transcription factor, developmental protein, DNA binding	TGF-β/BMP signaling	Rare pathogenic variants in the *DLX6* gene have been identified with autism (Nakashima et al. 2010)	Nakashima et al. (2010)
EN2	Human	7q36.3	3	Rare single-gene pathogenic variant, genetic association, functional	Developmental protein, DNA binding	SHH	Gain of function pathogenic variant leads to an elevation in SHH expression	Choi et al. (2014)

(continued)

Table 1.1 (continued)

ASD causal genes	Species	Chromosome	Genetic score[a]	Gene category[a]	Molecular function	Associated signaling pathway	Relevance to ASD	Ref
FGF22	Mouse	19p13.3-11 q14.2-	–	–	Growth factor	FGF	Synapse formation impairment	Terauchi et al. (2010)
FGF7	Mouse	15q21.2	–	–	Heparin binding, growth factor, mitogen	FGF	Synapse formation impairment	Terauchi et al. (2010)
FGFR1	–	8p11.23	2	Genetic association	Transferase, tyrosine-protein kinase, heparin binding, kinase, receptor	FGF	An intronic polymorphism in the FGFR1 gene was the index SNP for a loci that reached genome-wide significance with ASD	Matoba et al. (2020)
FMRP	Monkey	Xq27.3	–	–	RNA binding	BMP	Increase in BMPR2 and activation of LIMK1, stimulation of actin reorganization for promoting neurite outgrowth and synapse formation	Kashima et al. (2016)
FOXN1	Human	17q11.2	–	–	Developmental protein, DNA binding	RA		Moreno-Ramos et al. (2015)
mGluR5	Mouse	11q14.2-q14.3	3	Rare single-gene pathogenic variant, genetic	G-protein-coupled receptor,	FGF	Increased mRNA levels of both *FGF10* and *NGF*	Huang and Lu (2017)

				association, functional	receptor, transducer			
PRICKLE2	Mouse	3p14.1	2	Rare single-gene pathogenic variant, functional	Zinc ion binding	WNT (noncanonical)	Altered social interactions, learning abnormalities, and behavioral inflexibility	Sowers et al. (2013)
PTCHD1	Mouse	Xp22.11	1	Rare single-gene pathogenic variant, genetic association	Transmembrane protein	SHH (hypothetical)	Disrupted synaptic transmission, SHH independent	Tora et al. (2017)
PTGS2	Humans	1q31.1	3	Rare single-gene pathogenic variant, genetic association, functional	Oxidoreductase, dioxygenase, peroxidase	WNT (canonical)	There are significant associations between the genotypes and the specific symptom domain scores of ADOS and ADI-R	Yoo et al. (2008)
RERE	Human	1p36.23	1	Rare single-gene pathogenic variant, syndromic, genetic association	Transcriptional repressor during development	RA		Fregeau et al. (2016)
RORA	Human	15q22.2	S	Rare single-gene pathogenic variant, syndromic, genetic association, functional	A member of the nuclear hormone-receptor superfamily	RA	Reduction in the levels of *BCL-2* and *RORA* in the brain, as well as aberrant methylation	Nguyen et al. (2010)

(continued)

Table 1.1 (continued)

ASD causal genes	Species	Chromosome	Genetic score[a]	Gene category[a]	Molecular function	Associated signaling pathway	Relevance to ASD	Ref
TCF7L2	Human and mouse	10q25.2-q25.3	1	Rare single-gene pathogenic variant, syndromic		WNT (canonical)	Required for thalamocortical axonal projection formation	De Rubeis et al. (2014); Iossifov et al. (2014); Zhou et al. (2004); Lee et al. (2017)
UBE3A	Human	15q11.2	1	Rare single-gene pathogenic variant, syndromic, genetic association	Transferase	WNT (canonical)	Nuclear β-catenin is stabilized, and canonical WNT signaling is stimulated	Yi et al. (2017)
UBE3A	Mouse	15q11.2	1	Rare single-gene pathogenic variant, syndromic, genetic association	Transferase	RA	ALDH1A2 negative regulation and impaired synaptic plasticity	Xu et al. (2018)
UBE3A	Drosophila	15q11.2	1	Rare single-gene pathogenic variant, syndromic, genetic association	Transferase	BMP	Upregulation of BMP signaling in the nervous system and compromising endocytosis in the NMJs	Li et al. (2016)
WNT1	Human	12q13.12	3	Genetic association	Developmental protein	WNT (canonical)	Increased expression	Martin et al. (2013)

[a]Genetic scores and categories that were retrieved from SFARI Gene database, S: syndromic. The genetic scores rank the strength of association of each gene with ASD from the strongest association to the weakest association as follows: (1) high confidence; (2) strong candidate; (3) suggestive evidence

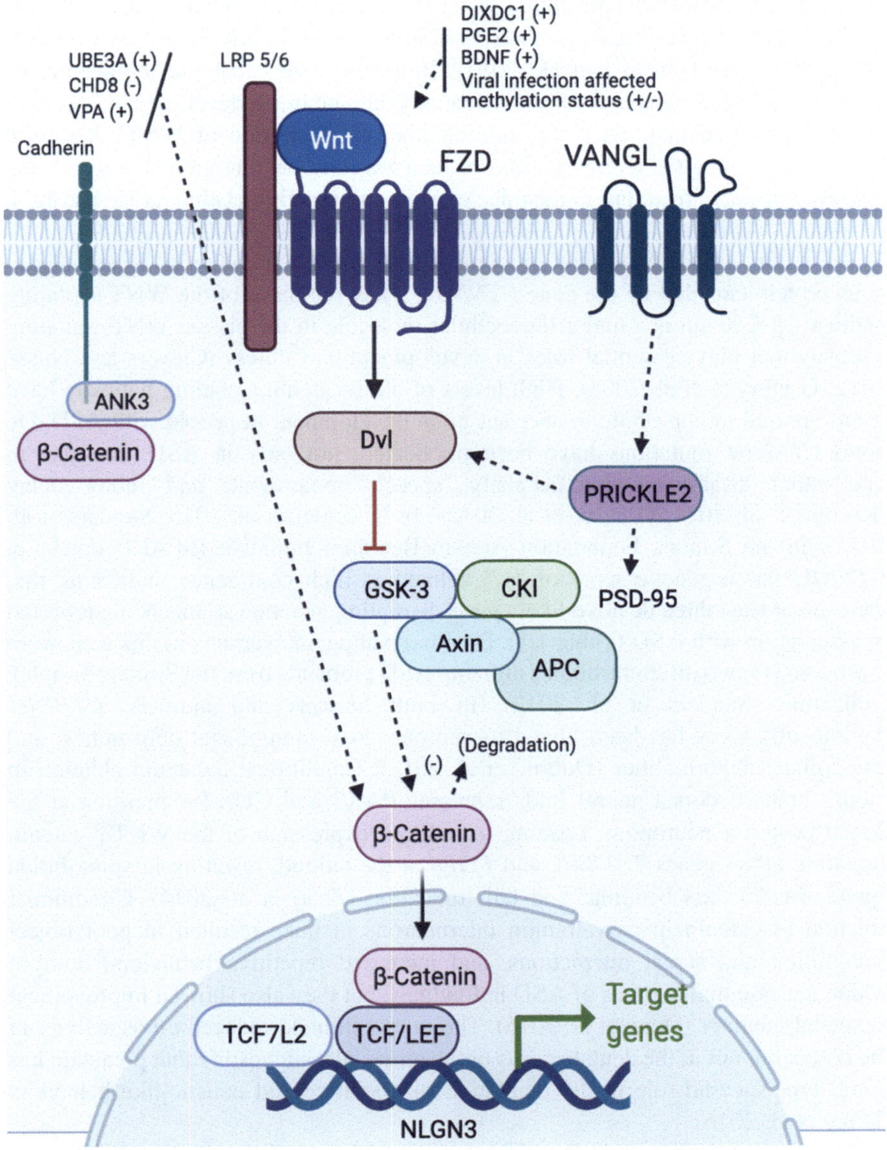

Fig. 1.1 Plausible interactions between the ASD-related genes and WNT signaling. The majority of molecules encoded by ASD-related genes either play a key role in WNT signaling pathways or are modulators. A plus sign signifies upregulation and a negative sign implies downregulation. The figure was created using BioRender (https://biorender.com/)

The functions of WNT receptors, in addition to the other WNT ligands (such as FZD1-FZD10 protein family), in ASD etiology remain poorly understood (Onishi et al. 2014; MacDonald and He 2012). Research shows that the duplication or deletion in FZD9 receptors might potentially impair brain development and thus cause ASD (Kalkman 2012). In addition, the administration of WNT2 has been found to stimulate the overexpansion of neurons in dopaminergic pathways in the midbrain of mice, resulting in the mice engaging in repetitive behavior (Sousa et al. 2010).

β-Catenin, an adherent junction component linked to E-cadherin, is an endogenous protein encoded by the gene *CTNNB1*, a key regulator of the WNT signaling pathway. β-Catenin is a major intracellular molecule in the classic WNT signaling pathway that plays essential roles in development and illness (Clevers and Nusse 2012; Grigoryan et al. 2008). High levels of the β-catenin signaling pathway have been reported to contribute to aberrant brain development in people with ASD. De novo *CTNNB1* mutations have been implicated, not only in ASD, but also in intellectual disabilities, microcephaly, speech impairment, and motor delay (Krumm et al. 2014; O'Roak et al. 2012a, b; Kuechler et al. 2015; Sanders et al. 2012). In the Simons Foundation Autism Research Initiative (SFARI) database, *CTNNB1* has a genetic score of "1," defined as high confidence, indicating that there are at least three de novo likely gene-disrupting genetic variants being reported in association with ASD (Table 1.1). De novo pathogenic variants in this gene were discovered in two different studies utilizing ASD probands from the Simons Simplex Collection (Sanders et al. 2012). In both humans and animals, *CTNNB1* haploinsufficiency has been linked to neuronal loss, craniofacial deformities, and hair follicle abnormalities (Dubruc et al. 2014). Conditional β-catenin ablation in mouse embryo dorsal neural folds represses PAX3 and CDX2 expression at the dorsal posterior neuropore. Leading to reduced expression of the WNT/β-catenin signaling target genes *T, TBX6*, and *FGF8* at the tailbud, resulting in spina bifida aperta, caudal axis bending, and tail truncation (Zhao et al. 2014). Conditional ablation of catenin in parvalbumin interneurons in mice resulted in poor object recognition and social interactions, and increased repetitive behaviors, both of which are essential features of ASD individuals, but they also showed improvement in spatial memory (Dong et al. 2016). The mice exhibited reduced c-Fos activity in the cortex but not in the dentate gyrus or the amygdala, suggesting that β-catenin has a cell type-specific role in the regulation of cognitive and autistic-like behaviors (Dong et al. 2016).

There are a number of ASD-linked genes, besides *CTNNB1*, that are believed to affect β-catenin functions; among them is adenomatous polyposis coli (*APC*) (Zhou et al. 2007). APC is a tumor suppressor that functions as a negative regulator of β-catenin and plays a significant role in the β-catenin-destruction complex (MacDonald et al. 2009). APC-inactivating gene variations in humans have been linked to ASD (Zhou et al. 2007; Barber et al. 1994). APC protein deletion mutation in mice was found to lead to autism-like disabilities, as well as learning and memory deficits in APC conditional knockout compared to wild-type littermates (Mohn et al. 2014). In addition, APC knockout forebrain neurons had higher levels of β-catenin and increased transcript levels of the canonical WNT target genes *Dkkl, Sp5, Neurogl,* and *SYN2* (Mohn et al. 2014). Furthermore, APC knockout mouse lysates from the hippocampus, cortical, and striatal areas were shown to have greater β-catenin levels than control mice (Mohn et al. 2014), along with seizures, behavioral traits, and cognitive deficits (Pirone et al. 2017). These findings imply that WNT/β-catenin signaling activity might be associated with ASD. Besides being an important regulator of the β-catenin level, APC has other roles that are critical for neurogenesis and function, including its role in regulating microtubule and actin cytoskeleton dynamics (Zumbrunn et al. 2001; Akiyama and Kawasaki 2006) and as an mRNA-binding protein with several of its targets involved in brain development (Preitner et al. 2014).

Another gene that plays an important role in the WNT pathway is the transcription factor 7-like 2 (*TCF7L2*), also known as *TCF4*. *TCF7L2* is one of the TCF/LEF1 transcription factors in the WNT/β-catenin signaling pathway that aid in the initiation of gene transcriptional responses when WNT ligands engage their receptors on the membrane and the signal is transduced to the nucleus (Grant et al. 2006) (Fig. 1.1). *TCF7L2* has been implicated in developmental delays (DDs), intellectual disabilities (IDs), neurodevelopmental disorders (NDDs), and attention-deficit hyperactivity disorder (ADHD) (De Rubeis et al. 2014; Iossifov et al. 2014; Dias et al. 2021).

Recently, two de novo loss-of-function mutations in the *TCF7L2* gene have been identified in ASD cases (De Rubeis et al. 2014; Iossifov et al. 2014). In addition, 11 people with de novo *TCF7L2* mutations who had a syndromic neurodevelopmental condition were identified, four of which had ASD (Dias et al. 2021). It has been shown that TCF7L2, like the key WNT co-receptor Lrp6, is crucial for the development of thalamocortical axonal projections in mice (Zhou et al. 2004; Lee et al. 2017) implying that abnormal thalamocortical axonal inputs might be playing a vital role in the development of the disorder. Nevertheless, there is still a paucity of knowledge about the connection of other transcription factors from the TCF/Lef1 family with ASD. Furthermore, the significance of TCF4 in brain development remains unknown and will need to be investigated further in future research.

Core elements of the noncanonical WNT signaling Whole-exome sequencing (WES) conducted on ASD-affected families has identified genetic variations of the WNT/PCP pathway genes *PRICKLE1* and *PRICKLE2*. *PRICKLE1* acts as a nuclear receptor that negatively regulates the WNT/β-catenin signaling pathway. Moreover, it is involved in the planar cell polarity route, which has several roles, such as convergent extension during gastrulation and neural tube closure, whereas the function of *PRICKLE2* remains unclear. Pathogenic variants in *PRICKLE1* and *PRICKLE2* have previously been linked to epilepsy (EP) (Bosoi et al. 2011; Tao et al. 2011), a disease that is often associated with ASD (Buckley and Holmes 2016). According to an in vitro study, *PRICKLE1* and *PRICKLE2* promote neurite outgrowth through a dishevelled-dependent mechanism (Fujimura et al. 2009). *PRICKLE1* +/− mice were reported to have ASD-like characteristics, such as abnormal social interactions and disturbed circadian rhythms. In addition, *PRICKLE1* is required for synapsin1 (Syn1) function in the pre-synapse (Paemka et al. 2013), while *PRICKLE2* interacts with postsynaptic density protein-95 (PSD95) and NMDA receptors in the postsynapse (Hida et al. 2011). Mice with disrupted *PRICKLE2* were reported exhibiting impaired social behavior, learning disabilities, and behavioral rigidity. Moreover, mouse models with ASD were shown to exhibit behavioral and physiological abnormalities that are similar to those of *PRICKLE2* mice (Sowers et al. 2013). Furthermore, rare on-synonymous *PRICKLE2* genetic variants (p.E8Q and p.V153I) have been identified in individuals with ASD (Sowers et al. 2013). Dendrite branching, synapse number, and PSD size were all reduced in *PRICKLE2*-deficient mice's hippocampal neurons (Sowers et al. 2013). The discovery of a *PRICKLE2*-containing 3p interstitial deletion in identical twins with ASD adds to the evidence that *PRICKLE2* plays a role in ASD (Okumura et al. 2014). *PRICKLE2*'s interaction with PSD-95 is improved by Vangl2, a key component in the noncanonical WNT/PCP pathway (Nagaoka et al. 2015). The role of other PCP genes in ASD etiology, and the signaling interaction between the PCP and WNT/β-catenin pathways, should be investigated in the future.

1.4 Modulators and Effectors of WNT Signaling in ASD Etiology

Several genes have been implicated in ASD, such as chromodomain helicase DNA-binding protein 8 (*CHD8*) (Willsey et al. 2013; Cotney et al. 2015), ankyrin-G (*ANK3*) (Shi et al. 2013; Iqbal et al. 2013; Bi et al. 2012), DIX domain-containing 1 (*DIXDC1*) (Kwan et al. 2016b), prostaglandin E2 (*PGE2*) (Wong et al. 2016), and HECT domain E3 ubiquitin ligase (*UBE3A*) (Yi et al. 2017) (Fig. 1.1). In addition, recent research finds that neuroligin 3 (*NLGN3*), an ASD-associated gene, is a direct downstream target of WNT/β-catenin signaling during synaptogenesis (Medina et al. 2018) (Fig. 1.1). Furthermore, due to the identification of multiple people with genetic variants in phosphatase and tensin homolog (*PTEN*), it has been described as a high-risk candidate ASD-associated gene that plays a role in WNT signaling (O'Roak et al. 2012b; Spinelli et al. 2015; Frazier et al. 2015; McBride et al. 2010).

Canonical WNT signaling is one of the key pathways controlled by CHD8 (Thompson et al. 2008; Nishiyama et al. 2012). CHD8 acts as a transcription repressor by altering the structure of chromatin (Kwan et al. 2016a). It binds β-catenin and inhibits the WNT signaling pathway, which is vital in vertebrates' early development and morphogenesis. Alternatively, spliced transcript variants encoding various isoforms have been discovered in CHD8 (Thompson et al. 2008). Due to its presence at active transcription sites with H3K4me3 or H3K27ac histone modifications, it is believed that CHD8 directly activates genes by binding to the transcriptional start site and boosting transcription factor activity or recruitment, according to the theory. It might also have an indirect effect on transcription by interacting with altered histone sites and other co-regulators to make chromatin more accessible to transcription factors (Barnard et al. 2015; Sugathan et al. 2014; Wilkinson et al. 2015; Cotney et al. 2015). CHD8 binding to p53 causes the development of a trimeric complex on chromatin with histone H1, which reduces p53-dependent transactivation and death during early embryogenesis (Nishiyama et al. 2009). CHD8 is also necessary for the expression of E2 adenovirus promoter-binding factor target genes during the cell cycle's G1/S transition (Subtil-Rodríguez et al. 2014). In mice, the *CHD8* gene deletion causes embryonic lethality (Nishiyama et al. 2009), whereas its heterozygous loss-of-function variants are associated with macrocephaly, craniofacial deformities, and behavioral impairments (Platt et al. 2017). Its knockdown in human neural progenitor cells changes the expression of neuronal development genes (Wilkinson et al. 2015). In the nucleus accumbens (NAc) region of the brain in CHD8+/− mice, WNT signaling is upregulated, suggesting the crucial function CHD8 plays in WNT signaling regulation in the NAc (Platt et al. 2017).

CHD8 is considered the most potential single candidate gene for non-syndromic ASDs (O'Roak et al. 2012a, b; Barnard et al. 2015; Krumm et al. 2014, 2015; Sanders 2015; Bernier et al. 2014). Multiple de novo, truncating, or missense mutations in *CHD8* have been found in people with ASDs (O'Roak et al. 2012a, b; Neale et al. 2012; Sugathan et al. 2014; Bernier et al. 2014; Talkowski et al. 2012; McCarthy et al. 2014). Novel risk loci as Balanced chromosomal abnormalities (BCAs) in the *CHD8* was found in people with ASD or other neurodevelopmental problems (Talkowski et al. 2012). An analysis of rare coding variation in 3871 ASD cases and 9937 ancestry-matched or paternal controls identified *CHD8* as a gene with high statistical significance with an FDR of 0.01, indicating that this gene had a 99% chance of being a true autism gene (De Rubeis et al. 2014). The gene *CHD8* has been identified as an antagonist to the signaling pathway that regulates canonical WNT signaling. This finding is consistent with the notion that increased canonical WNT signaling activity causes excessive proliferation of embryonic neural progenitor cells in the brain, which might help to explain in part the macrocephaly (or "big brain") phenotype observed in cases (Bernier et al. 2014). In Bernier and colleagues study (loss of funcation variants in *CHD8* was identified in children with developmental delay and ASD; a phenotypic comparison of patients with CHD8 variants in this report revealed recurrent phenotypes and dysmorphic facial features suggestive of a syndromic form of ASD (Bernier et al. 2014). According to in vivo research, *CHD8* loss-of-function pathogenic variant might activate more canonical WNT signals, resulting in macrocephaly, and ASD-like symptoms (Platt et al. 2017). In addition, recent research found that multiple *CHD8*-controlled genes were implicated in abnormal head size (Sugathan et al. 2014; Wang et al. 2015; Merner et al. 2016). It has been revealed that CHD8 is a positive regulator of WNT/β-catenin signaling neural progenitor cells while also negatively regulating the pathway in non-neuronal cells (Durak et al. 2016). This finding suggests that CHD8 modulates WNT signaling in a cell-specific manner and that some CHD8 mutations might not be as straightforward as WNT signaling loss-of-function mutations. Further studies are needed to understand how CHD8 modulates WNT signaling in diverse brain cell types, and how patients with

CHD8 pathogenic variants develop macrocephaly. It is also worth noting that WNT signaling is only one of several neurodevelopmental pathways regulated by CHD8, and recent research has revealed several other mechanisms, such as chromatin remodeling. Thus, more studies are required to explore the involvement of CHD8 in WNT signaling at various stages in the development of brain. This is critical to gain deeper insights into how CHD8 mutations might impair embryonic brain development.

Another gene that has been implicated in ASD is the *ANK3* gene, which encodes a scaffolding protein referred to as ankyrin-G (Shi et al. 2013; Iqbal et al. 2013; Bi et al. 2012). The protein was first discovered in the axonal initial segment and nodes of Ranvier of neurons, where it plays a role in axonal initial segment assembly and neuronal polarity (Hedstrom et al. 2008; Kordeli et al. 1995). In general, the ankyrin family of proteins are thought to aid in anchoring integral membrane proteins to the cytoskeleton. Furthermore, they are involved in a wide range of activities, including cell motility, activation, proliferation, contact, and maintenance of specialized membrane domains, among other things. Ankyrin-G promotes cell-cell contact by binding to E-cadherin at a conserved location separate from that of β-catenin and transporting it to the cell adhesion site with 2-spectrin in early embryos and cultured epithelial cells (Kizhatil et al. 2007). Ankyrin-G is abundant in the embryonic brain's ventricular zone, where it governs neural progenitor cell growth (Durak et al. 2015). Because of alternative splicing and different beginning exons, there are multiple protein isoforms of ankyrin-G13,17. The isoforms have distinct roles and tissue distributions, with some being expressed exclusively in the brain. Rare polymorphisms identified in ASD patients are mostly found in the brain-specific exons 371, 2, 3, 4, 5, 6, 7, and 8 (Bi et al. 2012; Ferreira et al. 2008; Schulze et al. 2009; Psychiatric GWAS Consortium Bipolar Disorder Working Group 2011; Tesli et al. 2011; Baum et al. 2008; Iqbal et al. 2013; Kosmicki et al. 2017). In ASD cases, whole-genome and whole-exome sequencing investigations have documented several genetic variants in *ANK3* (Shi et al. 2013; Iqbal et al. 2013; Bi et al. 2012). Loss of function in ANK3 promotes neural progenitor cell proliferation and nuclear β-catenin expression, most likely by disrupting the β-catenin/cadherin connection (Durak et al. 2015).

Several missense mutations in *DIXDC1* have been identified in individuals with ASD (Kwan et al. 2016b). These mutations hinder DIXDC1 isoform 1 phosphorylation, resulting in dendritic and spine development defects (Kwan et al. 2016b). DIXDC1 encodes for a protein that acts as a positive regulator of the WNT signaling pathway that modulates excitatory neuron dendrite formation and synapse function in the mouse cortex (Kwan et al. 2016b). *MARK1*, which has also been associated with ASD, phosphorylates DIXDC1 to regulate dendritic and spine formation via isoform-specific cytoskeletal network regulation (Kwan et al. 2016b). DIXDC1-deficient animals were found to exhibit behavioral abnormalities, including decreased social interaction, which can be relieved by pharmacological inhibition of glycogen synthase kinase 3 (GSK3) to increase WNT/β-catenin signaling (Martin et al. 2018; Kivimäe et al. 2011). These findings point to a possible approach to ASD treatment, including modification of WNT/β-catenin signaling activity. Exome sequencing of the *DIXDC1* gene has shown a higher burden of rare sequencing-disrupting SNVs among ASD cases in comparison to controls (Martin et al. 2018). In DIXDC1 knockout neurons, it has been reported that rare *DIXDC1* missense variants in ASD cases failed to rescue deficits in glutamatergic synapse density and spine density, with a subset of *DIXDC1* missense variants displaying hyperactivity in WNT/β-catenin signaling activity as opposed to dominant-negative effects on spine density and glutamatergic synapse density in wild-type neurons. The existence of functionally relevant ASD missense pathogenic variants in controls, on the other hand, complicates the genetic evidence connecting *DIXDC1* to ASD, with reports of high functionally relevant ASD missense variants in controls. Lack of information regarding the mode of inheritance and segregation of variants in ASD cases, along with the presence of sequence-disrupting *DIXDC1* variants in controls, all make it very challenging to understand the link between *DIXDC1* and ASD.

PGE2, an endogenous lipid molecule, has been shown to alter the expression of downstream WNT pathway genes previously linked to neurodevelopmental problems (Wong et al. 2016). The primary regulator of PGE2 synthesis is cyclooxygenase-2 (COX2). COX2/PGE2 signaling abnormalities have been linked to ASD (Wong et al. 2016). In addition, a growing body of evidence shows that a variety of environmental risk factors, such as inadequate dietary supplementation, infections, misoprostol use during pregnancy, air pollutants, or chemicals, can have a negative effect on PGE2 levels and can be indirectly linked to ASD (Tamiji and Crawford 2010; Wong et al. 2015; Bandim et al. 2003; Landrigan 2010). Furthermore, PGE2 has been shown to downregulate *PTGS2* while upregulating *MMP9* and *CCND1* in undifferentiated stem cells. On the other hand, in differentiating neuronal cells, it upregulates *WNT3*, *TCF4*, and *CCND1* expression (Wong et al. 2016).

Another gene that has been implicated in ASD etiology is *UBE3A* which encodes an E3 ubiquitin-protein ligase, which is part of the ubiquitin protein degradation mechanism. This imprinted gene is expressed maternally in the brain and biallelically in other organs. Several studies have shown genetic associations and rare polymorphisms in the *UBE3A* gene that are linked to ASD. A link was discovered in the collaborative linkage study of autism families and rare variations were discovered in instances of European ancestry (Nurmi et al. 2001). Another polymorphism identified in *UBE3A* was T485A which is a de novo autism-linked *UBE3A* pathogenic variants that was reported to be involved in ubiquitinating numerous proteasome subunits, decreasing their number and activity, stabilizing nuclear β-catenin, and activating the canonical WNT pathway more efficiently in comparison to wild-type *UBE3A* (Yi et al. 2017). Dysfunction of *UBE3A* has been associated with cancer, Angelman syndrome, ADHD, DD/NDD, extrapyramidal symptoms (EPS), ID, and EP (Yi et al. 2017).

Pathogenic variants in the neurexin and neuroligin families have also been implicated in the development of autism spectrum disorders (Südhof 2008). Neurexins and neuroligins interact cooperatively to regulate the synaptic activity, both excitatory and inhibitory, in the brain (Knight et al. 2011). *NRXN1, NLGN1, NLGN3, CNTN4, CNTN6, and CNTNAP2* are examples of superfamily genes that have a function in autism spectrum disorder (Berg and Geschwind 2012). Knockout

ASD mouse models of genes such as *NRXN1, SHANK3, FMR1,* and *CNTNAP2* were found to have imbalances in excitation and inhibition in brain regions, and these knockout models exhibited social interaction deficits and reduced ultrasonic vocalizations, which overlapped with behavioral endophenotypes relevant to ASD (Rosti et al. 2014). Imbalances in excitation and inhibition in brain regions were discovered in knockout ASD mouse models of genes such as *NRXN1, CNTNAP2, FMR1,* and *SHANK3*. It has been shown that ASD-affected individuals have variants in the neuroligins *NLGN3* and *NLGN4* (Jamain et al. 2003). These type I transmembrane proteins serve as adhesion molecules for brain cells and are required to establish and develop synaptic connections in the brain (Zhang et al. 2017). According to chromatin immunoprecipitation and promoter luciferase experiments, WNT/β-catenin signaling directly controls the production of NLGN3, a transcription factor (Medina et al. 2018). It would however be essential to establish whether or not WNT/β-catenin signaling influences the expression of additional ASD-associated genes. It is noteworthy that neuronal activity is regulated by genes such as *GRIN2B, SCN1A,* and *SCN2A*, and these genes are thought to be involved in the mediation of synaptic plasticity. In addition, they encode for ion channels (O'Roak et al. 2012a). It is noteworthy to mention that UBE3A, PCDH10, DIA1, and NHE9/SLC9A9 are similarly affected by neuronal activity that regulates transcription factors (Morrow et al. 2008; Flavell et al. 2008).

Given the finding of several individuals with phosphatase and tensin homolog (*PTEN*) mutations, *PTEN* is considered a high-risk autism candidate gene that plays a role in WNT signaling (O'Roak et al. 2012b; Spinelli et al. 2015; Frazier et al. 2015; McBride et al. 2010). Furthermore, individuals with heterozygous *PTEN* pathogenic variants are also at risk of developing macrocephaly, implying that PTEN regulates brain size, which is thought to affect specific ASD instances (Page et al. 2009; Kwon et al. 2006; Chen et al. 2015; Vogt et al. 2015; Clipperton-Allen and Page 2014, 2015; Takeuchi et al. 2013; Tilot et al. 2015; Zhou and Parada 2012).

1.5 Altered Sonic Hedgehog (SHH) Signaling in ASD

SHH signaling is one such pathway that regulates neurogenesis and neuronal processes during CNS development (Choy and Cheng 2012). SMO-SHH signaling is implicated in various neurological activities, including neuronal cell proliferation and survival (Álvarez-Buylla and Ihrie 2014). While the function of primary neural cilia in CNS patterning during embryonic development is well understood, their relevance in adult CNS plasticity has only recently been discovered (Kirschen and Xiong 2017). SHH signaling at the main cilium has been investigated and defined, as shown in Fig. 1.2 (Seppala et al. 2017). In the absence of SHH activity, PTCH suppresses SMO effectively. This causes Gli (glioma-associated oncogene/transcription factor) proteins to be phosphorylated, after which they are proteolytically truncated into repressor forms that inhibit transcriptional activity. SHH binding to PTCH, on the other hand, induces internalization followed by breakdown, resulting in SMO buildup and protein phosphorylation. As a consequence, Gli is carried to the cytoplasm and enters the nucleus in its complete form, boosting target transcription even more. In children with ASD, pathological functions for SHH, Indian hedgehog (IHH), and BDNF have been proposed (Halepoto et al. 2015). SHH signaling influences both neurogenesis and neuronal patterning during the development of the central nervous system during pregnancy. Neurological diseases such as ASD are caused by an imbalance in SHH signaling in the brain (Patel et al. 2017). In the context of ASD, SHH has also been linked to oxidative stress (Ghanizadeh 2012). Autistic children exhibited substantially greater levels of oxygen free radicals (OFR) and serum SHH protein, suggesting that oxidative stress and SHH play a role in the development of ASD (Al-Ayadhi 2012). As shown in Fig. 1.2, the connection between ASD-associated genes and SHH signaling is illustrated.

The SHH pathway has also been implicated in ASD, and several pathogenic variants in patched domain-containing 1 (*PTCHD1*) have been identified. A pathogenic variant in this gene has been found to lead to problems with synaptic connections and irregularities in neuronal transmissions in male mice, resulting in hyperactivity and cognitive deficits (Tora et al. 2017; Ung et al. 2018) and behavioral abnormalities (Tora et al. 2017). However, while the PTCHD1 protein does not seem to play a role in SHH-dependent signaling when tested using a loss-of-function approach, it does appear to affect synaptic transmission in the mouse dentate gyrus (Tora et al. 2017). In addition it has been shown that PTCHD1 interacts with PSD95 and SAP102, two postsynaptic proteins (Ung et al. 2018). PTCHD1 deficiency ($PTCHD1^{-/y}$) in male mice causes widespread changes in synaptic gene expression, including changes in the expression of the immediate-early expression genes *EGR*1 and *NPAS4*, and disruptions in excitatory synaptic structure and neuronal excitatory activity in the hippocampus, which results in cognitive dysfunction, motor impairments, and hyperactivity (Ung et al. 2018).

Pathogenic variants in the 7-dehydrocholesterol reductase (*DHCR7*) gene have also been linked to ASD. Increased cholesterol production was found to have a negative effect on the activation of the transmembrane protein smoothened (SMO), which is responsible for transmitting SHH signals, and its localization to the major cilium (Blassberg et al. 2016). The transcription factor engrailed 2 (*EN2*) has also been implicated in the development of ASD (Brune et al. 2008; Sen et al. 2010; Wang et al. 2008; Yang et al. 2008, 2010). The higher levels of *EN2* seen in individuals with the EN2 ASD-associated haplotype (rs1861972-rs1861973 A-C) indicated that EN2 is associated with ASD susceptibility (Choi et al. 2014; Bi et al. 2012; Gharani et al. 2004). According to the evidence from postmortem samples, rising *EN2* levels have been associated with increased *SHH* expression (Choi et al. 2014). During brain development, *SHH* is one of the genes that coexpress *EN2* and other genes (Wechsler-Reya and Scott 1999; Simon et al. 2005). Furthermore, both mice with EN2 pathogenic variants and autistic humans had cerebellar structural abnormalities that were similar (Gharani et al. 2004).

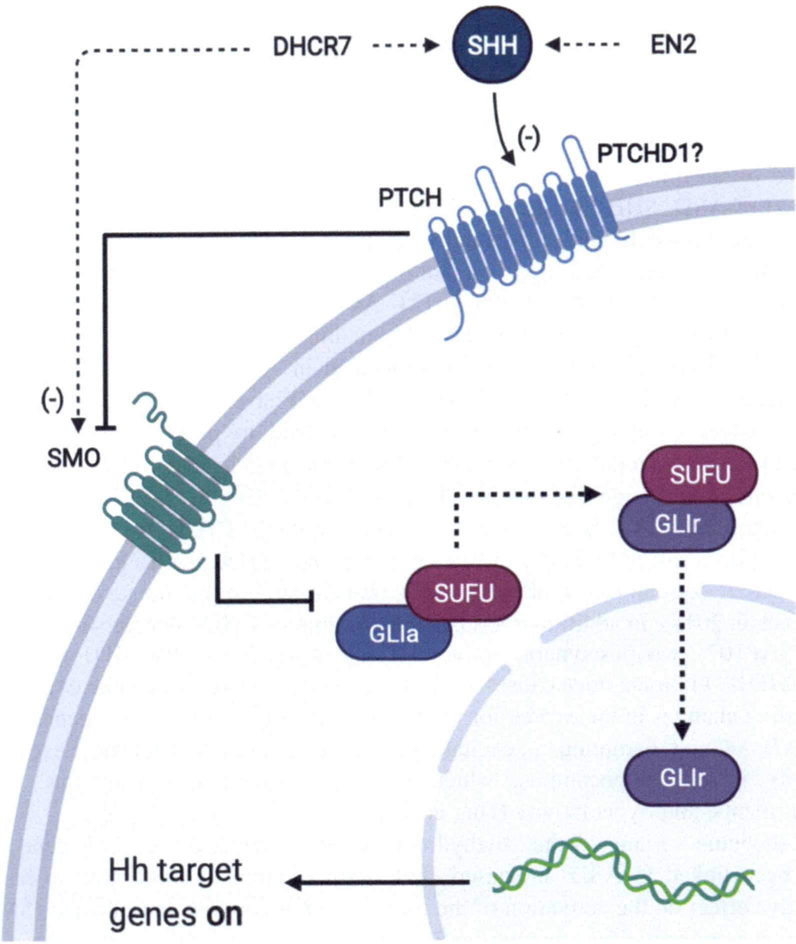

Fig. 1.2 Possible associations between the ASD-linked genes and SHH signaling. PTCHD1, EN2, and DHCR7 genes are all potential ASD-associated genes. A negative sign denotes downregulating activity. The figure was created using BioRender (https://biorender.com/)

1.6 Altered Retinoic Acid (RA) Signaling in ASD

RA, a functional metabolite of vitamin A, is a necessary morphogen during vertebrate development (Niederreither and Dollé 2008; Duester 2008). RA can influence several developmental genes that include RA response elements (RARE) and their regulatory areas. By binding to nuclear receptors known as retinoic acid receptors (RARs) and retinoid X receptors (RXRs), RA mediates both genomic transcriptional effects and non-genomic effects such as retinoylation (RA acylation), a posttranslational modification of proteins (Rhinn and Dollé 2012; Das et al. 2014). A number of co-activators and co-repressors have been identified as modulators of RA signaling activity (Das et al. 2014). RA is necessary for neural patterning, differentiation, proliferation, and establishment of neurotransmitter systems in the developing CNS (Zieger and Schubert 2017). Cortical neuron generation is regulated by RA from the meninges (Siegenthaler et al. 2009). During embryonic development, RA aids in the regulation of a group of HOX (homeobox) genes that form the upper-body pattern, both anteriorly and posteriorly, and are implicated in brain patterning. Dopaminergic and GABAergic neurons are involved in brain cell development in RA. It is also a critical inducing factor for the development of motor neurons from pluripotent stem cells (Faravelli et al. 2014).

Furthermore, RA is necessary for the appropriate functioning of motor neurons. RA has also been linked to neuronal migration and neurogenesis in the granular zone of the hippocampus, the sub-ventricular zone, and the olfactory bulb (Ghyselinck and Duester 2019). The above mentioned functions of RA indicate that neurodevelopmental diseases and RA signaling pathways are linked.

A sufficient quantity of retinol is required for the regular functioning of the RA signaling system, assuming that all of the enzymes involved in the RA pathway and nuclear factors operate properly. One of the key reasons for reduced intracellular RA signaling is retinol insufficiency. In China, some autistic people had lower levels of retinol than the normal control group, which might be a synergistic component in the development of ASD symptoms (Cheng et al. 2021). Retinol supplements have been shown to increase RAR expression and reduce ASD symptoms. Some investigations have shown that retinol deficiency in rats during pregnancy reduces RA receptor expression (RAR, beta isoform) in the hypothalamus, resulting in autistic-like symptoms in the neonates (Lai et al. 2018). In rats, a lack of vitamin A has been linked to ASD-like behavior (Lai et al. 2018). It has also been claimed that ASD is caused by an abnormality in the interaction of RA and sex hormones (Niculae and Pavăl 2016).

The nuclear receptors for RA have also been implicated in the pathophysiology of ASD. RORs (retinoic acid-related orphan receptors) operate as transcriptional regulators upon RA binding, triggering transcription of numerous genes. Autistic people were shown to have reduced protein expression of the RA-related orphan receptor alpha (RORA) due to hypermethylation (Nguyen et al. 2010), and RORA mutations have been linked to ASD (Sayad et al. 2017). RORA has been shown to transcriptionally control numerous ASD-relevant genes, including *NLGN1* (Sarachana and Hu 2013). P89L missense variant of *NLGN1*, identified in ASD cases, has been linked to alterations in cellular localization, protein degradation, and spine formation impairment. Mice with heterozygous P89L in the *NLGN1* have also been found to exhibit aberrant social behavior, which is a crucial hallmark of ASD (Nakanishi et al. 2017).

Furthermore, an immunohistochemical examination of the postmortem brains of autistic people revealed a reduced quantity of ROR alpha protein (Nguyen et al. 2010). Disruption of the retinoic acid enzymatic production pathway was found to be associated with ASD phenotypes and retinoic acid nuclear receptors, which have also been implicated in the pathophysiology of ASD. Therefore, more research is needed to establish a link between ASD pathogenesis and involvement of RAR and ROR agonists in autism treatment.

Another approach for detecting ASD signals is quantitative electroencephalography (EEG) analysis. Details of individual characterization of EEG fluctuations in ASD subjects could aid in examining brain issues, which would be helpful in observing automatic groupings and random draws of the patient population when analyzing sensory-processing issues of the brain and peripheral system (Ryu et al. 2021).

The conversion of retinol to RA with the aid of the enzyme (ALDH) retinaldehyde dehydrogenase, which assures the concentration of RA in the cell, is an essential metabolic step in the RA pathway. In vitro, over-ubiquitinoylation by *UBE3A* has been shown to increase the breakdown rate of this enzyme's isoform aldehyde dehydrogenase 1 family member A2 (ALDH 1A2). Autistic characteristics were detected in mice with *UBE3A* overexpression (Xu et al. 2017). In addition, loss-of-function pathogenic variants in the *UBE3A* have been linked to Angelman syndrome, demonstrating the role of *UBE3A* in brain development (Khatri and Man 2019). Surprisingly, *UBE3A* overexpression suppresses ALDH1A2 and inhibits RA-mediated synaptic plasticity in ASD, which might be improved with RA supplementation (Xu et al. 2018). All-trans-RA can increase CD38 expression in ASD lymphoblastoid cell lines, but CD38-deficient animals display ASD-like behavior (Riebold et al. 2011; Kim et al. 2016). In BTBR mice, beta-carotene, a precursor of vitamin A, is a viable therapy for autistic-like behavior (Avraham et al. 2019). In a mouse model, a synthetic RORA/G agonist was investigated for its ability to ameliorate autistic symptoms (Wang et al. 2016). A South American cohort's WES has revealed a link between RA signaling genes, including a RA-synthesizing gene *ALDH1A3* and the RORA-regulated *FOXN1*, and ASD (Moreno-Ramos et al. 2015). A subpopulation of autistic people were shown to have low levels of *ALDH1A1* (Pavăl et al. 2017). ASD and other abnormalities linked with proximal 1p36 deletions might be caused by de novo variants in arginine-glutamic acid dipeptide repeats (RERE) that encode a nuclear receptor coregulator for RA signaling (Fregeau et al. 2016). These findings point to potential pharmacological methods for ASD treatment that include targeting RA and associated signaling pathways. Figure 1.3 depicts the potential connections between ASD-associated genes and RA signaling.

1.7 Altered Fibroblast Growth Factor (FGF) Signaling in ASD

FGF signaling is critical in brain patterning, and its disruption might result in a variety of neurological disorders (Turner et al. 2016). In the mammalian FGF family, there are 18 secreted FGFs and 4 tyrosine kinase FGF receptors (FGFRs) whose interaction is controlled by cofactors and external binding proteins (Ornitz and Itoh 2015). When FGFRs are activated, tyrosine residues are phosphorylated, resulting in interactions between cytosolic adaptor proteins and RAS-MAPK, PI3K-AKT, PLC, and STAT intracellular signaling pathways (Ornitz and Itoh 2015). It has been

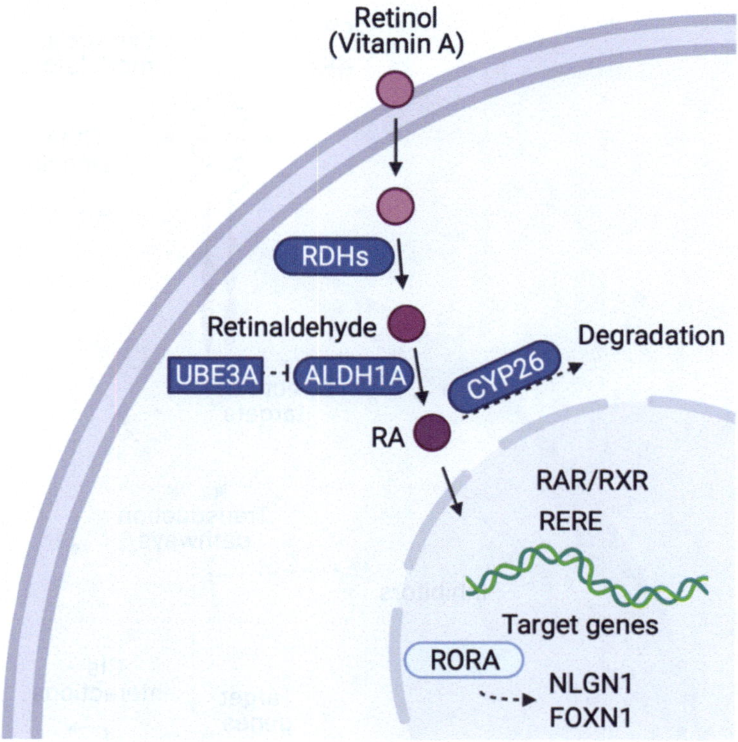

Fig. 1.3 Possible interactions between ASD-associated genes and RA signaling. UBE3A influences ALDH1A expression, which in turn affects the RA signaling pathway. RORA is linked to ASD, which impacts NLGN1. ASD is also linked to the RA signaling coregulator RERE. The figure was created using BioRender (https://biorender.com/)

proposed that dysregulation of FGF signaling plays a role in the etiology of ASD (Iwata and Hevner 2009). Cortical abnormalities in autistic brains, for example, have been linked to faulty FGF signaling (Turner et al. 2016; Amaral et al. 2008). ASD is thought to be caused by a disruption in the number of excitatory and inhibitory synapses (Terauchi et al. 2010). Mutant mice FGF22 or FGF7 knockout (KO) showed defective synapse formation in hippocampal CA3 pyramidal neurons (Terauchi et al. 2010), indicating the pathogenic significance of dysregulated FGF signaling in ASD. The metabotropic glutamate receptor 5 (mGluR5) loss-of-function pathogenic variant causes abnormal dendritogenesis in cortical neurons, which is one of the traits identified in autistic brains, by raising nerve growth factor (NGF) and *FGF10* mRNA levels (Huang and Lu 2017). Although several members of FGF gene subfamilies have been linked to CNS development and/or function, many of their roles remain largely unknown, and no direct evidence of altered FGF signaling in ASD has been reported. Nevertheless, it has been well documented that epigenetic mechanisms (Zhu et al. 2007), and posttranslational modifications of FGFs and FGFRs (Sarabipour and Hristova 2016; Triantis et al. 2010; Wheeler and

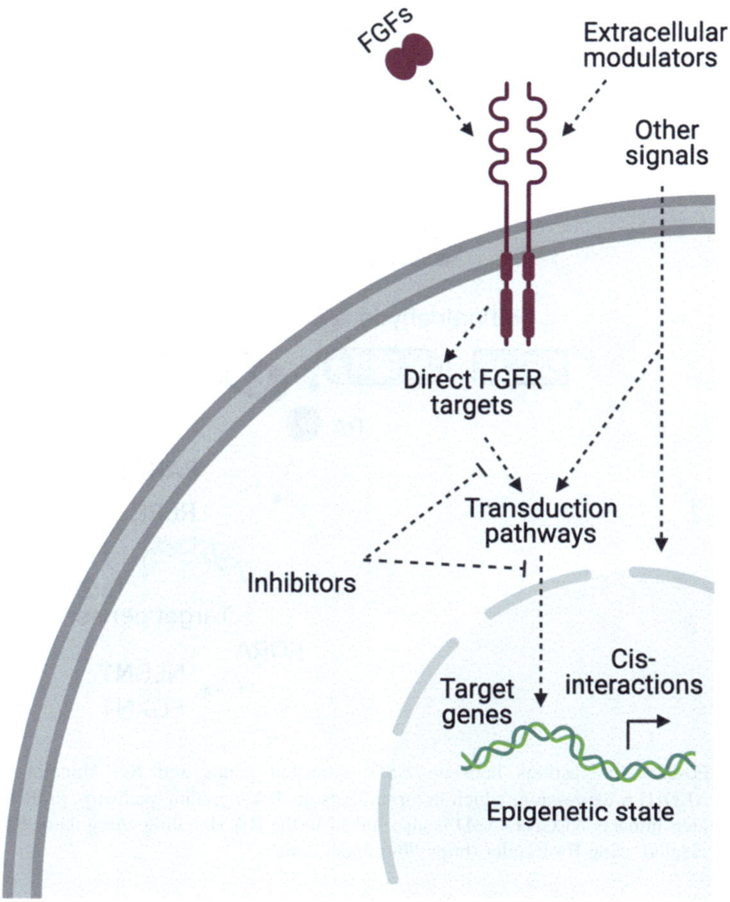

Fig. 1.4 Possible mechanisms of FGF signaling. A simplified depiction of the possible mechanisms through which an FGF signal might be modulated to influence cell fate. The figure was created using BioRender (https://biorender.com/)

Clinkenbeard 2019; Kucinska et al. 2019; Porebska et al. 2018), are involved in the regulation of the expressions of FGF/FGFR signaling (Xie et al. 2020). Figure 1.4 summarizes the plausible mechanisms by which FGF signaling modulates cell fate.

1.8 Altered TGF-β/BMP Signaling in ASD

TGF-β/activin and bone morphogenetic protein (BMP)/growth and differentiation factor (GDF) are the two subcategories of the TGF-β superfamily (Zi et al. 2012). BMPs are the most abundant in the TGF-β superfamily (Lory and Rosen 2018), and they play an essential role in nervous system development (Bond et al. 2012). Their signaling has been demonstrated to be dysregulated in ASD. BMPs influence gene expression via both canonical (Smad-dependent) and noncanonical pathways (such as the MAPK cascade) (Wang et al. 2014). The binding of BMPs to type I or type II serine/threonine kinase receptors results in a heterotetrameric complex in the canonical pathway. The type II receptor then transphosphorylates the type I receptor, following which the type I receptor phosphorylates the R-Smads (Smad1/5/8). Phosphorylated Smad1/5/8 and the co-Smad (Smad4) translocate to the nucleus and regulate gene expression. BMP signaling is modulated by various factors, including plasma membrane co-receptors and external and intracellular factors (Wang et al. 2014). The BTBR T + Itpr3tf/J (BTBR) mice are commonly utilized in ASD studies (Ansari et al. 2017). TGF-β levels have been found to be lower in BTBR mice than in B6 mice (Ansari et al. 2017). TGF-β expression levels in the spleen and brain tissues of BTBR mice were found to be significantly higher than in adenosine A2A receptor (A2AR) agonist CGS 21680 (CGS)-treated mice (Ansari et al. 2017). Components of the TGF-β pathway were identified as novel hyperserotonemia-related ASD genes in a network-based gene set enrichment analysis (NGSEA) based on loss-of-function and missense de novo variants (Chen et al. 2017). Figure 1.5 summarizes the interactions between ASD-associated genes and TGF-β/BMP signaling.

Fragile X syndrome (FXS) is the most frequent heritable form of ASD and intellectual impairment caused by *FMR1* silencing (Kashima et al. 2016). Depletion of the *FMR1* protein (FMRP) leads to increased bone morphogenetic protein type II receptor (BMPR2) and activation of a noncanonical BMP signaling component, LIM domain kinase 1 (LIMK1), which stimulates actin rearrangement to promote neurite outgrowth and synapse formation (Kashima et al. 2016). Increased BMPR2 and LIMK1 activity has been observed in the prefrontal cortex of FXS patients compared to healthy participants (Kashima et al. 2016).

UBE3A has been associated with the regulation of synapse growth and endocytosis by inhibiting BMP signaling (Li et al. 2016). The ubiquitin-proteasome pathway degrades the BMP receptor Tkv, a direct substrate of *UBE3A* (Li et al. 2016). Through the BMP signaling pathway, Drosophila *UBE3A* is known to modulate neuromuscular junction (NMJ) development in presynaptic neurons (Li et al. 2016). Drosophila *UBE3A* mutants have been found to be viable and productive. Nevertheless, they have been demonstrated to have impaired endocytosis in the NMJs and upregulated BMP signaling in the nervous system, owing to an increase in Tkv (Li et al. 2016).

The *DLX* genes, which encode homeodomain transcription factors, have also been linked to ASD (Liu et al. 2009; Hamilton et al. 2005; Rubenstein and Merzenich 2003). These genes regulate craniofacial patterning, differentiation, and survival of forebrain inhibitory neurons (Hamilton et al. 2005). The BMP-binding endothelial regulator (Bmper) is upregulated in a cell line overexpressing *DLX5* (Sajan et al. 2011), indicating that dysregulated *DLX* activity in individuals with ASD might lead to aberrant BMP signaling.

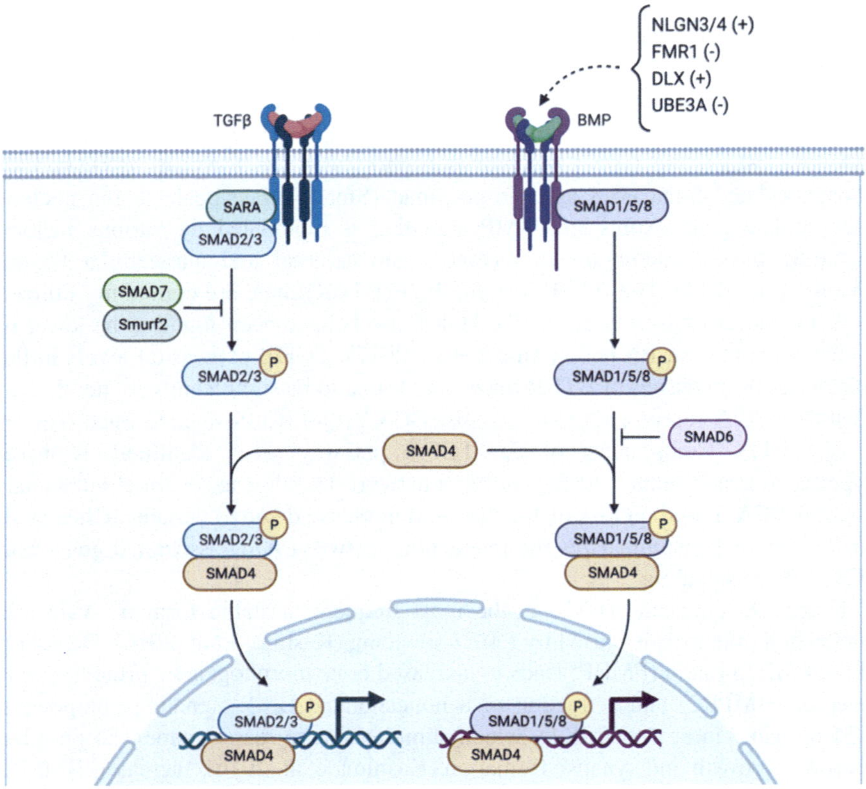

Fig. 1.5 Possible interactions between ASD-associated genes and TGF-β/BMP signaling. ASD-associated gene-encoded proteins, such as NLGN3/4, FMR1, DLX, and UBE3A, interact with BMP signaling. Plus sign indicates upregulation; minus sign indicates downregulation. The figure was created using BioRender (https://biorender.com/)

1.9 Signaling Crosstalk in ASD

While it is known that crosstalk occurs across signaling pathways in many developmental stages and illnesses, availability of evidence on autistic models is limited. Research has shown that the WNT signaling proteins might affect both developmental and inflammatory processes. It was shown that WNT signaling and SHH signaling are connected and influence each other at the transcriptional level (Chatterjee and Sil 2019). Cell-line studies have shown that Gli stimulates β-catenin nuclear translocation by E-cadherin and Snail (Li et al. 2007). Gli regulates WNT5a and WNT2b to elevate WNT signaling (Katoh and Katoh 2009). Casein kinase 1 (CK1), p53, PTEN, and GSK3 are protein modulators in these signaling pathways, and each serves a different purpose. They collaborate to boost β-catenin and Gli signaling (Song et al. 2014). Due to evidence of cross talk between WNT and SHH/Gli signaling in colon cancer, this signaling pathway is now a possible candidate for colon cancer therapy (Song et al. 2015). CK1 and GSK3 are known to have antagonistic functions in regulating β-catenin and Gli1 (Wang and Li 2006; Tempé et al. 2006; Hart et al. 1999), and a detrimental impact on TCF and downstream genes of metastatic colon cancer. Also, they were shown to have opposing roles (Varnat et al. 2010). Another problem is that the gene for a kinase inhibitor (Sufu), which represses Gli1 activity, has been shown to change the distribution of β-catenin between the nucleus and cytoplasm (Meng et al. 2001; Dunaeva et al. 2003; Tukachinsky et al. 2010). Loss of p53 or PTEN causes Gli1 and β-catenin to activate in colon cancer (Varnat et al. 2010; Rychahou et al. 2008). The authors have seen that the use of SMO inhibitors keeps Gli1, an upstream effector, from becoming active, which prevents the amount of active β-catenin from rising and increases the nuclear exclusion of β-catenin (Arimura et al. 2009). Furthermore, Gli1 prevents Gli3R, whereas Gli3R stops Gli1 (Varnat et al. 2010). Additionally, β-catenin activity has been shown to reduce with Gli3R (Ulloa et al. 2007). There is evidence that Gli1 boosts the expression of WNT4, WNT2b, and WNT7b (Li et al. 2007).

Research on the embryonic development of WNT signaling, TGF-β, RA signaling, and BMP revealed that interactions exist among them (Hayward et al. 2008; Boyle et al. 2011; Pelullo et al. 2019). There are a number of ways that WNT signaling might also interact with cytokine signaling and NF-B to support inflammatory responses (Koopmans et al. 2017).

The development of the zebrafish tailbud (Stulberg et al. 2012) and the mouse craniofacial area are thought to be influenced by the crosstalk between the FGF and WNT pathways (Wang et al. 2011). Researchers have shown that increased FGF and WNT signaling is positive (Stulberg et al. 2012). The gene Fgf8 was shown to play a key role in the induction of neural tissues in the face. It is particularly active in the anterior neural ridge and face ectoderm, with WNT/β-catenin signaling enhancing its role and β-catenin mutation leading to aberrant levels of Fgf8 expression in the face ectoderm (Glinka et al. 2011). WNT boosts FGF signaling by boosting Erk phosphorylation in the MAPK branch (Stulberg et al. 2012). FGF blocks WNT antagonists dkk1 and notum1a, which results in WNT signaling being raised (Stulberg et al. 2012). A mutated *UBE3A* gene, linked to ASD, affects both the WNT and BMP signaling pathways, indicating their possible connection (Li et al. 2016; Yi et al. 2017). Xu and her colleagues observed UBE3A as negatively regulating ALDH1A2, which means that this gene makes RA formation more difficult (Xu et al. 2018). The ASD-associated gene might additionally influence BMP signaling, even though it has been suggested that it was likely to be a direct target of WNT/β-catenin signaling (Medina et al. 2018). The above results indicate that ASD-connected genes enhance the signaling of cross talk via morphogenetic pathways, emphasizing the necessity for further study into the interconnections between signals in normal physiological and pathological circumstances. WNT and BMP signaling cross talk seems to be the result of a tissue-specific process (Itasaki and Hoppler 2010). Signaling interactions occur contextually in the eye, because WNT signaling positively controls RA signaling in the dorsal optic cup but suppresses RA signaling during orofacial development (Itasaki and Hoppler 2010).

Cyclosporine-enhanced cell proliferation has also been seen in human gingival fibroblasts (Chung and Fu 2013). In addition, interactions between TGF-β and SHH pathways have been found in cancer (Javelaud et al. 2012). TGF-β, a growth factor, binds to receptors found on cancer cell membranes, including the neuropilin-1 (NRP1). NRP1 has been demonstrated to promote the canonical SMAD2/3 signaling in response to TGF-β (Glinka et al. 2011). Also, NRP1 transduction is strengthened by HH signaling, and NRP1 has been shown to increase HH target gene activation by assisting the SMO/SUFU interconnection (Hillman et al. 2011; Hochman et al. 2006). While it is essential in the growth of SMO-associated cancers (Fan et al. 2010), it has been shown that TGF-β might increase GLI2 and GLI1 expression by blocking PKA activity (Pierrat et al. 2012). According to the hierarchical structure of cross talk, TGF-β upregulates SHH, resulting in increased SHH synthesis and cell proliferation in gingival fibroblasts. This occurs through cyclosporine, which enhances SHH production (Chung and Fu 2013).

1.10 Nongenetic ASD Etiologies

With ASD susceptibility estimated to be 40–80% hereditary, the risk is probably driven by environmental variables, which are most likely involved in ASD etiology via epigenetic regulation as the primary mechanism.

Hundreds of plausible environmental risk variables have been identified, including increased parental age, viral infections, and prenatal exposure to anticonvulsants, such as valproic acid (VPA). All the proposed risk factors can cause various neurodevelopmental diseases with disrupted WNT signaling (O'Roak et al. 2012a; Rasalam et al. 2005; Kong et al. 2012; Ohkawara et al. 2015).

1.11 Advancing Paternal Age

Since the 1970s, there has been speculation about a connection between paternal age and ASD (Allen et al. 1971; Treffert 1970). A study has reported that increasing paternal age was significantly associated with the susceptibility to ASD, even after adjusting for confounding factors, such as maternal age (Reichenberg et al. 2006). De novo germline pathogenic variants or changes in genomic imprinting might have a role, or at least in part, in the observed association. Nevertheless, further research is needed to validate these findings.

1.12 Viral Infection

New research suggests a link between viral infections and neurodegenerative and neurobehavioral disorders such as autism. Infection with a virus at key stages of early in utero neurodevelopment might increase the chance of autism in the child. Clinical and epidemiological research has revealed associations between congenital

CMV infection and autistic symptoms (Stubbs 1978; Stubbs et al. 1984; Markowitz 1983). An antibody response to the virus, positive viral culture from the urine, decreased hearing, and retinal inflammation revealed evidence of congenital CMV infection. During pregnancy or at birth, CMV infection is a significant cause of sensorineural hearing loss (Grosse et al. 2008; Fowler et al. 1999) and other neurological impairments (Dollard et al. 2007; Townsend et al. 2013). Earlier case studies found that children with congenital CMV infection had typical autistic traits such as inability to build strong interpersonal relationships, poor eye contact, delayed use of language, and nonthematic usage of items.

In 1990, there were numerous reports of severely impaired children with autism who also had congenital CMV infection (Ivarsson et al. 1990). Yamashita and colleagues (Sarabipour and Hristova 2016) demonstrated positive serum CMV-specific IgM antibodies and CMV-DNA in the urine of children with typical autistic disorders. The brain magnetic resonance imaging (MRI) findings indicated an unusually intense region in the periventricular white matter, indicating disrupted myelination. Seten et al. (Seten et al. 2004) discussed the potential function of congenital and perinatal CMV infection in inducing an altered immune response or autoimmune process. The infection or the resulting immune response might impair the development of brain regions or structures, leading to autism. Engman et al. (Engman et al. 2015) also examined the frequency of congenital CMV infection in a representative sample of children with ASD. Congenital CMV infection was found in 1 of the 33 infants with autistic disorder and intellectual deficits, corresponding to a 0.2% prevalence in the general Swedish newborn population. Sakamoto et al. (Sakamoto et al. 2015) hypothesized that congenital CMV infection was involved in a subset of children with ASD since the rate of CMV infection was greater than the incidence of congenital CMV infection in Nagasaki, Japan.

There has been growing evidence describing a connection between rubella infection in early pregnancy and ASD (Chess 1971). Recent studies have demonstrated that in utero viral immune activation results in persistent hyper- and hypomethylated CpGs at WNT signaling genomic regions (WNT3, WNT7B, WNT8A) (Richetto et al. 2017), leading to the disruption of the transcription of downstream target genes, which is thought to have a role in the development of ASD. Other viral infections that have been implicated in ASD include Zika (Nielsen-Saines et al. 2019; Abtibol-Bernardino et al. 2020), cytomegalovirus, and seasonal and pandemic influenza infections (Atladóttir et al. 2012; Deykin and Macmahon 1979).

1.13 Valproic Acid (VPA)

VPA is a medication used to treat EP and bipolar disorder. VPA usage during early pregnancy has been associated with ASD and autistic features in children (Moore et al. 2000). Prenatal VPA exposure in rats has been reported to be related to autism-like phenotypes (Zhang et al. 2012). Prenatally VPA treatment in rats has been shown to cause an imbalance in oxidative homeostasis, which increases vulnerability to autism (Zhang et al. 2012; Qin and Dai 2015). In addition, VPA-administered rats demonstrated reduced social contact, among other autism-like behavioral features. Sulindac, a small-molecule inhibitor of the WNT/β-catenin signaling pathway (Zhang et al. 2012; Qin and Dai 2015), has been reported to be capable of reversal of the VPA-induced autistic-like behaviors. It also caused a decrease in p-GSK3β expression and an increase in β-catenin expression in the hippocampus, prefrontal lobe, and cerebellum (Qin and Dai 2015).

In addition, GATA-3 is a transcription factor that is required for brain development (Tsarovina et al. 2010). It is implicated in the WNT (Notani et al. 2010) and TGF-β/BMP (Kim et al. 2018; Forsman et al. 2013) signaling pathways. It has been shown to be associated with ASD, after exposure to valproate, thalidomide, and alcohol, which might contribute to the development of ASD (Rout and Clausen 2009).

1.14 Precision Medicine Approaches in ASD

Various pathophysiological processes and etiologies characterize ASD and therefore precision medicine seems to be the most promising way of finding treatment options for ASD (Uddin et al. 2019; Al-Dewik and Alsharshani 2020; Styles et al. 2020). The field of precision medicine attempts to mix groundbreaking pathophysiologically based treatments (biomarker stratification) with testing to determine which therapeutic course of action might assist a particular person (Loth et al. 2016). Understanding the molecular and cellular pathways associated with ASD might enable the development of more effective treatment options. The current approach has more significant information about the molecular and cellular processes involved in ASD (Loth et al. 2016). However, following recent therapeutic research failures involving monogenic ASD variants, new challenges have emerged which include conceptual and methodological problems (i.e., the difficulty of translating from animal models to humans and then conducting suitable clinical trials). Other issues are the impact of placebo effects, clinical trial design, and the need to identify mechanistically plausible, measurable outcomes (Loth et al. 2016). Additionally, as many as 70% of individuals with ASD are likely to have a psychiatric or physical co-condition (Simonoff et al. 2008), which might complicate treatment, which could cause an impairment that will last far into adulthood (Shaltout et al. 2020). A transdisciplinary, multidisciplinary, and collaborative approach would be required to deal with the many issues that future research projects must cover.

1.15 Future Perspectives and Conclusions

Research and knowledge of ASD are growing, thanks to many efforts that have been rigorous and comprehensive. Nonetheless, a multi-domain expert collaboration is required to have a cohesive picture of ASD and adequately evaluate its increasing genetic, epidemiological, and environmental components. During embryogenesis, dysregulation of various signaling neurodevelopmental pathways such as WNT, BMP/TGF, SHH, FGF, or RA seems to induce ASD and alter neurogenesis. The signaling pathways might function upstream or downstream of ASD causative genes. Several studies have revealed that the pathways mentioned above play important roles in developing targeted therapies for ASD. However, comprehensive developmental investigations are needed to determine the time window during which disruption of these signals has the most significant impact on brain structure and function, and the subsequent impairment in behavior.

Furthermore, such research might aid in the understanding of the processes involved in ASD etiology, and the upstream and downstream signaling pathways involved. The interaction between different signaling pathways in neuronal and glial cells for ASD should be explored, as this might aid in devising therapy and addressing the disturbed signaling in a cell-specific manner. Only a few ASD risk genes have been investigated in the context of disrupted signaling pathways so far. The exploration of functions of additional ASD genes in neurodevelopment and the control of other signaling pathways might improve the existing knowledge of the processes underlying ASD etiology. Early identification is necessary for a successful therapy that treats the symptoms while reversing neurons to normal circumstances. In addition to the clinical value of a genetic diagnosis, early diagnosis has also been proven to enhance the understanding, give a feeling of empowerment, and improve the quality of life of the affected child and the parents. As a result, additional investigations are needed to comprehend the mechanistic picture of signaling pathways, communication in ASD, and causative gene interaction with these pathways.

Acknowledgement The authors want to thank their respective institutions for their continued support.

Declarations Ethics Approval and Consent to Participate: Not applicable.
Consent for Publication: Not applicable.
Availability of Data and Materials: Not applicable.
Competing/Conflict of Interest: The authors declare no conflicts of/or competing interests.
Funding: This research received no specific grant from any funding agency in the public, commercial, or not-for-profit sectors.

References

Abrahams BS, Geschwind DH (2008) Advances in autism genetics: on the threshold of a new neurobiology. Nat Rev Genet 9:341

Abtibol-Bernardino MR, de Almeida Peixoto LdFA, de Oliveira GA, de Almeida TF, Rodrigues GRI, Otani RH, Soares Chaves BC, de Souza Rodrigues C, de Andrade ABCA, de Fatima Redivo E et al (2020) Neurological findings in children without congenital microcephaly exposed to Zika virus in utero: a case series study. Viruses 12:1335. https://doi.org/10.3390/v12111335

Abu-Khalil A, Fu L, Grove EA, Zecevic N, Geschwind DH (2004) Wnt genes define distinct boundaries in the developing human brain: implications for human forebrain patterning. J Comp Neurol 474:276–288

Akiyama T, Kawasaki Y (2006) Wnt signaling and the actin cytoskeleton. Oncogene 25:7538–7544. https://doi.org/10.1038/sj.onc.1210063

Al-Ayadhi LY (2012) Relationship between sonic hedgehog protein, brain-derived neurotrophic factor and oxidative stress in autism spectrum disorders. Neurochem Res 37:394–400

Al-Dewik N, Alsharshani M (2020) New horizons for molecular genetics diagnostic and research in autism spectrum disorder. In: Essa M, Qoronfleh M (eds) Personalized food intervention and therapy for autism spectrum disorder management. Advances in neurobiology, vol 24. Springer, Cham. https://doi.org/10.1007/978-3-030-30402-7_2

Al-Dewik N et al (2020) Overview and introduction to autism spectrum disorder (ASD). In: Essa M, Qoronfleh M (eds) Personalized food intervention and therapy for autism spectrum disorder management. Advances in neurobiology, vol 24. Springer, Cham. https://doi.org/10.1007/978-3-030-30402-7_1

Allen J, DeMeyer MK, Norton JA, Pontius W, Yang E (1971) Intellectuality in parents of psychotic, subnormal, and normal children. J Autism Child Schizophr 1(3):1311–1326

Álvarez-Buylla A, Ihrie RA (2014) Sonic hedgehog signaling in the postnatal brain. Semin Cell Dev Biol 33:105–111

Amaral DG, Schumann CM, Nordahl CW (2008) Neuroanatomy of autism. Trends Neurosci 31: 137–145

Ansari MA, Attia SM, Nadeem A, Bakheet SA, Raish M, Khan TH et al (2017) Activation of adenosine A2A receptor signaling regulates the expression of cytokines associated with immunologic dysfunction in BTBR T(+) Itpr3(tf)/J mice. Mol Cell Neurosci 82:76–87

Arimura S, Matsunaga A, Kitamura T, Aoki K, Aoki M, Taketo MM (2009) Reduced level of smoothened suppresses intestinal tumorigenesis by down-regulation of Wnt signaling. Gastroenterology 137:629–638

Atladóttir H, Henriksen TB, Schendel DE, Parner ET (2012) Autism after infection, febrile episodes, and antibiotic use during pregnancy: an exploratory study. Pediatrics 130:e1447–e1454. https://doi.org/10.1542/peds.2012-1107

Avraham Y, Berry EM, Donskoy M, Ahmad WA, Vorobiev L, Albeck A et al (2019) Beta-carotene as a novel therapy for the treatment of "autistic like behavior" in animal models of autism. Behav Brain Res 364:469–479

Bae SM, Hong JY (2018) The Wnt signaling pathway and related therapeutic drugs in autism spectrum disorder. Clin Psychopharmacol Neurosci 16(2):129–135

Bandim JM, Ventura LO, Miller MT, Almeida HC, Costa AES (2003) Autism and Möbius sequence: an exploratory study of children in northeastern Brazil. Arq Neuropsiquiatr 61: 181–185

Barber JC, Ellis KH, Bowles LV, Delhanty JD, Ede RF, Male BM et al (1994) Adenomatous polyposis coli and a cytogenetic deletion of chromosome 5 resulting from a maternal intrachromosomal insertion. J Med Genet 31:312–316

Barnard RA, Pomaville MB, O'Roak BJ (2015) Mutations and modelling of the chromatin remodeler CHD8 define an emerging autism etiology. Front Neurosci 9:477. https://doi.org/10.3389/fnins.2015.00477

Baum AE et al (2008) A genome-wide association study implicates diacylglycerol kinase eta (DGKH) and several other genes in the etiology of bipolar disorder. Mol Psychiatry 13:197–207

Bengoa-Vergniory N, Kypta RM (2015) Canonical and noncanonical Wnt signaling in neural stem/progenitor cells. Cell Mol Life Sci 72:4157–4172

Berg JM, Geschwind DH (2012) Autism genetics: searching for specificity and convergence. Genome Biol 13:247–247. https://doi.org/10.1186/gb-2012-13-7-247

Bernier R, Golzio C, Xiong B, Stessman HA, Coe BP, Penn O, Witherspoon K, Gerdts J, Baker C, Vulto-van Silfhout AT et al (2014) Disruptive CHD8 mutations define a subtype of autism early in development. Cell 158(2):263–276. https://doi.org/10.1016/j.cell.2014.06.017

Bi C, Wu J, Jiang T, Liu Q, Cai W, Yu P et al (2012) Mutations of ANK3 identified by exome sequencing are associated with autism susceptibility. Hum Mutat 33:1635–1638
Bielen H, Houart C (2014) The Wnt cries many: Wnt regulation of neurogenesis through tissue patterning, proliferation, and asymmetric cell division. Dev Neurobiol 74:772–780
Blassberg R, Macrae JI, Briscoe J, Jacob J (2016) Reduced cholesterol levels impair Smoothened activation in Smith-Lemli-Opitz syndrome. Hum Mol Genet 25(4):693–705
Bond AM, Bhalala OG, Kessler JA (2012) The dynamic role of bone morphogenetic proteins in neural stem cell fate and maturation. Dev Neurobiol 72:1068–1084
Bosoi CM, Capra V, Allache R, Trinh VQ, De Marco P, Merello E, Drapeau P, Bassuk AG, Kibar Z (2011) Identification and characterization of novel rare mutations in the planar cell polarity gene PRICKLE1 in human neural tube defects. Hum Mutat 32:1371–1375
Boyle SC, Kim M, Valerius MT, McMahon AP, Kopan R (2011) Notch pathway activation can replace the requirement for Wnt4 and Wnt9b in mesenchymal-to-epithelial transition of nephron stem cells. Development 138:4245–4254. https://doi.org/10.1242/dev.070433
Brune CW, Korvatska E, Allen-Brady K, Cook EH, Dawson G, Devlin B et al (2008) Heterogeneous association between engrailed-2 and autism in the CPEA network. Am J Med Genet B Neuropsychiatr Genet 147B:187–193
Buckley AW, Holmes GL (2016) Epilepsy and autism. Cold Spring Harb Perspect Med 6:a022749
Burden SJ (2000) Wnts as retrograde signals for axon and growth cone differentiation. Cell 100: 495–497
Chatterjee S, Sil PC (2019) Targeting the cross talks of Wnt pathway with Hedgehog and Notch for cancer therapy. Pharmacol Res 142:251–261. https://doi.org/10.1016/j.phrs.2019.02.027
Chen Y, Huang WC, Sejourne J, Clipperton-Allen AE, Page DT (2015) Pten mutations alter brain growth trajectory and allocation of cell types through elevated beta-catenin signaling. J Neurosci 35(28):10252–10267. https://doi.org/10.1523/JNEUROSCI.5272-14.2015
Chen R, Davis LK, Guter S, Wei Q, Jacob S, Potter MH et al (2017) Leveraging blood serotonin as an endophenotype to identify de novo and rare variants involved in autism. Mol Autism 8:14
Cheng B, Zhu J, Yang T, Guo M, Lai X, Li Q, Chen J, Li T (2021) Vitamin A deficiency increases the risk of gastrointestinal comorbidity and exacerbates core symptoms in children with autism spectrum disorder. Pediatr Res 89:211–216. https://doi.org/10.1038/s41390-020-0865-y
Chess S (1971) Autism in children with congenital rubella. J Autism Child Schizophr 1:33–47
Choi J, Ababon MR, Soliman M, Lin Y, Brzustowicz LM, Matteson PG et al (2014) Autism associated gene, engrailed2, and flanking gene levels are altered in post-mortem cerebellum. PLoS One 9:e87208
Chow ML, Pramparo T, Winn ME, Barnes CC, Li HR, Weiss L et al (2012) Age-dependent brain gene expression and copy number anomalies in autism suggest distinct pathological processes at young versus mature ages. PLoS Genet 8:e1002592
Choy SW, Cheng SH (2012) Hedgehog signaling. Vitam Horm 88:1–23
Chung Y, Fu E (2013) Crosstalk between Shh and TGF-β signaling in cyclosporine-enhanced cell proliferation in human gingival fibroblasts. PLoS One 8:e70128
Clevers H, Nusse R (2012) Wnt/β-catenin signaling and disease. Cell 149:1192–1205
Clipperton-Allen AE, Page DT (2014) Pten haploinsufficient mice show broad brain overgrowth but selective impairments in autism-relevant behavioral tests. Hum Mol Genet 23(13): 3490–3505. https://doi.org/10.1093/hmg/ddu057
Clipperton-Allen AE, Page DT (2015) Decreased aggression and increased repetitive behavior in Pten haploinsufficient mice. Genes Brain Behav 14(2):145–157. https://doi.org/10.1111/gbb.12192
Cotney J, Muhle RA, Sanders SJ, Liu L, Willsey AJ, Niu W, Liu W, Klei L, Lei J, Yin J et al (2015) The autism-associated chromatin modifier CHD8 regulates other autism risk genes during human neurodevelopment. Nat Commun 6:6404. https://doi.org/10.1038/ncomms7404
Das BC, Thapa P, Karki R, Das S, Mahapatra S, Liu TC et al (2014) Retinoic acid signaling pathways in development and diseases. Bioorg Med Chem 22:673–683

De Rubeis S, He X, Goldberg AP, Poultney CS, Samocha K, Cicek AE et al (2014) Synaptic, transcriptional and chromatin genes disrupted in autism. Nature 515:209–215
Deykin EY, Macmahon B (1979) Viral exposure and autism. Am J Epidemiol 109:628–638. https://doi.org/10.1093/oxfordjournals.aje.a112726
Dias C, Pfundt R, Kleefstra T, Shuurs-Hoeijmakers J, Boon EMJ, van Hagen JM et al (2021) De novo variants in TCF7L2 are associated with a syndromic neurodevelopmental disorder. Am J Med Genet A 185(8):2384–2390
Dollard SC, Grosse SD, Ross DS (2007) New estimates of the prevalence of neurological and sensory sequelae and mortality associated with congenital cytomegalovirus infection. Rev Med Virol 17:355–363. https://doi.org/10.1002/rmv.544
Dong F, Jiang J, McSweeney C, Zou D, Liu L, Mao Y (2016) Deletion of CTNNB1 in inhibitory circuitry contributes to autism-associated behavioral defects. Hum Mol Genet 25:2738–2751
Dubruc E, Putoux A, Labalme A, Rougeot C, Sanlaville D, Edery P (2014) A new intellectual disability syndrome caused by CTNNB1 haploinsufficiency. Am J Med Genet A 164A:1571–1575
Duester G (2008) Retinoic acid synthesis and signaling during early organogenesis. Cell 134:921–931
Dunaeva M, Michelson P, Kogerman P, Toftgard R (2003) Characterization of the physical interaction of Gli proteins with SUFU proteins. J Biol Chem 278:5116–5122
Durak O, de Anda FC, Singh KK, Leussis MP, Petryshen TL, Sklar P et al (2015) Ankyrin-G regulates neurogenesis and Wnt signaling by altering the subcellular localization of β-catenin. Mol Psychiatry 20:388–397
Durak O, Gao F, Kaeser-Woo YJ, Rueda R, Martorell AJ, Nott A, Liu CY, Watson LA, Tsai LH (2016) Chd8 mediates cortical neurogenesis via transcriptional regulation of cell cycle and Wnt signaling. Nat Neurosci 19(11):1477–1488
Engman ML, Sundin M, Miniscalco C, Westerlund J, Lensohn-Fuchs I, Gillberg C, Fernell E (2015) Prenatal acquired cytomegalovirus infection should be considered in children with autism. Acta Paediatr 104:792–795. https://doi.org/10.1111/apa.13032
Fan Q, He M, Sheng T, Zhang X, Sinha M, Luxon B et al (2010) Requirement of TGFbeta signaling for SMO-mediated carcinogenesis. J Biol Chem 285:36570–36576
Faravelli I, Bucchia M, Rinchetti P, Nizzardo M, Simone C, Frattini E, Corti S (2014) Motor neuron derivation from human embryonic and induced pluripotent stem cells: experimental approaches and clinical perspectives. Stem Cell Res Ther 5:1–13. https://doi.org/10.1186/scrt476
Fazel Darbandi S, Robinson Schwartz SE, Qi Q, Catta-Preta R, Pai EL, Mandell JD et al (2018) Neonatal Tbr1 dosage controls cortical layer 6 connectivity. Neuron 100:831–845.e7
Ferreira MA et al (2008) Collaborative genome-wide association analysis supports a role for ANK3 and CACNA1C in bipolar disorder. Nat Genet 40:1056–1058
Flavell SW, Kim TK, Gray JM, Harmin DA, Hemberg M, Hong EJ, Markenscoff-Papadimitriou E, Bear DM, Greenberg ME (2008) Genome-wide analysis of MEF2 transcriptional program reveals synaptic target genes and neuronal activity-dependent polyadenylation site selection. Neuron 60:1022–1038. https://doi.org/10.1016/j.neuron.2008.11.029
Forsman CL, Ng BC, Heinze RK, Kuo C, Sergi C, Gopalakrishnan R et al (2013) BMP-binding protein twisted gastrulation is required in mammary gland epithelium for normal ductal elongation and myoepithelial compartmentalization. Dev Biol 373:95–106
Fowler KB, Dahle AJ, Boppana SB, Pass RF (1999) Newborn hearing screening: will children with hearing loss caused by congenital cytomegalovirus infection be missed? J Pediatr 135:60–64. https://doi.org/10.1016/S0022-3476(99)70328-8
Frazier TW, Embacher R, Tilot AK, Koenig K, Mester J, Eng C (2015) Molecular and phenotypic abnormalities in individuals with germline heterozygous PTEN mutations and autism. Mol Psychiatry 20(9):1132–1138. https://doi.org/10.1038/mp.2014.125
Fregeau B, Kim BJ, Hernández-García A, Jordan VK, Cho MT, Schnur RE et al (2016) De novo mutations of RERE cause a genetic syndrome with features that overlap those associated with proximal 1p36 deletions. Am J Hum Genet 98:963–970

Fujimura L, Watanabe-Takano H, Sato Y, Tokuhisa T, Hatano M (2009) Prickle promotes neurite outgrowth via the Dishevelled dependent pathway in C1300 cells. Neurosci Lett 467:6–10
Geschwind DH (2008) Autism: many genes, common pathways? Cell 135:391
Ghanizadeh A (2012) Malondialdehyde, Bcl-2, superoxide dismutase and glutathione peroxidase may mediate the association of sonic hedgehog protein and oxidative stress in autism. Neurochem Res 37:899–901
Gharani N, Benayed R, Mancuso V, Brzustowicz LM, Millonig JH (2004) Association of the homeobox transcription factor, ENGRAILED 2, 3, with autism spectrum disorder. Mol Psychiatry 9:474–484. https://doi.org/10.1038/sj.mp.4001498
Ghyselinck NB, Duester G (2019) Retinoic acid signaling pathways. Development 146:dev167502. https://doi.org/10.1242/dev.167502
Glinka Y, Stoilova S, Mohammed N, Prud'homme GJ (2011) Neuropilin-1 exerts co-receptor function for TGF-beta-1 on the membrane of cancer cells and enhances responses to both latent and active TGF-beta. Carcinogenesis 32:613–621
Golden CE, Buxbaum JD, De Rubeis S (2017) Disrupted circuits in mouse models of autism spectrum disorder and intellectual disability. Curr Opin Neurobiol 48:106–112
Grainger S, Willert K (2018) Mechanisms of Wnt signaling and control. Wiley Interdiscip Rev Syst Biol Med e1422
Grant SFA, Thorleifsson G, Reynisdottir I, Benediktsson R, Manolescu A, Sainz J et al (2006) Variant of transcription factor 7-like 2 (TCF7L2) gene confers risk of type 2 diabetes. Nat Genet 38:320–323
Grigoryan T, Wend P, Klaus A, Birchmeier W (2008) Deciphering the function of canonical Wnt signals in development and disease: conditional loss- and gain-of-function mutations of beta-catenin in mice. Genes Dev 22:2308–2341
Grosse SD, Ross DS, Dollard SC (2008) Congenital cytomegalovirus (CMV) infection as a cause of permanent bilateral hearing loss: a quantitative assessment. J Clin Virol 41:57–62. https://doi.org/10.1016/j.jcv.2007.09.004
Halepoto DM, Bashir S, Zeina R, Al-Ayadhi LY (2015) Correlation between hedgehog (Hh) protein family and brain-derived neurotrophic factor (BDNF) in autism spectrum disorder (ASD). J Coll Physicians Surg Pak 25(12):882–885
Hamilton SP, Woo JM, Carlson EJ, Ghanem N, Ekker M, Rubenstein JLR (2005) Analysis of four DLX homeobox genes in autistic probands. BMC Genet 6:52
Hart M, Concordet JP, Lassot I, Albert I, del los Santos R, Durand H et al (1999) The F-box protein beta-TrCP associates with phosphorylated beta-catenin and regulates its activity in the cell. Curr Biol 9:207–210
Hayward P, Kalmar T, Arias AM (2008) Wnt/Notch signaling and information processing during development. Development 135:411–424. https://doi.org/10.1242/dev.000505
Hedstrom KL, Ogawa Y, Rasband MN (2008) AnkyrinG is required for maintenance of the axon initial segment and neuronal polarity. J Cell Biol 183:635–640
Hida Y, Fukaya M, Hagiwara A, Deguchi-Tawarada M, Yoshioka T, Kitajima I, Inoue E, Watanabe M, Ohtsuka T (2011) Prickle2 is localized in the postsynaptic density and interacts with PSD-95 and NMDA receptors in the brain. J Biochem 149:693–700
Hiester BG, Galati DF, Salinas PC, Jones KR (2013) Neurotrophin and Wnt signaling cooperatively regulate dendritic spine formation. Mol Cell Neurosci 56:115–127
Hillman RT, Feng BY, Ni J, Woo WM, Milenkovic L, Hayden Gephart MG et al (2011) Neuropilins are positive regulators of hedgehog signal transduction. Genes Dev 25:2333–2346
Hochman E, Castiel A, Jacob-Hirsch J, Amariglio N, Izraeli S (2006) Molecular pathways regulating pro-migratory effects of hedgehog signaling. J Biol Chem 281:33860–33870
Hormozdiari F, Penn O, Borenstein E, Eichler EE (2015) The discovery of integrated gene networks for autism and related disorders. Genome Res 25:142–154
Huang JY, Lu HC (2017) mGluR5 tunes NGF/TrkA signaling to orient spiny stellate neuron dendrites toward thalamocortical axons during whisker-barrel map formation. Cereb Cortex 28(6):1991–2006

Inestrosa NC, Varela-Nallar L (2015) Wnt signaling in neuronal differentiation and development. Cell Tissue Res 359:215–223

Iossifov I, O'Roak BJ, Sanders SJ, Ronemus M, Krumm N, Levy D et al (2014) The contribution of de novo coding mutations to autism spectrum disorder. Nature 515:216–221

Iqbal Z, Vandeyer G, van der Voet M, Waryah AM, Zahoor MY, Besseling JA et al (2013) Homozygous and heterozygous disruptions of ANK3: at the crossroads of neurodevelopmental and psychiatric disorders. Hum Mol Genet 22:1960–1970

Itasaki N, Hoppler S (2010) Crosstalk between Wnt and bone morphogenic protein signaling: a turbulent relationship. Dev Dyn 239:16–33

Ivarsson S, Bjerre I, Vegfors P, Ahlfors K (1990) Autism as one of several disabilities in two children with congenital cytomegalovirus infection. Neuropediatrics 21:102–103. https://doi.org/10.1055/s-2008-1071471

Iwata T, Hevner RF (2009) Fibroblast growth factor signaling in development of the cerebral cortex. Develop Growth Differ 51:299–323

Jamain S, Quach H, Betancur C, Råstam M, Colineaux C, Gillberg IC et al (2003) Mutations of the X-linked genes encoding neuroligins NLGN3 and NLGN4 are associated with autism. Nat Genet 34:27–29

Javelaud D, Pierrat MJ, Mauviel A (2012) Crosstalk beten TGF-β and hedgehog signaling in cancer. FEBS Lett 586:2016–2025

Jin YR, Han XH, Taketo MM, Yoon JK (2012) Wnt9b-dependent FGF signaling is crucial for outgrowth of the nasal and maxillary processes during upper jaw and lip development. Development 139:1821–1830

Juriloff DM, Harris MJ, McMahon AP, Carroll TJ, Lidral AC (2006) Wnt9b is the mutated gene involved in multifactorial nonsyndromic cleft lip with or without cleft palate in A/WySn mice, as confirmed by a genetic complementation test. Birth Defects Res A Clin Mol Teratol 76:574–579

Kalkman HO (2012) A review of the evidence for the canonical Wnt pathway in autism spectrum disorders. Mol Autism 3:10. https://doi.org/10.1186/2040-2392-3-10

Kashima R, Roy S, Ascano M, Martinez-Cerdeno V, Ariza-Torres J, Kim S et al (2016) Augmented noncanonical BMP type II receptor signaling mediates the synaptic abnormality of fragile X syndrome. Sci Signal 9:ra58

Katoh Y, Katoh M (2009) Hedgehog target genes: mechanisms of carcinogenesis induced by aberrant hedgehog signaling activation. Curr Mol Med 9:873–886. https://doi.org/10.2174/156652409789105570

Khatri N, Man H-Y (2019) The autism and Angelman syndrome protein Ube3A/E6AP: the gene, E3 ligase ubiquitination targets and neurobiological functions. Front Mol Neurosci 12:109. https://doi.org/10.3389/fnmol.2019.00109

Kim S, Kim T, Lee HR, Jang EH, Ryu HH, Kang M et al (2016) Impaired learning and memory in CD38 null mutant mice. Mol Brain 9:16

Kim HW, Hong R, Choi EY, Yu K, Kim N, Hyeon JY et al (2018) A probiotic mixture regulates T cell balance and reduces atopic dermatitis symptoms in mice. Front Microbiol 9:2414

Kirschen GW, Xiong Q (2017) Primary cilia as a novel horizon between neuron and environment. Neural Regen Res 12:1225–1230

Kivimäe S, Martin PM, Kapfhamer D, Ruan Y, Heberlein U, Rubenstein JLR et al (2011) Abnormal behavior in mice mutant for the Disc1 binding partner, Dixdc1. Transl Psychiatry 1:e43

Kizhatil K, Davis JQ, Davis L, Hoffman J, Hogan BLM, Bennett V (2007) Ankyrin-G is a molecular partner of E-cadherin in epithelial cells and early embryos. J Biol Chem 282:26552–26561

Klaus A, Birchmeier W (2008) Wnt signaling and its impact on development and cancer. Nat Rev Cancer 8:387–398. https://doi.org/10.1038/nrc2389

Knight D, Xie W, Boulianne GL (2011) Neurexins and neuroligins: recent insights from invertebrates. Mol Neurobiol 44:426–440. https://doi.org/10.1007/s12035-011-8213-1

Komiya Y, Habas R (2008) Wnt signal transduction pathways. Organogenesis 4:68–75. https://doi.org/10.4161/org.4.2.5851

Kong A, Frigge ML, Masson G, Besenbacher S, Sulem P, Magnusson G et al (2012) Rate of de novo mutations and the importance of father's age to disease risk. Nature 488:471–475. https://doi.org/10.1038/nature11396

Koopmans T, Eilers R, Menzen M, Halayko A, Gosens R (2017) β-catenin directs nuclear factor-κB p65 output via CREB-binding protein/p300 in human airway smooth muscle. Front Immunol 8: 1086. https://doi.org/10.3389/fimmu.2017.01086

Kordeli E, Lambert S, Bennett V (1995) AnkyrinG. A new ankyrin gene with neural-specific isoforms localized at the axonal initial segment and node of Ranvier. J Biol Chem 270:2352–2359

Kosmicki JA et al (2017) Refining the role of de novo protein-truncating variants in neurodevelopmental disorders by using population reference samples. Nat Genet 49:504–510

Krumm N, O'Roak BJ, Shendure J, Eichler EE (2014) A de novo convergence of autism genetics and molecular neuroscience. Trends Neurosci 37(2):95–105. https://doi.org/10.1016/j.tins.2013.11.005

Krumm N, Turner TN, Baker C, Vives L, Mohajeri K, Witherspoon K, Raja A, Coe BP, Stessman HA, He ZX et al (2015) Excess of rare, inherited truncating mutations in autism. Nat Genet 47(6):582–588. https://doi.org/10.1038/ng.3303

Kucinska M et al (2019) Differential regulation of fibroblast growth factor receptor 1 trafficking and function by extracellular galectins. Cell Commun Signal 17:65

Kuechler A, Willemsen MH, Albrecht B, Bacino CA, Bartholomew DW, van Bokhoven H et al (2015) De novo mutations in beta-catenin (CTNNB1) appear to be a frequent cause of intellectual disability: expanding the mutational and clinical spectrum. Hum Genet 134:97–109

Kwan V, Unda BK, Singh KK (2016a) Wnt signaling networks in autism spectrum disorder and intellectual disability. J Neurodev Disord 8:45. https://doi.org/10.1186/s11689-016-9176-3

Kwan V, Meka DP, White SH, Hung CL, Holzapfel NT, Walker S et al (2016b) DIXDC1 phosphorylation and control of dendritic morphology are impaired by rare genetic variants. Cell Rep 17:1892–1904

Kwon CH, Luikart BW, Powell CM, Zhou J, Matheny SA, Zhang W, Li Y, Baker SJ, Parada LF (2006) Pten regulates neuronal arborization and social interaction in mice. Neuron 50(3): 377–388. https://doi.org/10.1016/j.neuron.2006.03.023

Lai X, Wu X, Hou N, Liu S, Li Q, Yang T, Miao J, Dong Z, Chen J, Li T (2018) Vitamin A deficiency induces autistic-like behaviors in rats by regulating the RARβ-CD38-oxytocin axis in the hypothalamus. Mol Nutr Food Res 62:1700754. https://doi.org/10.1002/mnfr.201700754

Landrigan PJ (2010) What causes autism? Exploring the environmental contribution. Curr Opin Pediatr 22:219–225

Lee M, Yoon J, Song H, Lee B, Lam DT, Yoon J et al (2017) Tcf7l2 plays crucial roles in forebrain development through regulation of thalamic and habenular neuron identity and connectivity. Dev Biol 424:62–76

Levy D, Ronemus M, Yamrom B, Lee Y, Leotta A, Kendall J et al (2011) Rare de novo and transmitted copy-number variation in autistic spectrum disorders. Neuron 70:886–897

Li X, Deng W, Lobo-Ruppert SM, Ruppert JM (2007) Gli1 acts through Snail and E-cadherin to promote nuclear signaling by beta-catenin. Oncogene 26:4489–4498. https://doi.org/10.1038/sj.onc.1210241

Li W, Yao A, Zhi H, Kaur K, Zhu YC, Jia M et al (2016) Angelman syndrome protein Ube3a regulates synaptic growth and endocytosis by inhibiting BMP signaling in Drosophila. PLoS Genet 12:e1006062

Lie DC, Colamarino SA, Song HJ, Désiré L, Mira H, Consiglio A et al (2005) Wnt signaling regulates adult hippocampal neurogenesis. Nature 437:1370–1375

Lin PI, Chien YL, Wu YY, Chen CH, Gau SS, Huang YS et al (2012) The WNT2 gene polymorphism associated with speech delay inherent to autism. Res Dev Disabil 33:1533–1540

Liu P, Wakamiya M, Shea MJ, Albrecht U, Behringer RR, Bradley A (1999) Requirement for Wnt3 in vertebrate axis formation. Nat Genet 22:361–365

Liu X, Novosedlik N, Wang A, Hudson ML, Cohen IL, Chudley AE et al (2009) The DLX1and DLX2 genes and susceptibility to autism spectrum disorders. Eur J Hum Genet 17:228–235

Lory JW, Rosen V (2018) Bone morphogenetic protein-based therapeutic approaches. Cold Spring Harb Perspect Biol 10:a022327

Loth E, Murphy DG, Spooren W (2016) Defining precision medicine approaches to autism spectrum disorders: concepts and challenges. Front Psychiatry 7:188

MacDonald BT, He X (2012) Frizzled and LRP5/6 receptors for Wnt/β-catenin signaling. Cold Spring Harb Perspect Biol 4:a007880

MacDonald BT, Tamai K, He X (2009) Wnt/beta-catenin signaling: components, mechanisms, and diseases. Dev Cell 17:9–26

Markowitz PI (1983) Autism in a child with congenital cytomegalovirus infection. J Autism Dev Disord 13:249–253. https://doi.org/10.1007/BF01531564

Martin PM, Yang X, Robin N, Lam E, Rabinowitz JS, Erdman CA et al (2013) A rare WNT1 missense variant overrepresented in ASD leads to increased WNT signal pathway activation. Transl Psychiatry 3:e301

Martin PM, Stanley RE, Ross AP, Freitas AE, Moyer CE, Brumback AC, Iafrati J, Stapornwongkul KS, Dominguez S, Kivimäe S, Mulligan KA, Pirooznia M, McCombie WR, Potash JB, Zandi PP, Purcell SM, Sanders SJ, Zuo Y, Sohal VS, Cheyette B (2018) DIXDC1 contributes to psychiatric susceptibility by regulating dendritic spine and glutamatergic synapse density via GSK3 and Wnt/β-catenin signaling. Mol Psychiatry 23(2):467–475. https://doi.org/10.1038/mp.2016.184

Marui T, Funatogawa I, Koishi S, Yamamoto K, Matsumoto H, Hashimoto O et al (2010) Association between autism and variants in the wingless-type MMTV integration site family member 2 (WNT2) gene. Int J Neuropsychopharmacol 13:443–449

Matoba N, Liang D, Sun H, Aygün N, McAfee JC, Davis JE et al (2020) Common genetic risk variants identified in the SPARK cohort support DDHD2 as a candidate risk gene for autism. Transl Psychiatry 10(1):265

McBride KL, Varga EA, Pastore MT, Prior TW, Manickam K, Atkin JF, Herman GE (2010) Confirmation study of PTEN mutations among individuals with autism or developmental delays/mental retardation and macrocephaly. Autism Res 3(3):137–141. https://doi.org/10.1002/aur.132

McCarthy SE, Gillis J, Kramer M, Lihm J, Yoon S, Berstein Y, Mistry M, Pavlidis P, Solomon R, Ghiban E et al (2014) De novo mutations in schizophrenia implicate chromatin remodeling and support a genetic overlap with autism and intellectual disability. Mol Psychiatry 19(6):652–658. https://doi.org/10.1038/mp.2014.29

McMahon AP, Bradley A (1990) The Wnt-1 (int-1) proto-oncogene is required for the development of a large region of the mouse brain. Cell 62:1073–1085

McMahon AP, Joyner AL, Bradley A, McMahon JA (1992) The midbrain-hindbrain phenotype of Wnt-1-/Wnt-1-mice results from stepwise deletion of engrailed-expressing cells by 9.5 days postcoitum. Cell 69:581–595

Medina MA, Andrade VM, Caracci MO, Avila ME, Verdugo DA, Vargas MF et al (2018) Wnt/β-catenin signaling stimulates the expression and synaptic clustering of the autism-associated Neuroligin 3 gene. Transl Psychiatry 8:45

Meng X, Poon R, Zhang X, Cheah A, Ding Q, Hui CC et al (2001) Suppressor of fused negatively regulates beta-catenin signaling. J Biol Chem 276:40113–40119

Merner N, Forgeot d'Arc B, Bell SC, Maussion G, Peng H, Gauthier J, Crapper L, Hamdan FF, Michaud JL, Mottron L et al (2016) A de novo frameshift mutation in chromodomain helicase DNA-binding domain 8 (CHD8): a case report and literature review. Am J Med Genet A 170(5): 1225–1235. https://doi.org/10.1002/ajmg.a.37566

Mohn JL, Alexander J, Pirone A, Palka CD, Lee SY, Mebane L et al (2014) Adenomatous polyposis coli protein deletion leads to cognitive and autism-like disabilities. Mol Psychiatry 19:1133–1142

Moore SJ, Turnpenny P, Quinn A, Glover S, Lloyd DJ, Montgomery T et al (2000) A clinical study of 57 children with fetal anticonvulsant syndromes. J Med Genet 37:489–497

Moreno-Ramos OA, Olivares AM, Haider NB, de Autismo LC, Lattig MC (2015) Whole-exome sequencing in a South American cohort links ALDH1A3, FOXN1 and retinoic acid regulation pathways to autism spectrum disorders. PLoS One 10:e0135927

Morrow EM, Yoo SY, Flavell SW, Kim TK, Lin Y, Hill RS, Mukaddes NM, Balkhy S, Gascon G, Hashmi A, Al-Saad S, Ware J, Joseph RM et al (2008) Identifying autism loci and genes by tracing recent shared ancestry. Science 321:218–223. https://doi.org/10.1126/science.1157657

Mulligan KA, Cheyette BN (2017) Neurodevelopmental perspectives on Wnt signaling in psychiatry. Mol Neuropsychiatry 2(4):219–246

Nagaoka T, Tabuchi K, Kishi M (2015) PDZ interaction of Vangl2 links PSD-95 and Prickle2 but plays only a limited role in the synaptic localisation of Vangl2. Sci Rep 5:12916

Nakanishi M, Nomura J, Ji X, Tamada K, Arai T, Takahashi E et al (2017) Functional significance of rare neuroligin 1 variants found in autism. PLoS Genet 13(8):e1006940

Nakashima N, Yamagata T, Mori M, Kuwajima M, Suwa K, Momoi MY (2010) Expression analysis and mutation detection of DLX5 and DLX6 in autism. Brain Dev 32(2):98–104

Neale BM, Kou Y, Liu L, Ma'ayan A, Samocha KE, Sabo A, Lin CF, Stevens C, Wang LS, Makarov V et al (2012) Patterns and rates of exonic de novo mutations in autism spectrum disorders. Nature 485(7397):242–245. https://doi.org/10.1038/nature11011

Nguyen A, Rauch TA, Pfeifer GP, Hu VW (2010) Global methylation profiling of lymphoblastoid cell lines reveals epigenetic contributions to autism spectrum disorders and a novel autism candidate gene, RORA, whose protein product is reduced in autistic brain. FASEB J 24:3036–3051. https://doi.org/10.1096/f.10-154484

Niculae AS, Pavăl D (2016) From molecules to behavior: an integrative theory of autism spectrum disorder. Med Hypotheses 97:74–84

Niederreither K, Dollé P (2008) Retinoic acid in development: towards an integrated view. Nat Rev Genet 9:541–553

Nielsen-Saines K, Brasil P, Kerin T, Vasconcelos Z, Gabaglia CR, Damasceno L, Pone M, de Carvalho LMA, Pone SM, Zin AA (2019) Delayed childhood neurodevelopment and neurosensory alterations in the second year of life in a prospective cohort of ZIKV-exposed children. Nat Med 25:1213–1217. https://doi.org/10.1038/s41591-019-0496-1

Nishiyama M, Oshikawa K, Tsukada Y, Nakagawa T, Iemura S, Natsume T et al (2009) CHD8 suppresses p53-mediated apoptosis through histone H1 recruitment during early embryogenesis. Nat Cell Biol 11:172–182

Nishiyama M, Skoultchi AI, Nakayama KI (2012) Histone H1 recruitment by CHD8 is essential for suppression of the Wnt-β-catenin signaling pathway. Mol Cell Biol 32(2):501–512. https://doi.org/10.1128/MCB.06409-11

Notani D, Gottimukkala KP, Jayani RS, Limaye AS, Damle MV, Mehta S et al (2010) Global regulator SATB1 recruits beta-catenin and regulates T(H)2 differentiation in Wnt-dependent manner. PLoS Biol 8:e1000296

Nurmi EL, Bradford Y, Chen Y-h, Hall J, Arnone B, Gardiner MB et al (2001) Linkage disequilibrium at the Angelman syndrome gene UBE3A in autism families. Genomics 77(1):105–113

O'Roak BJ, Vives L, Girirajan S, Karakoc E, Krumm N, Coe BP et al (2012a) Sporadic autism exomes reveal a highly interconnected protein network of de novo mutations. Nature 485:246–250

O'Roak BJ, Vives L, Fu W, Egertson JD, Stanaway IB, Phelps IG, Carvill G, Kumar A, Lee C, Ankenman K et al (2012b) Multiplex targeted sequencing identifies recurrently mutated genes in autism spectrum disorders. Science 338(6114):1619–1622. https://doi.org/10.1126/science.1227764

Ohkawara T, Katsuyama T, Ida-Eto M, Narita N, Narita M (2015) Maternal viral infection during pregnancy impairs the development of fetal serotonergic neurons. Brain Dev 37:88–93. https://doi.org/10.1016/j.braindev.2014.03.007

Okerlund ND, Cheyette BNR (2011) Synaptic Wnt signaling—a contributor to major psychiatric disorders. J Neurodev Disord 3:162–174

Okumura A, Yamamoto T, Miyajima M, Shimojima K, Kondo S, Abe S et al (2014) 3p interstitial deletion including PRICKLE2 in identical twins with autistic features. Pediatr Neurol 51:730–733

Onishi K, Hollis E, Zou Y (2014) Axon guidance and injury-lessons from Wnts and Wnt signaling. Curr Opin Neurobiol 27:232–240

Ornitz DM, Itoh N (2015) The fibroblast growth factor signaling pathway. Wiley Interdiscip Rev Dev Biol 4:215–266

Oron O, Elliott E (2017) Delineating the common biological pathways perturbed by ASD's genetic etiology: lessons from network-based studies. Int J Mol Sci 18:828

Paemka L, Mahajan VB, Skeie JM, Sors LP, Ehaideb SN, Gonzalez-Alegre P, Sasaoka T, Tao H, Miyagi A, Ueno N, Takao K, Miyakawa T, Wu S, Darbro BW, Ferguson PJ, Pieper AA, Britt JK, Wemmie JA, Rudd DS, Wassink T, El-Shanti H, Mefford HC, Carvill GL, Manak JR, Bassuk AG (2013) PRICKLE1 interaction with SYNAPSIN I reveals a role in autism spectrum disorders. PLoS One 8:e80737

Page DT, Kuti OJ, Prestia C, Sur M (2009) Haploinsufficiency for Pten and Serotonin transporter cooperatively influences brain size and social behavior. Proc Natl Acad Sci U S A 106(6): 1989–1994. https://doi.org/10.1073/pnas.0804428106

Patel SS, Tomar S, Sharma D, Mahindroo N, Udayabanu M (2017) Targeting sonic hedgehog signaling in neurological disorders. Neurosci Biobehav Rev 74:76–97

Pavăl D, Rad F, Rusu R, Niculae A, Colosi HA, Dobrescu I et al (2017) Low retinal dehydrogenase 1 (RALDH1) level in prepubertal boys with autism spectrum disorder: a possible link to dopamine dysfunction? Clin Psychopharmacol Neurosci 15(3):229–236

Pelullo M, Zema S, Nardozza F, Checquolo S, Screpanti I, Bellavia D (2019) Wnt, notch, and TGF-β pathways impinge on hedgehog signaling complexity: an open window on cancer. Front Genet 10:711. https://doi.org/10.3389/fgene.2019.00711

Pierrat MJ, Marsaud V, Mauviel A, Javelaud D (2012) Expression of microphthalmia-associated transcription factor (MITF), which is critical for melanoma progression, is inhibited by both transcription factor GLI2 and transforming growth factor-β. J Biol Chem 287:17996–18004

Pilarowski GO, Vernon HJ, Applegate CD, Boukas L, Cho MT, Gurnett CA et al (2018) Missense variants in the chromatin remodeler CHD1 are associated with neurodevelopmental disability. J Med Genet 55(8):561

Pirone A, Alexander J, Lau LA, Hampton D, Zayachkivsky A, Yee A et al (2017) APC conditional knock-out mouse is a model of infantile spasms with elevated neuronal β-catenin levels, neonatal spasms, and chronic seizures. Neurobiol Dis 98:149–157. https://doi.org/10.1016/j.nbd.2016.11.002

Platt RJ, Zhou Y, Slaymaker IM, Shetty AS, Weisbach NR, Kim JA, Sharma J, Desai M, Sood S, Kempton HR et al (2017) Chd8 mutation leads to autistic-like behaviors and impaired striatal circuits. Cell Rep 19:335–350. https://doi.org/10.1016/j.celrep.2017.03.052

Porebska N et al (2018) Targeting cellular trafficking of fibroblast growth factor receptors as a strategy for selective cancer treatment. J Clin Med 8:7

Preitner N, Quan J, Nowakowski DW, Hancock ML, Shi J, Tcherkezian J et al (2014) APC is an RNA-binding protein, and its interactome provides a link to neural development and microtubule assembly. Cell 158:368–382. https://doi.org/10.1016/j.cell.2014.05.042

Psychiatric GWAS Consortium Bipolar Disorder Working Group (2011) Large-scale genome-wide association analysis of bipolar disorder identifies a new susceptibility locus near ODZ4. Nat Genet 43:977–983

Qin L, Dai X (2015) Effect of sulindac on improving autistic behaviours in rats. Nan Fang Yi Ke Da Xue Xue Bao 35:1162–1165

Rasalam AD, Hailey H, Williams JHG, Moore SJ, Turnpenny PD, Lloyd DJ et al (2005) Characteristics of fetal anticonvulsant syndrome associated autistic disorder. Dev Med Child Neurol 47:551–555. https://doi.org/10.1017/s0012162205001076

Reichenberg A, Gross R, Weiser M, Bresnahan M, Silverman J, Harlap S et al (2006) Advancing paternal age and autism. Arch Gen Psychiatry 63(9):1026–1032

Rhinn M, Dollé P (2012) Retinoic acid signaling during development. Development 139:843–858

Richetto J, Massart R, Weber-Stadlbauer U, Szyf M, Riva MA, Meyer U (2017) Genome-wide DNA methylation changes in a mouse model of infection-mediated neurodevelopmental disorders. Biol Psychiatry 81:265–276

Riebold M, Mankuta D, Lerer E, Israel S, Zhong S, Nemanov L et al (2011) All-trans retinoic acid upregulates reduced CD38 transcription in lymphoblastoid cell lines from autism spectrum disorder. Mol Med 17:799–806

Rosso SB, Inestrosa NC (2013) WNT signaling in neuronal maturation and synaptogenesis. Front Cell Neurosci 7:103

Rosti RO, Sadek AA, Vaux KK, Gleeson JG (2014) The genetic landscape of autism spectrum disorders. Dev Med Child Neurol 56:12–18. https://doi.org/10.1111/dmcn.12278

Rout UK, Clausen P (2009) Common increase of GATA-3 level in PC-12 cells by three teratogens causing autism spectrum disorders. Neurosci Res 64:162–169

Rubenstein JLR, Merzenich MM (2003) Model of autism: increased ratio of excitation/inhibition in key neural systems. Genes Brain Behav 2:255–267

Rychahou PG, Kang J, Gulhati P, Doan HQ, Chen LA, Xiao SY et al (2008) Akt2 overexpression plays a critical role in the establishment of colorectal cancer metastasis. Proc Natl Acad Sci U S A 105:20315–20320

Ryu J, Bar-Shalita T, Granovsky Y, Weissman-Fogel I, Torres E (2021) Personalized biometrics of physical pain agree with psychophysics by participants with sensory over responsivity. J Pers Med 11:93. https://doi.org/10.3390/jpm11020093

Sajan SA, Rubenstein JLR, Warchol ME, Lovett M (2011) Identification of direct downstream targets of Dlx5 during early inner ear development. Hum Mol Genet 20:1262–1273

Sakamoto A, Moriuchi H, Matsuzaki J, Motoyama K, Moriuchi M (2015) Retrospective diagnosis of congenital cytomegalovirus infection in children with autism spectrum disorder but no other major neurologic deficit. Brain Dev 37:200–205. https://doi.org/10.1016/j.braindev.2014.03.016

Sanders SJ (2015) First glimpses of the neurobiology of autism spectrum disorder. Curr Opin Genet Dev 33:80–92. https://doi.org/10.1016/j.gde.2015.10.002

Sanders SJ, Murtha MT, Gupta AR, Murdoch JD, Raubeson MJ, Willsey AJ et al (2012) De novo mutations revealed by whole-exome sequencing are strongly associated with autism. Nature 485:237–241

Sarabipour S, Hristova K (2016) Mechanism of FGF receptor dimerization and activation. Nat Commun 7:10262

Sarachana T, Hu VW (2013) Genome-wide identification of transcriptional targets of RORA reveals direct regulation of multiple genes associated with autism spectrum disorder. Mol Autism 4(1):14

Sayad A, Noroozi R, Omrani MD, Taheri M, Ghafouri-Fard S (2017) Retinoic acid-related orphan receptor alpha (RORA) variants are associated with autism spectrum disorder. Metab Brain Dis 32:1595–1601

Schulze TG et al (2009) Two variants in Ankyrin 3 (ANK3) are independent genetic risk factors for bipolar disorder. Mol Psychiatry 14:487–491

Sen B, Singh AS, Sinha S, Chatterjee A, Ahmed S, Ghosh S et al (2010) Family-based studies indicate association of Engrailed 2 gene with autism in an Indian population. Genes Brain Behav 9:248–255

Seppala M, Fraser GJ, Birjandi AA, Xavier GM, Cobourne MT (2017) Sonic hedgehog signaling and development of the dentition. J Dev Biol 5:6

Seten TL, Posey DJ, McDougle CJ (2004) Brief report: autistic disorder in three children with cytomegalovirus infection. J Autism Dev Disord 34:583–586. https://doi.org/10.1007/s10803-004-2552-y

Shaltout E, Al-Dewik N, Samara M, Morsi H, Khattab A (2020) Psychological comorbidities in autism spectrum disorder. In: Essa M, Qoronfleh M (eds) Personalized food intervention and therapy for autism spectrum disorder management. Advances in neurobiology, vol 24. Springer, Cham. https://doi.org/10.1007/978-3-030-30402-7_6

Shi L, Zhang X, Golhar R, Otieno FG, He M, Hou C et al (2013) Whole-genome sequencing in an autism multiplex family. Mol Autism 4:8

Siegenthaler JA, Ashique AM, Zarbalis K, Patterson KP, Hecht JH, Kane MA et al (2009) Retinoic acid from the meninges regulates cortical neuron generation. Cell 139:597–609

Simon HH, Scholz C, O'Leary DDM (2005) Engrailed genes control the developmental fate of serotonergic and noradrenergic neurons in mid-and hindbrain in a gene dose-dependent manner. Mol Cell Neurosci 28:96–105

Simonoff E, Pickles A, Charman T, Chandler S, Loucas T, Baird G (2008) Psychiatric disorders in children with autism spectrum disorders: prevalence, comorbidity, and associated factors in a population-derived sample. J Am Acad Child Adolesc Psychiatry 47(8):921–929. https://doi.org/10.1097/CHI.0b013e318179964f

Song L, Li Z-Y, Liu W-P, Zhao M-R (2014) Crosstalk beten Wnt/β-catenin and Hedgehog/Gli signaling pathways in colon cancer and implications for therapy. Cancer Biol Ther 16:1–7. https://doi.org/10.4161/15384047.2014.972215

Song L, Li ZY, Liu WP, Zhao MR (2015) Crosstalk beten Wnt/β-catenin and hedgehog/Gli signaling pathways in colon cancer and implications for therapy. Cancer Biol Ther 16:1–7

Sousa KM, Villaescusa JC, Cajanek L, Ondr JK, Castelo-Branco G, Hofstra W, Bryja V, Palmberg C, Bergman T, Wainwright B et al (2010) Wnt2 regulates progenitor proliferation in the developing ventral midbrain. J Biol Chem 285:7246–7253. https://doi.org/10.1074/jbc.M109.079822

Sowers LP, Loo L, Wu Y, Campbell E, Ulrich JD, Wu S et al (2013) Disruption of the non-canonical Wnt gene PRICKLE2 leads to autism-like behaviors with evidence for hippocampal synaptic dysfunction. Mol Psychiatry 18:1077–1089

Spinelli L, Black FM, Berg JN, Eickholt BJ, Leslie NR (2015) Functionally distinct groups of inherited PTEN mutations in autism and tumour syndromes. J Med Genet 52(2):128–134. https://doi.org/10.1136/jmedgenet-2014-102803

Stubbs EG (1978) Autistic symptoms in a child with congenital cytomegalovirus infection. J Autism Child Schizophr 8:37–43. https://doi.org/10.1007/BF01550276

Stubbs EG, Ash E, Williams CP (1984) Autism and congenital cytomegalovirus. J Autism Dev Disord 14:183–189. https://doi.org/10.1007/BF02409660

Stulberg MJ, Lin A, Zhao H, Holley SA (2012) Crosstalk beten Fgf and Wnt signaling in the zebrafish tailbud. Dev Biol 369:298–307

Styles M, Alsharshani D, Samara M, Alsharshani M, Khattab A, Qoronfleh MW et al (2020) Risk factors, diagnosis, prognosis and treatment of autism. Front Biosci (Landmark Ed) 25:1682–1717

Subtil-Rodríguez A, Vázquez-Chávez E, Ceballos-Chávez M, Rodríguez-Paredes M, Martín-Subero JI, Esteller M et al (2014) The chromatin remodeller CHD8 is required for E2F-dependent transcription activation of S-phase genes. Nucleic Acids Res 42:2185–2196

Sudhof TC (2008) Neuroligins and neurexins link synaptic function to cognitive disease. Nature 455:903

Südhof TC (2008) Neuroligins and neurexins link synaptic function to cognitive disease. Nature 455:903–911. https://doi.org/10.1038/nature07456

Sugathan A, Biagioli M, Golzio C, Erdin S, Blumenthal I, Manavalan P, Ragavendran A, Brand H, Lucente D, Miles J et al (2014) CHD8 regulates neurodevelopmental pathways associated with autism spectrum disorder in neural progenitors. Proc Natl Acad Sci U S A 111(42):E4468–E4477. https://doi.org/10.1073/pnas.1405266111

Takeuchi K, Gertner MJ, Zhou J, Parada LF, Bennett MV, Zukin RS (2013) Dysregulation of synaptic plasticity precedes appearance of morphological defects in a Pten conditional knockout mouse model of autism. Proc Natl Acad Sci U S A 110(12):4738–4743. https://doi.org/10.1073/pnas.1222803110

Talkowski ME, Rosenfeld JA, Blumenthal I, Pillalamarri V, Chiang C, Heilbut A, Ernst C, Hanscom C, Rossin E, Lindgren AM et al (2012) Sequencing chromosomal abnormalities reveals neurodevelopmental loci that confer risk across diagnostic boundaries. Cell 149(3): 525–537. https://doi.org/10.1016/j.cell.2012.03.028

Tamiji J, Crawford DA (2010) The neurobiology of lipid metabolism in autism spectrum disorders. Neurosignals 18:98–112

Tang SJ (2014) Synaptic activity-regulated Wnt signaling in synaptic plasticity, glial function and chronic pain. CNS Neurol Disord Drug Targets 13:737–744

Tao H, Manak JR, Sors L, Mei X, Kiyonari H, Abe T, Dahdaleh NS, Yang T, Wu S, Chen S, Fox MH, Gurnett C, Montine T, Bird T, Shaffer LG, Rosenfeld JA, McConnell J, Madan-Khetarpal-S, Berry-Kravis E, Griesbach H, Saneto RP, Scott MP, Antic D, Reed J, Boland R, Ehaideb SN, El-Shanti H, Mahajan VB, Ferguson PJ, Axelrod JD et al (2011) Mutations in prickle orthologs cause seizures in flies, mice, and humans. Am J Hum Genet 88:138–149

Tempé D, Casas M, Karaz S, Blanchet-Tournier M-F, Concordet JP (2006) Multisite protein kinase A and glycogen synthase kinase 3beta phosphorylation leads to Gli3 ubiquitination by SCFbetaTrCP. Mol Cell Biol 26:4316–4326

Terauchi A, Johnson-Venkatesh EM, Toth AB, Javed D, Sutton MA, Umemori H (2010) Distinct FGFs promote differentiation of excitatory and inhibitory synapses. Nature 465:783–787

Tesli M et al (2011) Association analysis of ANK3 gene variants in Nordic bipolar disorder and schizophrenia case-control samples. Am J Med Genet B Neuropsychiatr Genet 156B:969–974

Thomas KR, Capecchi MR (1990) Targeted disruption of the murine int-1 proto-oncogene resulting in severe abnormalities in midbrain and cerebellar development. Nature 346:847–850

Thompson BA, Tremblay V, Lin G, Bochar DA (2008) CHD8 is an ATP-dependent chromatin remodelling factor that regulates beta-catenin target genes. Mol Cell Biol 28:3894–3904

Tilot AK, Frazier TW 2nd, Eng C (2015) Balancing proliferation and connectivity in PTEN-associated autism spectrum disorder. Neurotherapeutics 12(3):609–619. https://doi.org/10.1007/s13311-015-0356-8

Tora D, Gomez AM, Michaud JF, Yam PT, Charron F, Scheiffele P (2017) Cellular functions of the autism risk factor PTCHD1 in mice. J Neurosci 37:11993–12005

Townsend CL, Forsgren M, Ahlfors K, Ivarsson S-A, Tookey PA, Peckham CS (2013) Long-term outcomes of congenital cytomegalovirus infection in Sden and the United Kingdom. Clin Infect Dis 56:1232–1239. https://doi.org/10.1093/cid/cit018

Treffert DA (1970) Epidemiology of infantile autism. Arch Gen Psychiatry 22(5):431–438

Triantis V et al (2010) Glycosylation of fibroblast growth factor receptor 4 is a key regulator of fibroblast growth factor 19-mediated down-regulation of cytochrome P450 7A1. Hepatology 52:656–666

Tsarovina K, Reiff T, Stubbusch J, Kurek D, Grosveld FG, Parlato R et al (2010) The Gata3 transcription factor is required for the survival of embryonic and adult sympathetic neurons. J Neurosci 30:10833–10843

Tukachinsky H, Lopez LV, Salic A (2010) A mechanism for vertebrate hedgehog signaling: recruitment to cilia and dissociation of SuFu-Gli protein complexes. J Cell Biol 191:415–428

Turner CA, Eren-Koçak E, Inui EG, Watson SJ, Akil H (2016) Dysregulated fibroblast growth factor (FGF) signaling in neurological and psychiatric disorders. Semin Cell Dev Biol 53:136–143

Uddin M, Wang Y, Woodbury-Smith M (2019) Artificial intelligence for precision medicine in neurodevelopmental disorders. NPJ Digit Med 2(1):112

Ulloa F, Itasaki N, Briscoe J (2007) Inhibitory Gli3 activity negatively regulates Wnt/beta-catenin signaling. Curr Biol 17:545–550

Ung DC, Iacono G, Méziane H, Blanchard E, Papon MA, Selten M et al (2018) Ptchd1 deficiency induces excitatory synaptic and cognitive dysfunctions in mouse. Mol Psychiatry 23(5): 1356–1367

Varnat F, Siegl-Cachedenier I, Malerba M, Gervaz P, Ruiz i Altaba A (2010) Loss of WNT-TCF addiction and enhancement of HH-GLI1 signaling define the metastatic transition of human colon carcinomas. EMBO Mol Med 2:440–457

Vogt D, Cho KK, Lee AT, Sohal VS, Rubenstein JL (2015) The parvalbumin/somatostatin ratio is increased in Pten mutant mice and by human PTEN ASD alleles. Cell Rep 11(6):944–956. https://doi.org/10.1016/j.celrep.2015.04.019

Wada H, Okamoto H (2009) Roles of noncanonical Wnt/PCP pathway genes in neuronal migration and neurulation in zebrafish. Zebrafish 6:3–8

Wang B, Li Y (2006) Evidence for the direct involvement of βTrCP in Gli3 protein processing. Proc Natl Acad Sci U S A 103:33–38

Wang L, Jia M, Yue W, Tang F, Qu M, Ruan Y et al (2008) Association of the ENGRAILED 2 (EN2) gene with autism in Chinese Han population. Am J Med Genet B Neuropsychiatr Genet 147B:434–438

Wang Y, Song L, Zhou CJ (2011) The canonical Wnt/β-catenin signaling pathway regulates Fgf signaling for early facial development. Dev Biol 349:250–260

Wang RN, Green J, Wang Z, Deng Y, Qiao M, Peabody M et al (2014) Bone morphogenetic protein (BMP) signaling in development and human diseases. Genes Dis 1:87–105

Wang P, Lin M, Pedrosa E, Hrabovsky A, Zhang Z, Guo W, Lachman HM, Zheng D (2015) CRISPR/Cas9-mediated heterozygous knockout of the autism gene CHD8 and characterization of its transcriptional networks in neurodevelopment. Mol Autism 6:55. https://doi.org/10.1186/s13229-015-0048-6

Wang Y, Billon C, Walker JK, Burris TP (2016) Therapeutic effect of a synthetic RORα/γ agonist in an animal model of autism. ACS Chem Neurosci 7:143–148

Wassink TH, Piven J, Vieland VJ, Huang J, Swiderski RE, Pietila J et al (2001) Evidence supporting WNT2 as an autism susceptibility gene. Am J Med Genet 105:406–413

Wechsler-Reya RJ, Scott MP (1999) Control of neuronal precursor proliferation in the cerebellum by sonic hedgehog. Neuron 22:103–114

Wheeler JA, Clinkenbeard EL (2019) Regulation of fibroblast growth factor 23 by iron, EPO, and HIF. Curr Mol Biol Rep 5:8–17

Wilkinson B, Grepo N, Thompson BL, Kim J, Wang K, Evgrafov OV, Lu W, Knowles JA, Campbell DB (2015) The autism-associated gene chromodomain helicase DNA-binding protein 8 (CHD8) regulates noncoding RNAs and autism-related genes. Transl Psychiatry 5(5):e568. https://doi.org/10.1038/tp.2015.62

Willsey AJ, Sanders SJ, Li M, Dong S, Tebbenkamp AT, Muhle RA et al (2013) Coexpression networks implicate human midfetal deep cortical projection neurons in the pathogenesis of autism. Cell 155:997–1007

Wodarz A, Nusse R (1998) Mechanisms of Wnt signaling in development. Annu Rev Cell Dev Biol 14:59–88

Wong CT, Wais J, Crawford DA (2015) Prenatal exposure to common environmental factors affects brain lipids and increases the risk of developing autism spectrum disorders. Eur J Neurosci 42:2742–2760

Wong CT, Ussyshkin N, Ahmad E, Rai-Bhogal R, Li H, Crawford DA (2016) Prostaglandin E2 promotes neural proliferation and differentiation and regulates Wnt target gene expression. J Neurosci Res 94:759–775. https://doi.org/10.1002/jnr.23759

Xie Y, Su N, Yang J et al (2020) FGF/FGFR signaling in health and disease. Signal Transduct Target Ther 5:181. https://doi.org/10.1038/s41392-020-00222-7

Xu X, Li C, Gao X, Xia K, Guo H, Li Y, Hao Z, Zhang L, Gao D, Xu C et al (2017) Excessive UBE3A dosage impairs retinoic acid signaling and synaptic plasticity in autism spectrum disorders. Cell Res 28:48–68. https://doi.org/10.1038/cr.2017.132

Xu X, Li C, Gao X, Xia K, Guo H, Li Y et al (2018) Excessive UBE3A dosage impairs retinoic acid signaling and synaptic plasticity in autism spectrum disorders. Cell Res 28:48–68

Yang P, Lung FW, Jong YJ, Hsieh HY, Liang CL, Juo SH (2008) Association of the homeobox transcription factor gene ENGRAILED 2 with autistic disorder in Chinese children. Neuropsychobiology 57:3–8

Yang P, Shu B-C, Hallmayer JF, Lung FW (2010) Intronic single nucleotide polymorphisms of engrailed homeobox 2 modulate the disease vulnerability of autism in a Han Chinese population. Neuropsychobiology 62:104–115

Yi JJ, Paranjape SR, Walker MP, Choudhury R, Wolter JM, Fragola G et al (2017) The autism-linked UBE3A T485A mutant E3 ubiquitin ligase activates the Wnt/β-catenin pathway by inhibiting the proteasome. J Biol Chem 292:12503–12515

Yoo HJ, Cho IH, Park M, Cho E, Cho SC, Kim BN, Kim JW, Kim SA (2008) Association between PTGS2 polymorphism and autism spectrum disorders in Korean trios. Neurosci Res 62(1): 66–69. https://doi.org/10.1016/j.neures.2008.05.008

Yuen RK, Merico D, Cao H, Pellecchia G, Alipanahi B, Thiruvahindrapuram B et al (2016) Genome-wide characteristics of de novo mutations in autism. NPJ Genom Med 1:160271–1602710

Yuen RKC, Merico D, Bookman M, Howe JL, Thiruvahindrapuram B, Patel RV et al (2017) Whole genome sequencing resource identifies 18 new candidate genes for autism spectrum disorder. Nat Neurosci 20(4):602–611

Zhang Y, Sun Y, Wang F, Wang Z, Peng Y, Li R (2012) Downregulating the canonical Wnt/β-catenin signaling pathway attenuates the susceptibility to autism-like phenotypes by decreasing oxidative stress. Neurochem Res 37:1409–1419

Zhang X, Rui M, Gan G, Huang C, Yi J, Lv H et al (2017) Neuroligin 4 regulates synaptic growth via the bone morphogenetic protein (BMP) signaling pathway at the Drosophila neuromuscular junction. J Biol Chem 292:17991–18005

Zhao T, Gan Q, Stokes A, Lassiter RN, Wang Y, Chan J, Han JX, Pleasure DE, Epstein JA, Zhou CJ (2014) β-catenin regulates Pax3 and Cdx2 for caudal neural tube closure and elongation. Development (Cambridge, England) 141(1):148–157. https://doi.org/10.1242/dev.101550

Zhou J, Parada LF (2012) PTEN signaling in autism spectrum disorders. Curr Opin Neurobiol 22(5):873–879. https://doi.org/10.1016/j.conb.2012.05.004

Zhou CJ, Pinson KI, Pleasure SJ (2004) Severe defects in dorsal thalamic development in low-density lipoprotein receptor-related protein-6 mutants. J Neurosci 24:7632–7639

Zhou XL, Giacobini M, Anderlid BM, Anckarsäter H, Omrani D, Gillberg C et al (2007) Association of adenomatous polyposis coli (APC) gene polymorphisms with autism spectrum disorder (ASD). Am J Med Genet B Neuropsychiatr Genet 144B(3):351–354

Zhu X, Lee K, Asa SL, Ezzat S (2007) Epigenetic silencing through DNA and histone methylation of fibroblast growth factor receptor 2 in neoplastic pituitary cells. Am J Pathol 170:1618–1628

Zi Z, Chapnick DA, Liu X (2012) Dynamics of TGF-β/Smad signaling. FEBS Lett 586:1921–1928

Zieger E, Schubert M (2017) New insights into the roles of retinoic acid signaling in nervous system development and the establishment of neurotransmitter systems. Int Rev Cell Mol Biol 330:1–84

Zoghbi HY (2003) Postnatal neurodevelopmental disorders: meeting at the synapse? Science 302: 826

Zumbrunn J, Kinoshita K, Hyman AA, Näthke IS (2001) Binding of the adenomatous polyposis coli protein to microtubules increases microtubule stability and is regulated by GSK3 β phosphorylation. Curr Biol 11:44–49. https://doi.org/10.1016/s0960-9822(01)00002-1

Genes and Specific (Related) Proteins in Neurodevelopmental Disorders

2

Sabah Nisar, Mohammad Haris, and Khalid A. Fakhro

Abstract

Neurodevelopmental disorders (NDDs) represent a group of heterogeneous disabilities characterized by impaired cognition, memory, learning abilities, self-control, and psychomotor skills. NDDs encompass many rare genetic syndromes and heritable genetic conditions such as attention-deficit hyperactivity disorder (ADHD), autism spectrum disorder (ASD), and intellectual disability (ID). Understanding the molecular etiology and pathophysiological mechanisms of NDDs is challenging, mainly owing to the genetic and phenotypic heterogeneity of these conditions and the fact that both heritable and environmental factors influence NDD risk. Despite the complex genetic architecture of NDDs, emerging evidence suggests that proteins involved in NDDs converge on common pathways, including synaptic function, chromatin remodeling, and mTOR. In-depth insight of the recent advancements in genomic technologies has enabled the identification of genetic mutations underlying NDDs and uncovered mechanisms behind these pathways that point towards treatment options. In this chapter, the authors review the available literature to compile a comprehensive set of genes and proteins underlying NDDs and identify signaling pathways involved in various NDDs.

S. Nisar · M. Haris
Laboratory of Molecular and Metabolic Imaging, Sidra Medicine, Doha, Qatar

K. A. Fakhro (✉)
Department of Human Genetics, Sidra Medicine, Doha, Qatar

Department of Genetic Medicine, Weill Cornell Medical College, Doha, Qatar
e-mail: kfakhro@sidra.org

M. W. Qoronfleh et al. (eds.), *Proteins Associated with Neurodevelopmental Disorders*, Nutritional Neurosciences, https://doi.org/10.1007/978-981-15-9781-7_2

Keywords

Neurodevelopmental disorders · Attention-deficit hyperactivity disorder · Autism spectrum disorder · Intellectual disability · Genetic mutations · Whole-exome sequencing · Chromatin remodeling · Synaptic function · Transcriptional pathways · Signaling pathways

2.1 Background

Neurodevelopmental disorders (NDDs) are conditions affecting brain development and function, characterized by the inability to develop emotional, cognitive, learning, and motor skills. As of 2016, it has been estimated that 52.9 million children under the age of 5 have developmental disabilities globally (Olusanya et al. 2018). The prevalence of these conditions continues to grow in present days with better ascertainment and reporting. The heterogeneous etiology of NDDs could lead to impairment in cognition, behavior, adaptive behavior, and psychomotor skills which are described in the Diagnostic and Statistical Manual of Mental Disorders (DSM) 5 category (Morris-Rosendahl and Crocq 2020). The most prevalent NDDs with behavioral disorders are autism spectrum disorder (ASD), attention-deficit hyperactivity disorder (ADHD), intellectual disability (ID), learning disorders such as dyslexia, motor disorders, and speech or language disorders (Parenti et al. 2020). Though the exact underlying cause of NDDs is complex, recent studies revealed an interplay/interaction between environmental, genetic, and traumatic factors causing altered transcription, translational, and presynaptic dysfunction, all considered to be predisposing factors for the development of NDDs (van Loo and Martens 2007) (Fig. 2.1). Such a network of interacting players has been believed to share various molecular pathways (Mitchell 2014; D'Gama and Walsh 2018), creating difficulties in establishing genotype-phenotype correlations, and influence the clinical presentation of NDDs. Thus, delineating the phenotype-genotype correlation is necessary for individualized characterization of symptoms, monitoring the progression, future complications, and clinical management of patients with NDDs.

Identification of genes or genetic networks has traditionally been through forward genetic screens of patient cohorts with NDDs, to pinpoint causative mutations and altered pathways. To date, many different types of mutations have been associated with NDDs, including copy number variations (CNVs), chromosomal rearrangements, and point mutations. However, despite the advancements in technologies to facilitate these screens and the identification of hundreds of genes and loci, many patients still do not get a molecular diagnosis. In many cases, it has been suggested that NDDs might have a polygenic nature, which could explain phenotypic variability and missing heritability (Parenti et al. 2020; Mitchell 2014). Further, while most genetic mutations associated with the NDDs are considered to be inherited or de novo germline mutations, recent advancements in sequencing technologies have shown that somatic mutations also contribute to the impaired neurodevelopment (D'Gama and Walsh 2018).

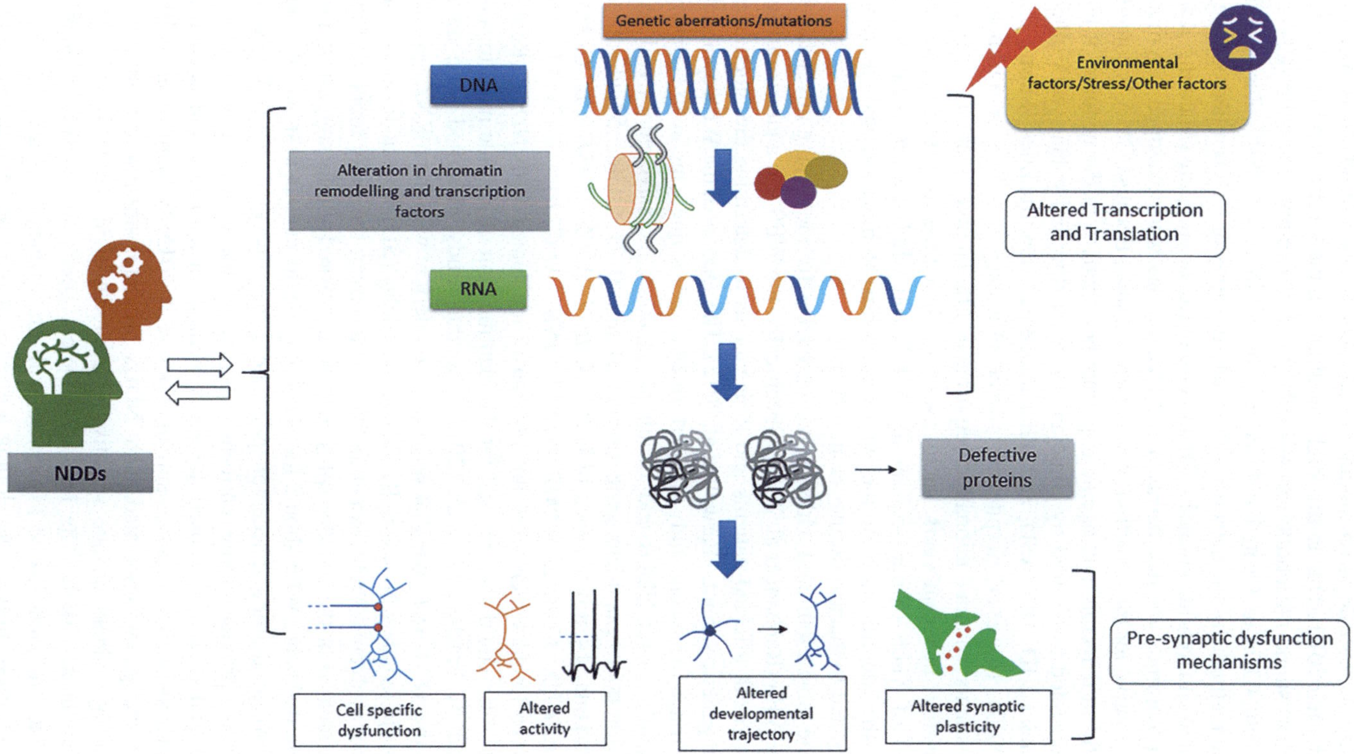

Fig. 2.1 Mechanisms in NDDs. Genetic mutations/aberrations due to environmental and other factors alter the transcriptional and translational processes, which result in the production of defective proteins. These defective proteins contribute to different presynaptic dysfunction mechanisms and subsequently lead to the development of NDDs

It was estimated that a newborn acquires 50–100 new mutations at the diploid level resulting in 0.86 de novo mutations, also known as non-inherited sporadic mutations (Lynch 2010). Due to this high variability in the human genome, several variants such as single-nucleotide variants (SNVs) or copy number variants (CNVs) associated with NDDs are located in different genes (Wilfert et al. 2017; Zarrei et al. 2019). Initial methods for the detection of the heterogeneity of NDDs relied on using fluorescence in situ hybridization (FISH) and G-banded karyotyping (Manning et al. 2010). Later, chromosome microarray (CMA) technology was developed as a more accurate alternative for the detection of structural variants. Because CMA is considered powerful in detecting pathogenic copy number changes in the genome, it has been recommended to be used by clinicians for evaluating individuals with otherwise unexplained ID, DD, ASD, and other congenital abnormalities (Martin and Ledbetter 2017). However, despite the proven efficiency of CMA, a significant fraction of patients remained idiopathic, requiring additional investigation to identify genetic causality. In recent years, such cases have benefited immensely from the development of whole-exome sequencing (WES) and whole-genome sequencing (WGS) for the identification of the entire spectrum of inherited and de novo mutations related to disease (Reuter et al. 2017).

One of the most useful and widely used approaches for identifying novel causal mutations is WES (Rabbani et al. 2014). Specifically, the WES-trio analysis utilizes samples from both affected patients and their parents to determine the inheritance pattern of all genetic variants (Farwell et al. 2015). The success of the WES approach was demonstrated in a recent study that performed exome sequencing in 152 consanguineous families with NDDs and identified 52 novel recessive genes and one autosomal dominant candidate gene (Reuter et al. 2017). Of the 52 identified recessive variants, 14 novel genes (*TRAP1, NARG2, AMZ2, FAM234B, CLMN, GALNT2, EEF1D, CCAR2, GRM7, STX1A, INIP, SEC23IP, SLC44A1,* and *LRRIQ3*) were found which were not previously associated with NDDs (Reuter et al. 2017). WES could also be used for the identification of intronic mutations as demonstrated by Prachalova et al. (2017). Another study identified a de novo mutation in the *SYNGAP1* gene that was found to interfere with the mRNA splicing in a 31-year-old female with severe ID, DD, speech delay (SD), autistic features, and sleep and gait problems (Prchalova et al. 2017). Moreover, the study found that the variant was located in the broader splice region of intron 10, thus suggesting that variants identified using WES should focus not only on the canonical splice sites but also on the broader donor and acceptor splice sites (Prchalova et al. 2017). WGS, on the other hand, provides more uniform coverage of the coding regions in the genome. Several studies have used the WGS method for delineation of genic and nongenic variants in NDDs such as autism (Yuen et al. 2015; Williams et al. 2019).

Some of the genes associated with NDDs are haplo-insufficient and were characterized by evident dosage sensitivity (Courchet et al. 2018; Khalil et al. 2018; Uehara et al. 2019). The haplo-insufficient genes are considered highly vulnerable and are found to be associated with an increased disease risk. Also, there is a high probability of developing a disease phenotype upon the disruption of these genes even in the absence of other disease-causing events (Parenti et al.

2020). Thus, mutations in the haplo-insufficient genes contribute towards negative selective pressure. On the other hand, genes that are less sensitive to disruptive mutations are not under negative selective pressure and are thus transgenerational (Niemi et al. 2018). Moreover, some of the studies have shown that common genetic variants contribute to the polygenic nature of NDDs, such as autism, developmental delay (DD), and ID (Niemi et al. 2018; Kurki et al. 2019). Also, the monoallelic expression influences the expression-level variance within and across populations and is an important epigenetic regulatory mechanism that might be useful in predicting the pathogenicity of NDDs (Savova et al. 2017). Epistatic interactions and dosage sensitivity are important determinants of phenotypical outcomes (Mitra et al. 2017) as they strongly correlate with the genetic load as high genetic load (more disruptive mutations) could result in reduced fitness at the population level which could contribute to the development of more complex disease phenotypes (Parenti et al. 2020). This finding was supported by a study showing that monoallelic (heterozygous) mutations in the *CACNA1A* gene resulted in milder cognitive deficits. In contrast, biallelic mutations in *CACNA1A* resulted in progressive cerebral, cerebellar, and optic nerve atrophy (Reinson et al. 2016). Similarly, in the *SCN8A* gene, monoallelic variants were found to be associated with mild conditions such as benign familial infantile seizures or movement disorders, while the biallelic variants were found to be severely affected by developmental and epileptic encephalopathy (DEE) (Wengert et al. 2019). Nevertheless, it is suggested that most of the NDDs are polygenic or multifactorial in nature that contribute to the heterogenous and complex nature of NDDs.

2.2 Genes Involved in Different NDDs

Currently, the mutational spectrum of NDDs includes genes that encode proteins involved in chromatin remodeling, synaptic function, and transcriptional pathways (De Rubeis et al. 2014) (Table 2.1). In the upcoming sections, we will explore the genetic landscape and the proteins associated with different NDDs.

2.2.1 Autism Spectrum Disorder (ASD)

Initial studies relied on the candidate gene approach. The genes known to play a role in neurodevelopment (such as the homeobox (HOX) and Wnt genes) were tested for their association with ASD. An initial impetus for examining the role of HOXA1 in autism susceptibility was published in a study in the year 2000 (Persico et al. 2001). It was followed shortly by another study by Conciatori et al. that showed the association of *HOXA1 A218G* polymorphism with an increased head circumference in autistic patients (Conciatori et al. 2004). Further, it has also been shown that the HOXB1 gene contributed to minor changes in the head circumference in autistic patients, while *HOXA1* showed more pronounced effects (Muscarella et al. 2010). On the other hand, there are contrasting studies that reported no association or

Table 2.1 Synaptic genes and associated proteins involved in various NDDs

Gene	Protein	Mutation	Suggested affected brain region	Effect	Type of study	References
SHANK3	SH3 and multiple ankyrin repeat domains protein 3	Conditional deletion	Cerebral cortex, striatum	Repetitive behaviors	Preclinical	Bey et al. (2018)
SHANK2	SH3 and multiple ankyrin repeat domains protein 2	Loss of function	Striatum	Hyperactivity, hyper-motivation	Preclinical	Modi et al. (2018)
		Deletion	Prefrontal cortex (positive parvalbumin-positive neurons)	Hyperactivity, enhanced-self grooming, suppressed brain excitation	Preclinical	Lee et al. (2018)
GRIN2B	Glutamate [NMDA] receptor subunit epsilon 2	Interstitial 12p13 deletions	Posterior parietal cortex	Intellectual disability, developmental delay, autistic-like behaviors, central hypotonia	Clinical	Mishra et al. (2016)
SYNGAP1	Synaptic Ras GTPase-activating protein	Loss of function	Cortex	Intellectual disability, developmental delay (language impairment), hypotonia	Clinical	Mignot et al. (2016)
CTNNB1	Catenin (cadherin-associated protein), beta 1	Loss of function	Frontal lobe, corpus callosum, ventricles	Intellectual disability, microcephaly, limited speech	Clinical	Kuechler et al. (2015)
		Nonsense mutation	Cerebral cortex (motor cortex)	Delayed speech and psychomotor development, hyperekplexia	Clinical	Winczewska-Wiktor et al. (2016)
		Conditional knockout mice in parvalbumin interneurons	Cerebral cortex	Impaired object recognition and social interaction, repetitive behaviors	Preclinical	Dong et al. (2016)

NRXN-1α	Neurexin-1-alpha	Deletion	Cerebral cortex	Altered social approach and reduced social investigation and locomotor activity	Preclinical	Grayton et al. (2013)
NRXN-2α	Neurexin-2-alpha	Deletion	Frontal cortex, hippocampus	Social memory and sociability deficits	Preclinical	Dachtler et al. (2014)
NLGN3	Neuroligin-3	NL3 (R451C) mutant mice	Amygdala	Aggressive behavior	Preclinical	Hosie et al. (2018)
		Nlgn3 dysfunction (Nlgn3−/y and Nlgn3R451C) mouse models	Prefrontal cortex, hippocampus	Impaired behavioral flexibility	Preclinical	Norris et al. (2019)
NLGN4	Neuroligin-4 protein	Loss of function	Somatosensory cortex	Decreased excitation and inhibition	Preclinical	Delattre et al. (2013)
CNTNAP2	Contactin-associated protein 2	*CNTNAP2* knockout rat model	Brainstem	Altered auditory processing, filtering, and reactivity	Preclinical	Scott et al. (2018)
DLG4	Postsynaptic density-95 (PSD-95)	Deletion	Cortico-amygdala	Increased repetitive and social behaviors, abnormal communication, impaired motor coordination	Preclinical	Feyder et al. (2010)
CDH11	Cadherin-11	*CDH11* knockout mice	Frontal cortex, cerebellum, striatum	Autistic-like behavioral deficits	Preclinical	Wu et al. (2021)
CDH2	Cadherin-2/N-cadherin	Missense, frameshift	Corpus callosum	Developmental delay, intellectual disability	Clinical	Accogli et al. (2019)
TAOK2	TAO kinase 2	*TAOK2* knockout mice	Hindbrain, midbrain, hypothalamus, thalamus, cerebellum, hippocampus, corpus callosum	Anxiety, impaired cognition, and social interaction	Preclinical	Richter et al. (2019)

linkage of *HOXA1* and *HOXB1* genes in autism (Devlin et al. 2002; Romano et al. 2003; Talebizadeh et al. 2002). Given the contradictory nature of the previous studies, the role of *HOX* genes in ASD remains inconclusive, and, therefore, further studies are required for a detailed understanding of these genes in ASD. Altogether, the candidate approach was successful in identifying the key genes involved in the etiology of ASD such as Reelin (*RELN*) (Persico et al. 2001), neuroligins (*NLGN3* and *NLGN4*) (Jamain et al. 2003), polyphosphate-1-phosphatase (*INPP1*), γ catalytic subunit of phosphatidyl 3-OH-kinase (PIK3CG), tuberous sclerosis complex 2 (*TSC2*) (Serajee et al. 2003), ubiquitin-protein ligase E3A (*UBE3A*) (Jiang et al. 2004), and aristaless-related homeobox (*ARX*) genes (Strømme et al. 2002).

In recent years, the advent of sequencing technologies allowed for a broader, unbiased approach to gene identification. In the above context, several key observations arose. First, evidence emerged that ASD could have a polygenic basis which was supported by the high genetic and phenotypic heterogeneity of ASD. Only a few disorders associated with ASD appeared to be truly monogenic, e.g., fragile X syndrome (FXS) (*FMR1* gene), Rett syndrome (*MECP2* gene), tuberous sclerosis (*TSC1* or *TSC2*), Timothy syndrome (*CACNA1C* gene), neurofibromatosis (*NF1* gene), Duchenne muscular dystrophy (*DMD* gene), and Schuurs–Hoeijmakers syndrome (*PACS1* gene) (Wiśniowiecka-Kowalnik and Nowakowska 2019; Artuso et al. 2011; Stern et al. 2017; Woodbury-Smith and Scherer 2018). Since then, a large number of genetic studies have been conducted to identify ASD-related risk genes. Most of the genes implicated in the pathology of ASD were found to be involved in transcriptional regulation, synaptic formation, and chromatin remodeling pathways (De Rubeis et al. 2014). The synaptic genes involved in ASD include synapsins (*SYN1* and *SYN2*), calcium channels (*CACNA1E, CACNA1D, CACNA2D3, CACNB2*), voltage-regulated sodium channels (*SCN1A, SCN2A, SCN3A*), potassium calcium-activated channels (*KCNMA1, KCNMB4*), potassium voltage-gated channels (*KCNQ3, KCNQ5, KCND2*), neurexins (*NRXN1*), neuroligins (*NLGN3, NLGN4*), contactin-associated protein-like 2 (*CNTNAP2*), cadherins (*CDH5, CDH8, CDH9, CDH10, CDH11, CDH13, CDH15*), protocadherins (*PCDHB4, PCDH10, PCDH19*), contactins (*CNTN4, CNTN5, CNTN6*), SH3 and multiple ankyrin domain proteins (*SHANK1, SHANK2, SHANK3*), synaptic Ras GTPase-activating protein 1 (*SYNGAP1*), and gamma 3 subunit of GABA-A receptor (*GABRG3*) (Giovedí et al. 2014; Sanders et al. 2012). It has also been suggested that even slight changes in *NRXN1* could contribute to ASD susceptibility (Kim et al. 2008).

Of the 33 ASD-risk genes identified by De Rubeis et al. (2014), seven genes (*MLL3, VIL1, ASH1L, ETFB, MYO9B, NAA15, MIB1*) were reported to be novel, and two of these genes (*ASH1L, MLL3*) were found to be involved in chromatin remodeling (De Rubeis et al. 2014). Moreover, multiple studies have found that mutations in genes such as *SHANK, NRXN, NLGN, MECP2, FMR1*, and *TSC1/2* converge on common cellular pathways associated with synapsis (Guang et al. 2018). The results confirmed the previous observations by Iossifov et al. (2012), who reported that most of the disrupted autism susceptibility genes associated with the fragile X protein (FMRP) demonstrated a link between autism and synaptic plasticity

(Iossifov et al. 2012). The chromatin genes implicated in ASD are methyl CpG-binding protein 2 (*MECP2*); core histone macro-H2A.1 (*H2AFY*); structural maintenance of chromosomes 1A (*SMC1A*); mono-ADP ribosylhydrolase 2 (*MACROD2*); lysine-specific demethylase 5C (*KDM5C*); methyl-CpG-binding domain protein 1 (*MBD1*); AT-rich interactive domain-containing protein 1B (*ARID1B*); SWI/SNF-related, matrix-associated, actin-dependent regulator of chromatin subfamily C members (*SMARCC1, SMARCC2*); jumonji domain containing 1C (*JMJD1C*); and chromodomain helicase DNA-binding protein 8 (*CHD8*) (Lasalle 2013). For determining the transcriptional organization of ASD and its connection to ASD-risk genes, a study by Gazestani et al. (2018) analyzed the transcriptomics data in ASD toddlers. The results demonstrated that the differentially expressed ASD network was enriched with the genes regulated by two high-confidence ASD-risk genes, namely *CHD8* and *FMR1* (Gazestani et al. 2018). The results also showed that most ASD-risk genes impact differentially expressed ASD networks through the PI3K/AKT, RAS/ERK, and Wnt/β-catenin signaling pathways (Gazestani et al. 2018). Moreover, a heterozygous mutation of a single copy of *SHANK3* (a scaffolding protein found at the excitatory synapsis) on the 22q13 chromosome resulted in language or communication disorders (Durand et al. 2007).

In recent years, exome sequencing studies, particularly in trios, have reported an avalanche of de novo protein-altering mutations disrupting genes in ASD patients. In particular, as cohorts grew, recurrent mutations observed in separate patients strengthened the evidence for causality. They revealed a role for genes such as forkhead box protein P1 (*FOXP1*), glutamate ionotropic receptor NMDA-type subunit 2B (*GRIN2B*), laminin subunit gamma 3 (*LAMC3*), and *SCN1A* and recurrent protein-altering mutations in *CHD8* and netrin G1 (*NTNG1*) genes in sporadic ASDs (O'Roak et al. 2011, 2012a). A study using WGS identified novel ASD-risk genes (*BIRC6, CD163L1, UTP6, SLCO1B3, DNAH10, ZNF774, KRT24, ZNF559, PCOLCE*) and also identified genes that were functionally implicated in ASD such as *MAP2K2, CHRNA2, KMT2D, KAT6B, SCN9A, NLGN1, LRRC7, KIF11, TENM1*, and *CACNB2* (Yuen et al. 2015).

In addition to common genetic variants, rare CNVs and single-nucleotide variants (SNVs) are important role players in ASD etiology. A genome-wide CNV study reported ASD candidate genes that were altered by inherited, de novo, or X-linked CNVs such as *HDAC9, SETD5*, and *miR37* and genes altered by both de novo CNVs and loss-of-function SNVs such as *DLL1, RIMS1*, and *TRIP12* (Pinto et al. 2014). Moreover, the genes altered by de novo CNVs and loss-of-function SNVs were found to converge on networks associated with neuronal signaling, chromatin regulation, and synapse function (Pinto et al. 2014). Another recent study identified rare CNVs; especially deletions in the Italian ASD cohort found that genes such as *WWOX, CNTNAP2, MACROD2, PARK2, RNF113A, GPHN, RBFOX1, VPS13B*, and *APBA2* had a significant role in ASD susceptibility (Bacchelli et al. 2020). In a genome-wide CNV study, novel candidate ASD genes associated with the GABAergic signaling and neural development pathways were identified. These genes include solute carrier family six member 11 (*SLC6A11*), which is a postsynaptic GABA transporter, GABA type a receptor-associated protein-like

1 (*GABARAPL1*), and diazepam-binding inhibitor (*DBI*). In addition, the study also identified CNVs in the cordon-bleu protein-encoding gene (*COBL*), which was found to be involved in neuronal morphogenesis, and the delta/notch-like EGF repeat-containing (*DNER*) gene, which is a neuron-specific notch ligand essential for the development of the cerebellum (Griswold et al. 2012). Targeted large-scale sequencing studies confirmed the association of the genes CHD8, dual-specificity tyrosine phosphorylation-regulated kinase 1A (*DYRK1A*), *GRIN2B*, T-box brain gene 1 (*TBR1*), phosphatase and tensin homolog (*PTEN*), and transducin (beta)-like 1 X-linked receptor 1 (*TBL1XR1*) with ASD pathology (Krumm et al. 2014; O'Roak et al. 2012b). Also, the gene expression profiling of lymphoblastoid cell lines derived from monozygotic twins revealed high expression of argininosuccinate synthetase (*ASS*), cell adhesion molecule L1 (*CHL1*), and 5-lipoxygenase-activating protein (*FLAP*) in the autistic twin compared to the normal sibling (Hu et al. 2006). A study of peripheral leukocytes obtained from ASD patients found 18 genes to be upregulated (*UBL4A, ITGA2B, PLCXD2, NRN1, KLHDC7A, MYOG, NUMBL, WDTC2, SDK2, NOVA2, SLC22A18AS, C21orf58, LZTR2, RKHD1, BC018095, FAM124A, LHB,* and *TAOK2*) and 1 (*UTS2*) downregulated (Kuwano et al. 2011). The altered genes were involved in cellular morphology, organization, and nerve development (Kuwano et al. 2011). A study reported adenosine A2A receptor signaling pathway to be the most dysregulated in the young autistic brain by utilizing the whole-genome analysis of mRNA levels and CNVs for the identification of abnormal gene expression in autistic individuals (Chow et al. 2012).

There is also increasing evidence that neurotrophins and their receptors play an important role in the pathophysiology of ASD. In support of this, a study evaluated the mRNA expression of neurotrophins and their receptors from the blood of patients with ASD using qRT-PCR (Segura et al. 2015). The study found lower mRNA expression of neurotrophin-3 (*NT3*) and neurotrophin-4 (*NT4*) and higher mRNA expression of p75 in ASD individuals as compared to healthy controls (Segura et al. 2015). Moreover, *NT3* and *NT4* are crucial for the development of the climbing fiber system of the Purkinje cells, which have been reported to be significantly reduced in number and size in the ASD postmortem brain (Bauman and Kemper 2005), thus suggesting a link between neurotrophins and ASD pathophysiology. Also, the expression of the molybdenum cofactor sulfurase (*MOCOS*) gene was found to be downregulated in the nasal olfactory stem cells of ASD individuals and was suggested to likely have an impact on neurodevelopment and neurotransmission (Féron et al. 2016). The comparative gene expression profiling of the lymphoblastoid cell lines revealed elevated expression of *FOXP1* in ASD patients compared to the healthy controls, thus suggesting its role in the pathogenesis of ASD (Chien et al. 2013). Altered mGLUR signaling was also found to contribute to the pathogenesis of non-syndromic autism (Kelleher 3rd et al. 2012). A study identified homer scaffold protein 1 (*HOMER1*), a novel ASD-risk gene that regulated mGLUR activity by interacting with *SHANK3* (Kelleher 3rd et al. 2012). The same study also identified rare coding variants that also affected the components of the RAS/MAPK cascade (Kelleher 3rd et al. 2012). Another study showed various de novo and inherited CNVs implicating ASD genes such as *SHANK2, DLGAP2, SYNGAP1,* and

DDX53-PTCHD1 locus. The study also found that disruptive CNVs of these gene sets were involved in cellular proliferation and Ras/GTPase signaling pathway (Pinto et al. 2010).

2.2.2 Attention-Deficit Hyperactivity Disorder (ADHD)

Attention-deficit hyperactivity disorder (ADHD) is a neurodevelopmental condition that occurs in childhood and adolescence and is characterized by symptoms such as inattention, hyperactivity, and impulsivity. Besides, there are many psychiatric comorbidities such as bipolar disorder, depression, anxiety disorders, substance-use disorder, and personality disorders that co-occur with ADHD. A 2009 meta-analysis study reported the pooled prevalence estimate of 2.5% (95% confidence interval) ADHD in adults (Simon et al. 2009). When diagnosed via the DSM-IV criteria, ADHD is found to affect 5–7% of children (Willcutt 2012; Katzman et al. 2017). Family studies suggest an increased familial liability for adult ADHD compared with childhood ADHD. However, multiple sources reported that the heritability of both clinically diagnosed adult and childhood ADHD is similar. Candidate gene and genome-wide studies suggested similar genes to be involved in adult and childhood ADHD. In few cases, different genes and alleles are responsible for adult and childhood ADHD.

Moreover, linkage and rare genetic variant studies in ADHD have led to the identification of novel genes and causative mutations (Franke et al. 2012). A 2018 study identified several ADHD candidate genes, namely, *DIRAS2, GRIN2B, GRM1, NOS1, PARK2, SNAP25, SYT2, ATM, CAMK2D, CAMK2G, PDE4D, ADRA1A, ADRA1B*, and *SLC6A9*, using gene network analysis (Hayman and Fernandez 2018). In a meta-analysis study, serotonin transporter gene (*5HTT*), serotonin receptor gene (*HTR1B*), dopamine transporter gene (*DAT1*), dopamine receptor genes (*DRD4, DRD5*), and a gene encoding for synaptosomal-associated protein (*SNAP25*) were found to be significantly associated with ADHD (Gizer et al. 2009). Another meta-analysis study found an association between 10 repeat (R) alleles of a variable number tandem repeat (VNTR) polymorphism in the 3′-untranslated region of the dopamine transporter gene *SLC6A3* and ADHD (Yang et al. 2007). Another study found the association of 9R allele with adult ADHD (Brown et al. 2011). In contrast, a recent study found that *SLC6A3* polymorphism influenced attentional/cognitive functions but it was not related to ADHD (Kuc et al. 2020). Another meta-analysis study in adult ADHD confirmed the association of brain-specific angiogenesis inhibitor 1-associated protein 2 (BAIAP2), which played an important role in neuronal proliferation and survival (Bonvicini et al. 2016). The study found a positive association of SNP rs8079781 in the *BAIPA2* gene with ADHD symptoms and showed that *BAIPA2* was highly expressed in the left human cerebral cortex of ADHD adults (Bonvicini et al. 2016). Forkhead box 2 (*FOXP2*) gene, which was found to encode a transcriptional factor involved in speech and language disorders, was also found to be implicated in ADHD (Ribasés et al. 2012). A FOXP2 alteration in a mouse model was shown to affect the cortico-basal ganglia circuits which are involved in speech and language regions (Lai et al. 2003; Enard et al. 2009). A large number of CNVs could be detected by common variant genotyping arrays used in genome-wide association (GWAS) studies. The data obtained from GWAS and meta-analysis study identified the 12 risk loci that were associated with several genes including *FOXP2*, dual-specificity phosphatase 6 (*DUSP6*), and sortilin-related vps10 domain-containing receptor 3 (*SORCS3*) in ADHD (Demontis et al. 2019).

Another GWAS study identified a higher prevalence of rare CNVs within the Parkinson protein 2 (*PARK2*) gene in ADHD patients compared to healthy controls (Jarick et al. 2014). A recent study by Palladino et al. (2020) suggests energy impairment in cell models derived from adult ADHD patients carrying *PARK2* CNV deletions or duplications (Palladino et al. 2020). The role of F-box-only protein 33 (*FBXO33*) and E3 ubiquitin ligase (*RNF122*) as novel susceptibility genes for ADHD was confirmed by other GWAS studies (Sánchez-Mora et al. 2015; Garcia-Martínez et al. 2017). Also, the investigators have reported duplications in the alpha-7 nicotinic acetylcholine receptor (*CHRNA7*) gene at chromosome 15q13.3 (Williams et al. 2012). The CNVs of behavior-related genes such as neuropeptide Y (*NPY*) were also found to be implicated in the pathogenesis of ADHD (Lesch et al. 2011). The results of the study showed duplication on chromosome 7p15.2-15.3 harboring NPY and a link between dose-dependent increase in NPY with reward and emotion processing in the ADHD duplication carriers (Lesch et al. 2011). Another GWAS study identified CNVs affecting metabotropic glutamate receptor genes in ADHD (Elia et al. 2011). The study reported deletions in metabotropic glutamate receptor genes (*GRM7* and *GRM8*) and duplications in *GRM1* in ADHD cohorts (Elia et al. 2011). Another study confirmed the involvement of glutamatergic genes in the pathophysiology of ADHD (Akutagava-Martins et al. 2014). Although the data showed no significant difference in the number of CNVs in ADHD and control samples, the study reported that CNVs in the *GRM5* gene are associated with anxiety disorders in ADHD patients (Akutagava-Martins et al. 2014). In an interesting research, authors studied the response of methylphenidate in ADHD children using GWAS and identified SNPs in *GRM7* and norepinephrine transporter gene *SLC6A2* (Mick et al. 2008). An increased burden of rare functional and disruptive variants in candidate risk genes in individuals with persistent ADHD was also documented by the WES study (Demontis et al. 2016). Using exome sequencing, Kim et al. identified six de novo missense variants in genes expressed in the brain (*DAGLA, QARS, TRPM2, WDR83, CSMD2,* and *TBC1D9*) in individuals with sporadic ADHD (Kim et al. 2017). Another exome chip study revealed that four significant loci (*NT5DC1, SEC23IP, PSD,* and *ZCCHC4*) that are found to be involved in signal transduction are important for cellular communication in ADHD (Zayats et al. 2016). Moreover, BDNF, which is a key protein involved in learning and memory processes, has also been found to be implicated in ADHD (Hawi et al. 2017). An NGS study identified rare DNA variants in *the BDNF* gene and reported that BDNF acts as a genetic risk factor for ADHD (Hawi et al. 2017).

2.2.3 Intellectual Disability (ID)

Intellectual disability (ID) replaces the older term of "mental retardation" and is characterized by a general learning disability in which an individual's cognitive abilities, functioning, and skills are limited. It is the most common neurodevelopmental disorder and represents an important socioeconomic problem as many healthcare professionals do not perceive it as a health condition. The global prevalence of ID is estimated to be lower than 1% (McKenzie et al. 2016). In the human genome, the X-chromosome comprises about 5% but interestingly accounts for 15% of the genes known to be associated with ID (Neri et al. 2018). The X-linked ID was specifically associated with X-linked recessive inheritance. More than 140 X-linked ID genes were identified so far by using high-resolution microarrays and NGS (Neri et al. 2018). Earlier, autosomal dominant forms of ID genes were considered to be more prevalent in Western countries, while in the Middle East regions, where inbreeding is common, autosomal recessive forms of ID genes were found to be prevalent (Hu et al. 2019; Kahrizi et al. 2019). Cytogenetic analysis such as karyotyping revealed trisomy 21 (also known as Down's syndrome) as the most common chromosomal aneuploidy in ID (Fukushi et al. 2012). Most of the genes involved in ID are either X-linked or autosomal (dominant or recessive) but the latter was found to be more common (Ropers 2008).

As identified by the WES study, a pathogenic de novo missense mutation in the neurodevelopmental gene-trio Rho guanine nucleotide exchange factor (*TRIO*) gene is associated with ID, microcephaly, and dysmorphism (Pengelly et al. 2016). Moreover, the study found that the mutations are specific to the Rac-guanine nucleotide exchange factor (GEF) domain of the *TRIO* gene and they were found to be inherited in an autosomal dominant manner (Pengelly et al. 2016). By performing SNP array and exome sequencing, another study identified an autosomal recessive mutation in trafficking protein particle complex 9 (*TRAPPC9*) in two Italian sisters born to non-consanguineous parents (Marangi et al. 2013). The detailed phenotypic analysis revealed that the homozygous *TRAPPC9* loss-of-function mutations resulted in a phenotype characteristic of ID, hypotonia, obesity, and peculiar facial appearance (Marangi et al. 2013). In a similar study, autosomal recessive ID candidate genes in Pakistani and Irani consanguineous families were identified (Harripaul et al. 2018). The novel candidate genes identified were *ABI2, MAPK8, MPDZ, PIDD1, SLAIN1, TBC1D23, TRAPPC6B, UBA7, and USP44,* and genes with missense mutations identified were *BDNF* or ten-eleven translocation methylcytosine dioxygenase 1 (*TET1*) (Harripaul et al. 2018). A study by Riazuddin et al. identified 30 novel candidate genes (Ka et al. 2016) and another study by Santos-Cortez et al. identified 7 novel genes (Fisher and DeFries 2002) for autosomal recessive ID in Pakistani consanguineous families. The study also identified homozygous variants in two ID candidate genes, *GRAMD1B* and *TBRG1* (Santos-Cortez et al. 2018). Another exome sequencing study reported 52 novel recessive genes and 1 autosomal dominant candidate gene in 152 consanguineous families with at least 1 offspring with ID (Reuter et al. 2017). By exploiting genomic tools such as molecular karyotyping, multigene panel, and exome sequencing, a study identified pathogenic variants in three candidate genes *DENND5A, NEMF,* and *DNHD1* harboring homozygous mutations in the ID cohort (Anazi et al. 2017). Additionally, recessive and de novo variants in 32 genes and causal variants in candidate genes (*ASTN1, HELZ, THOC6, WDR45B, ADRA2B,* and *CLIP1*) associated with the pathogenesis of ID were also identified (Anazi et al. 2017). In a study, 77 novel autosomal recessive ID candidate genes, 21 X-linked genes, and 9 genes implicated in other disorders were identified in consanguineous Iranian families by using WES and WGS (Hu et al. 2019). In addition, another exome sequencing in Turkish families with non-syndromic ID identified a novel gene *FAM183A* and seven causative variants in *MCPH1, WDR62, CC2D1A, TUSC3, ZNF335, RARS,* and *ASPM* that were previously associated with autosomal recessive ID (McSherry et al. 2018).

Autosomal dominant mutations caused due to heterozygous mutations in different genes and CNVs are more common as compared to the autosomal recessive mutations in ID. Although there are up to 650 autosomal dominant ID genes that have been reported using WES and WGS technologies, the most common autosomal dominant ID genes mutated are *SYNGAP1, DYRK1A, CTNNB1, STXB1, PACS1, FOXP1, SMARCA2, MED13L, ARID1B, and KCNQ2* (Wieczorek 2018; Deciphering Developmental Disorders Study 2015). Furthermore, some of the mutated genes such as *ARID1B* are involved in dendritic arborization and play an important role in synaptic transmission (Ka et al. 2016). A homozygous missense mutation in *GPR126,* a member of the adhesion G-protein-coupled receptor (GPCR) family, was also reported in patients with severe ID accompanied with speech impairment, seizures, microcephaly, and cerebellar hypoplasia. Moreover, the role of *GPR126* in myelination and radial sorting in Schwann cells suggested its role in the pathogenesis of ID (Hosseini et al. 2019).

2.2.4 Dyslexia

Developmental dyslexia (DD) is a neurodevelopmental disorder that is characterized by the inability to decode written text resulting in impaired reading acquisition despite having normal intelligence. The prevalence rate of dyslexia ranges from 5 to 17.5% among school children (Shaywitz 1998).

Dyslexia is found to be inherited, and a strong familial aggregation of the disorder has been reported (Fisher and DeFries 2002). Dyslexia-susceptibility-1-candidate-1 (*DYX1C1*) was reported to be the first gene found to be associated with dyslexia (Taipale et al. 2003). According to a 2003 study, two SNPs, namely “-3GA” and “1249GT,” were associated with DD in the Finnish population. Since then, several reports have been published showing the role of these SNPs in dyslexia (Taipale et al. 2003; Marino et al. 2007; Dahdouh et al. 2009; Scerri et al. 2004). Among these, some studies reported the association of the *DYX1C1* gene with dyslexia in Italian and German populations (Marino et al. 2007; Dahdouh et al. 2009). In contrast, no association between putative functional alleles of the *DYX1C1* gene with dyslexia was observed in a large sample of sibling pairs from the United

Kingdom (Scerri et al. 2004). According to a 2006 GWAS study, nine genetic loci (*DYX1–DYX9*) and four candidate genes (*DCDC2, KIAA0319, ROBO1, and DYX1C1*) for dyslexia were identified (Schulte-Körne et al. 2006). Apart from these genes, different studies have reported the association of several other candidate genes associated with dyslexia, such as *MRPL19, C2ORF3, NRSN1, FAM176A, KIAA0319L, TTRAP, THEM2,* and *FMR1* (Scerri et al. 2004; Skeide et al. 2016; Francks et al. 2004; Becker et al. 2014; Nayar et al. 2019). Contrastingly, a study found associations of *DCDC2* and *KIAA0319* genes with reading abilities and their involvement in dyslexia but did not find any association of the *MRPL19/C2ORF3* locus in reading abilities, suggesting no association of these genes with dyslexia (Scerri et al. 2011). Other genes such as *CNTNAP2, CNTNAP5, DOCK4,* and *FOXP2* were also found to be associated with dyslexia (Peter et al. 2011; Pagnamenta et al. 2010). Genes encoding for glutamate receptor (*GRIN2B*) and glucose transporter (*SLC2A3*) were found to contribute to memory-related cognitive impairments and glucose deficits in dyslexia (Ludwig et al. 2010; Mascheretti et al. 2015; Roeske et al. 2011). Another study reported that a genetic variant in the disco interacting protein 2 homolog A (*DIP2A*) gene was found to be associated with DD in the Chinese population (Kong et al. 2016). Besides, a noncoding variant in the S100 calcium-binding protein B (S100B) gene was associated with spelling performance in families of German origin, thus suggesting its role in DD (Matsson et al. 2015).

Most of the DD-susceptible genes were found to be involved in neuronal migration, cilia functions, and estrogen signaling (Kere 2014). For instance, *DYX1C1, DCDC2,* and *KIAA0319* are found to be involved in neuronal migration during embryonic development (Kere 2014). Mutations in the *DCDC2* gene were found to contribute to deficits in rapid auditory processing, working, and reference memory in mice, suggesting the association of *DCDC2* genetic variants with impairments in memory ability and phonological processing (Truong et al. 2014). Similarly, another study showed that knockdown of *DCDC2* resulted in impaired speech sound discrimination in rats (Centanni et al. 2016). Another study reported that mutation in the *DCDC2* gene enhanced glutamatergic synaptic transmission between layer four neurons of the mouse neocortex (Che et al. 2016). Another gene implicated in dyslexia, *FOXP2* was previously associated with speech and language disorders and has a significant impact on neurite outgrowth in primary neurons and neuronal models (Vernes et al. 2011; Lai et al. 2001). Mice models with disruption in the *FOXP2* gene showed impaired ultrasonic vocalization and motor impairment (Fujita et al. 2008; Shu et al. 2005). *CNTNAP2* is another gene that is involved in a broad range of NDDs such as autism, ID, and language impairment, including dyslexia. The role of *CNTNAP2* in neuronal connectivity and migration and synaptic functioning has been discussed in many studies (Gdalyahu et al. 2015; Peñagarikano et al. 2011; Rodenas-Cuadrado et al. 2014). In knockout mice studies, anomalies in auditory processing, poor social interactions, and reduced vocalizations were observed, suggesting the role of *CNTNAP2* in neural systems associated with learning and auditory processes (Rendall et al. 2016; Truong et al. 2015). *ROBO1* is another gene that is implicated in regulating axonal and dendritic growth and connections between different brain hemispheres (Kere 2014; Andrews et al. 2006).

2.2.5 Tourette's Syndrome (TS)

Tourette's syndrome (TS) is an NDD characterized by involuntary, repetitive movements and vocalizations called tics, the characteristic feature of TS. It is a heritable condition and is influenced by both genetic and environmental factors. Family studies have shown a 10–100-fold increase in the rate of TS in first-degree relatives compared to healthy controls (Kidd et al. 1980; Pauls et al. 1981; Mathews and Grados 2011).

Genes involved in the dopaminergic pathways have been generally linked to the etiology of TS. Candidate gene studies in TS have mainly focused on the dopaminergic pathway as dopamine antagonists serve as effective medications for the suppression of tics. Early studies reported the association of *DRD2* Taq1 A polymorphism (SNP rs1800497) with TS (Comings et al. 1991). This finding was replicated, and *DRD2* polymorphisms such as Taq I and H313H were found to be associated with TS in Taiwanese children (Lee et al. 2005). Similarly, another study showed a positive association for three SNPs (rs6279, rs1079597, and rs4648318) and five SNP haplotypes across the *DRD2* gene in a South American population. It confirmed the implication of *DRD2* in TS etiology (Herzberg et al. 2010). In contrast, some studies reported no association of the *DRD2* gene with TS (Díaz-Anzaldúa et al. 2004; Gelernter et al. 1994; Nöthen et al. 1994). A recent study compared healthy controls with TS patients and showed hypermethylation of DRD2 gene that positively correlated with tic severity (Müller-Vahl et al. 2017), suggesting the involvement of altered epigenetic regulation of DRD2 gene in the pathophysiology of TS (Müller-Vahl et al. 2017).

Polymorphisms of the dopamine transporter (*DAT1/SLC6A3*) gene were also associated with the etiology of TS. The 10-repeat allele of a 40 bp VNTR in the 3′ untranslated region of the *DAT1* gene was frequently implicated in TS (Comings et al. 1996). The *DAT1* 40 bp VNTR was also associated with increased tic severity in a TS family study (Tarnok et al. 2007). Besides, another *DAT1* polymorphism (DAT1 Ddel) was found to be associated with TS (Yoon et al. 2007). The dopamine β-hydroxylase (*DBH*) gene is found to influence dopaminergic and adrenergic systems as the β-hydroxylase enzymes are involved in the conversion of dopamine to norepinephrine (Weinshenker 2007). An association study using samples from Taiwanese children found that polymorphisms in *DAT1* and the dopamine β-hydroxylase gene (*DBH*) could be used as susceptibility markers in TS (Chou et al. 2013). In contrast, Ozbay et al. did not find any association of *DBH* in Canadian and Turkish samples. They stated that the finding could be false positive due to reduced sample size and number of informative transmissions (Ozbay et al. 2006).

Dopamine is not the only neurotransmitter involved in TS pathogenesis. Genes in the serotonergic pathway are also implicated in TS. The tryptophan hydroxylase 2 (*TPH2*) gene is primarily expressed in the serotonergic neurons and acts as a rate-limiting enzyme in the synthesis of serotonin (Walther et al. 2003). An association between the two SNPs of the *TPH2* gene (rs4565946 and rs4570625) with TS has been previously demonstrated (Mössner et al. 2007). The variants of the serotonin receptor gene (*HTR2C*) were studied in Chinese Han patients with TS, but no significant association between *HTR2C* variants and TS was reported (Guo et al. 2012). Functional studies of the HTR2C promoter region revealed two polymorphisms (C-759T, G-697C) that are associated with TS in male patients (Dehning et al. 2010). On the other hand, serotonin transporter (*SLC6A4*) that has been implicated in the etiology of obsessive-compulsive disorder (OCD) (Voyiaziakis et al. 2011) has provided inconclusive results regarding its association with TS. So far, many studies have focused on the serotonin transporter-linked polymorphic region (5-HTTPLR) but did not find any significant association of the 5-HTTPLR polymorphism with TS (Liu et al. 2011; Cavallini et al. 2000). However, a study found high expression of *SLC6A4* variants in TS probands alone and a rare gain-of-function variant (SERT I425V) in three male siblings with TS and/or OCD (Moya et al. 2013). Moreover, a recent study showed the increased expression of *SLC6A4* in TS individuals compared to controls. The study also observed that the L_{AC}/L_{AC} genotype of the 5-HTTLPR/rs25531/rs25532 three-locus haplotype is associated with higher *SLC6A4* mRNA expression levels in TS individuals (Hildonen et al. 2021).

Another gene, known as the monoamine oxidase-A gene (*MAOA*), which encodes for the MAO enzyme required for the catalytic oxidation of dopamine, norepinephrine, and serotonin, has been implicated in the etiology of TS. A study found the VNTR polymorphism at the X-linked *MAOA* gene to be associated with behavioral phenotypes in both TS and drug abuse (Gade et al. 1998). Similarly, another study observed high-activity alleles of the *MAOA* VNTR polymorphism in French-Canadian TS populations (Díaz-Anzaldúa et al. 2004). Contrastingly, one study found no significant association of the *MAOA* VNTR polymorphism in the Chinese Han population and stated that the negative finding could be due to the small population size used in the study (Liu et al. 2015).

The role of glutamatergic system was also demonstrated by numerous studies in the pathogenesis of TS (Pogorelov et al. 2015; Adamczyk et al. 2011; Anderson et al. 1992). A study identified E219D as a functional glutamate aspartate transporter 1 (*SLC1A3*) variant in 11 heterozygous individuals with TS (Adamczyk et al. 2011). The postsynaptic scaffolding protein (*DLGAP3*), which is highly expressed in striatal glutamatergic synapses, was evaluated as a candidate TS susceptibility gene. Nominal significant associations were reported between rs11264126 and two haplotypes containing rs11264126 and rs12141243 and TS. However, these results were considered not significant after correction for multiple hypothesis testing (Crane et al. 2011).

The neuronal transmembrane protein member of the neurexin family *CNTNAP2* gene, which has been implicated in various NDDs such as ASD, ID, and dyslexia, was also found to be associated with TS. A study by Verkerk et al. showed that the father and two children affected with TS shared a chromosome 2p21–p23 insertion on chromosome 7q35–q36, which resulted in the disruption of the *CNTNAP2* gene

(Verkerk et al. 2003). Additional breaks and structural changes in the *CNTNAP2* gene were also reported in a boy with speech delay and ASD, along with features of TS (Poot et al. 2010). Another study reported a familial balanced reciprocal translocation t(7;15)(q35;q26.1) of the *CNTNAP2* gene in phenotypically normal individuals and pointed that the truncation of the *CNTNAP2* gene does not necessarily lead to the symptoms of TS (Belloso et al. 2007). CNV analysis also revealed deletions (~400 kb deletions) and genomic rearrangements involving the *NRXN1* gene in Latin American TS populations (Nag et al. 2013). Similarly, another study identified *NRXN1* deletions in European TS samples, thus showing that rare structural variation significantly contributes to the genetic architecture of TS (Huang et al. 2017). Furthermore, in a study carried out in a family with varied neuropsychiatric illnesses, the latter were associated with a deletion of exons 4, 5, and 6 of NLGN4. The autistic boy (proband) has a TS- and ADHD-affected brother and a mother with a learning disorder, anxiety, and depression (Lawson-Yuen et al. 2008).

There have been reports on the association of sequence variants in slit and trk-like family member 1 (*SLITRK1*) with TS. Abelson and coworkers found functional variations (one frameshift mutation and sequence variants in the binding site for microRNA miR-189) in the *SLITRK1* gene in three subjects with TS (Abelson et al. 2005). The wild-type *SLITRK1* was found to enhance dendritic growth in neuronal cultures, thus suggesting that mutations in *SLITRK1* might impair dendritic activity and functioning (Abelson et al. 2005; Kang et al. 2016). In another association study, a rare 3′ UTR variant, var321 (*SLITRK1* var321), was identified in 5 (out of 515) Ashkenazi parents (probands) affected with TS and in 1 subject from an Ashkenazi control sample. The results of the present study did not support the previously reported association and suggested that var321 was overrepresented among the Ashkenazi population compared with other populations of European ancestry (Keen-Kim et al. 2006). The authors of the study believe that overrepresentation of rare variants in a specific ethnic group could complicate the interpretation of the association analyses of rare variants. The precisely matching cases and control population, thus, underscored the importance of association analyses of such variants (Keen-Kim et al. 2006). There are several studies which contradicted the findings and reported no association of the *SLITRK1* var321 in Caucasian patients (Lalli et al. 2016; Milani et al. 2015) and Taiwanese, Canadian, and Austrian populations (Zheng et al. 2016; Smith et al. 2006; Verpelli et al. 2012).

2.3 Proteins Associated with NDDs

Different types of proteins are associated with NDDs (Fig. 2.2), including scaffolding or postsynaptic density (PSD) proteins, synaptic proteins, bromodomain-containing proteins, and DNA-binding proteins. The main PSD proteins that were found to be implicated in ASD include SYNGAP1, CTNNB1, GRIN2B, NLGNs, and SHANKs (de la Torre-Ubieta et al. 2016; Bourgeron 2015). The loss of function of one copy of *SHANK3* gene causes Phelan-McDermid syndrome, an NDD characterized by ID and ASD. Reduced expression of SHANK3 protein leads to a

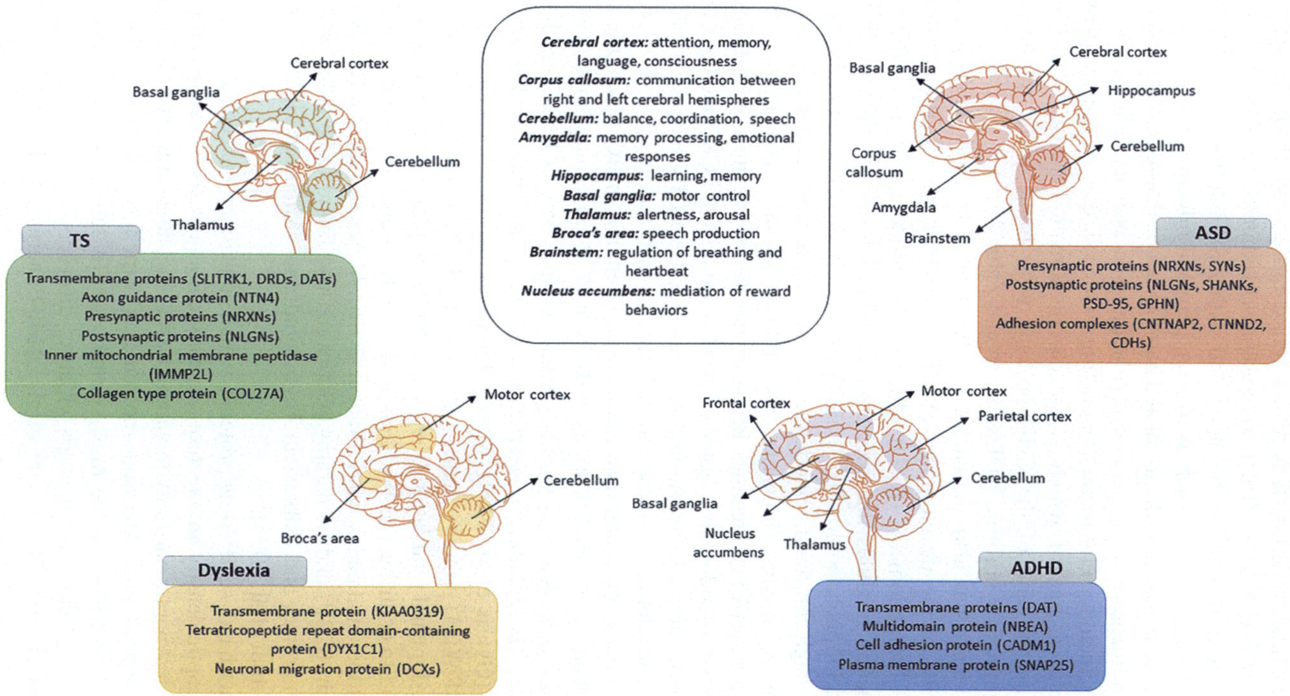

Fig. 2.2 Proteins associated with different NDDs. Mutation or disruption in proteins related to different NDDs that might contribute to the structural and functional impairment of the brain

reduced number of dendrites and impaired synaptic transmission and plasticity (Costales and Kolevzon 2015). Mutations in the fragile-X mental retardation 1 (*FMR1*) and cytoplasmic fragile-X mental retardation protein (FMRP)-interacting protein 1 (*CYFIP1*) (FMRP-CYFIP1 complex) involved in the regulation of various PSD proteins are also found to be associated with ASD (Clifton et al. 2020). As various environmental factors such as trauma, perinatal infection, and malnutrition could increase the risk of multifactorial and heterogeneous disorders, NDDs are also significantly influenced by epigenetic regulation. The primary epigenomic regulatory process includes chromatin remodeling, which is carried out by multiunit protein complexes, one of which is the chromodomain helicase DNA-binding (CHD) proteins. Several of the CHD proteins were found to be implicated in autosomal dominant syndromic NDDs (Yasin and Zahir 2020).

Recent evidence has also shown the involvement of bromodomain-containing proteins (BCPs) in the development of NDDs. BCPs have a particular domain known as the bromodomains (BRD) which are evolutionarily conserved protein-protein interaction modules that recognize acetylated lysine residues on histones (Fujisawa and Filippakopoulos 2017). BCPs are mainly involved in regulating gene expression through chromatin remodeling, histone modification and recognition, and transcriptional regulation (Fujisawa and Filippakopoulos 2017). All histone acetyltransferases (HATs), enzymes that acetylate lysine residues on histones, are found to contain bromodomains. Some of the BCPs such as CECR2 and BAZ1A are considered modulators of chromatin remodeling, while BCPs such as BRD2 and BRWD3 are considered as transcriptional regulators (Li et al. 2013). BCPs are critical for cellular events, and therefore genetic alteration in BCPs could result in complicated phenotypes. Modifications in BCPs or BCP-containing complexes have been shown to cause defect in the chromatin-targeting protein BRD2 resulting in neural tube defects including misexpression of the genes implicated in neuronal development (Gyuris et al. 2009). Mutant bromodomain and WD-repeat domain-containing (BRWD3) proteins altered various intracellular signaling pathways leading to X-linked mental retardation (Field et al. 2007). Haploinsufficiency of BAZ1B was also found to contribute to Williams syndrome by transcriptional dysregulation of neurodevelopmental pathways (Lalli et al. 2016). Moreover, mutant HAT CREB-binding protein (CBP) and its paralog p300 were also found to be associated with FXS, Rett syndrome, and Rubinstein-Taybi syndrome (Milani et al. 2015; Zheng et al. 2016; Smith et al. 2006).

On the other hand, scaffolding proteins, which constituted the central component of the postsynaptic density architecture, were critical regulators of various processes associated with synaptic plasticity including trafficking, anchoring, and clustering of glutamate receptors and adhesion molecules (Verpelli et al. 2012). DLGAP, SHANKs, and HOMER are the most common scaffolding proteins implicated in NDDs such as ASD and neuropsychiatric disorders such as schizophrenia (Rasmussen et al. 2017; Guilmatre et al. 2014; Banerjee et al. 2016; Spellmann et al. 2011).

2.4 Signaling Pathways Involved in NDDs

2.4.1 Autism Spectrum Disorder (ASD)

Any alteration or disruption of signaling pathways could contribute to the development of ASD and other NDDs. The causal genes of ASD might act upstream or downstream of different signaling pathways such as Wnt, sonic hedgehog (SHH), retinoic acid (RA), fibroblast growth factor (FGF), mTOR, and ERK1/2. Many of the ASD genes contribute to different ASD phenotypes through various signaling pathways and play an important role in the pathogenesis of ASD (Nisar et al. 2019). The presence of genetic variations in the voltage-gated calcium channels and other calcium signaling abnormalities showed the implication of calcium signaling in ASD (Liao and Li 2020). The calcium signals originated from the release of calcium through intracellular ion channels or extracellular calcium release suggests that most of the calciumopathies are due to ion channel defects (Gargus and Schmunk 2014). Several studies reported impaired inositol triphosphate (IP3)-mediated calcium signaling in sporadic and rare syndromic forms of ASD (Schmunk et al. 2017; Nguyen et al. 2018). Moreover, it was also suggested that the interaction of IP3 receptors with mitochondria might contribute to the mitochondrial dysfunction phenotypes observed in ASD (Nguyen et al. 2018). An imbalance of calcium/calmodulin-dependent kinase 4 (*CaMK4*) and specific isoforms of cytoplasmic fmr1-interacting protein 1 (*CYFIP1*) are found to contribute towards a higher ASD risk (Waltes et al. 2014). Similarly, a de novo mutation in the catalytic domain of calcium/calmodulin-dependent protein kinase type II subunit alpha (*CAMKIIα*), (Chang et al. 2019) decreased excitatory synaptic transmission and dendritic spine density in mice, thus suggesting the role of *CAMKIIα* in ASD-related behavioral phenotypes (Stephenson et al. 2017).

Wnt signaling is also implicated in different neurodevelopmental processes such as neurogenesis, neuronal migration and maturation, and synaptic plasticity (Rosso and Inestrosa 2013; Bielen and Houart 2014). Any perturbation in the Wnt signaling could lead to the development of NDDs. Moreover, three functional pathways, namely chromatin remodeling, mitochondrial dysfunction, and Wnt signaling, were found to be potentially involved in the pathology of ASD (Bae and Hong 2018). The Wnt signaling is broadly classified into canonical and noncanonical pathways, and both these pathways play significant roles in NDDs.

Sonic hedgehog (SHH) is another signaling pathway known to influence the neurogenesis mechanism during the development of the central nervous system (Mooney et al. 2016) and is implicated in ASD. The aberrant downregulation of the smoothened (SMO)-SHH signaling leads to the proteolytic cleavage of glioma-associated homolog (GLI) into GLI3 (repressor) which by suppressing target gene leads to increased oxidative stress and neuroinflammation (Rahi and Mehan 2020). Although the loss-of-function experiments showed that gene-patched domain-containing 1 (*PTCHD1*, an ASD-associated gene) is not required for the SHH-dependent neuronal precursor proliferation, it played an essential role in the synaptic transmission process in the mouse dentate gyrus (Tora et al. 2017).

Moreover, possible interactions between 7-dehydrocholesterol reductase (*DHCR7, another* ASD gene) and engrailed 2 (*EN2*) and SHH signaling were suggested (Kumar et al. 2019).

Another signaling that is found to be associated with ASD is the retinoic acid (RA) signaling. RA is a functional metabolite of vitamin A and is important for vertebrate cell growth, development, and organogenesis (Kin Ting Kam et al. 2012). RA mediates transcriptional effects by binding to retinoic acid receptors (RARs) and retinoid X receptors (RXRs) (Das et al. 2014). It has been found that the deficiency of vitamin A could increase autistic behaviors by decreasing the expression of hypothalamic RAR-beta (RARβ), CD38, and serum oxytocin levels in rat offspring (Lai et al. 2018). Moreover, aberrant methylation of RAR-related orphan receptor A (*RORA*) gene in the lymphoblastoid cell lines derived from autistic individuals and reduced RORA protein expression in the autistic brain suggested the role of the RA acid receptor genes in the pathophysiology of ASD (Nguyen et al. 2010). Also, *ALDH1A3*, an enzyme involved in RA synthesis, *FOXN1,* and RA regulatory pathways, was found to be linked to ASD (Moreno-Ramos et al. 2015). *RORA* was also found to transcriptionally regulate ASD-associated genes *NLGN1, NTRK2, A2BP1, ITPR1, CYP19A1,* and *HSD17B10* (Sarachana and Hu 2013). The excessive dosage of *UBE3A* was found to impair RA signaling and RA-mediated neuronal synaptic plasticity in ASD (Xu et al. 2018).

Fibroblast growth factor (FGF) signaling is implicated in the pathogenesis of various neurological disorders including ASD (Esnafoglu and Ayyıldız 2017). FGFs were involved in the patterning and neurogenesis during cortical development, and alteration or mutation in the FGF genes could lead to cortical abnormalities, hyperactivity, and social deficits and predispose to the development of ASD (Vaccarino et al. 2009). In support of this, a study found decreased serum levels of FGF-2 in autistic individuals as compared to the healthy controls (Esnafoglu and Ayyıldız 2017). Other signaling pathways that were found to be involved in ASD include the mTOR and ERK1/2 pathways (Subramanian et al. 2015). Moreover, GABA/glutamate abnormalities in the corticostriatal circuitry might be linked to the alteration of *NRXNS/NLGNS*, suggesting the role of GABA and glutamate signaling in the pathogenesis of ASD (Horder et al. 2018).

2.4.2 Attention-Deficit Hyperactivity Disorder (ADHD)

A study utilizing pathway analysis methods in two GWAS datasets identified several brain-relevant ADHD pathways such as transforming protein RhoA signaling pathway, fibroblast growth receptor activity, glycosaminoglycan biosynthesis, and signaling pathways associated with potassium channel genes (Mooney et al. 2016). Several studies also reported the association of genes involved in the Wnt pathway with ADHD. A GWAS study found α-catenin (*CTNNA2*), a regulator of synaptic plasticity, to be associated with adult ADHD (Lesch et al. 2008). Another study identified three canonical pathways, namely the inhibition of matrix metalloproteases (*ADAM10, ADAM12, MMP7*), axonal guidance signaling

(*ABLIM2, ADAM10, ADAM12, MMP7, PAK7, SLIT1*), and Wnt/β-catenin signaling pathway (*MMP7, RARB, SFRP4, SOX5*), that were enriched in genes implicated in psychiatric condition called oppositional defiant disorder (ODD) to be highly prevalent in ADHD (Aebi et al. 2016). A candidate gene association study found potassium voltage-gated channel-interacting protein 4 *KCNIP4* gene, which is found to be involved in the negative feedback loop in the Wnt/β-catenin signaling pathway to be associated with ADHD and personality disorders (PD) (Weißflog et al. 2013). Another evidence of the involvement of the Wnt signaling pathway in ADHD comes from a recent study that showed a significant correlation between low gray matter volume and miR-30e-5p, miR-126-5p, and miR-140-3p miRNAs that are involved in the axonal guidance and the Wnt signaling pathway (Wang et al. 2020). An interesting study showed that phosphatidylinositol 3-kinase-gamma (PI3Kγ) knockout mice exhibited ADHD-related symptoms such as impaired attention and mnemonic domains, social dysfunction, and hyperactivity (D'Andrea et al. 2015). The study also demonstrated that dysregulation in CREB signaling exerted by the interaction of PI3Kγ and phosphodiesterase 4D (PDE4D) in the noradrenergic neurons of the locus coeruleus contributed to the ADHD phenotype in mice (D'Andrea et al. 2015). Another study performed pathway and network analysis and found nicotine signaling, *N*-methyl-D-aspartate (NMDA) receptor trafficking, cannabinoid receptor signaling, nitric oxide synthase (NOS) signaling, and alpha-1 adrenergic receptor signaling to be associated with ADHD (Hayman and Fernandez 2018).

2.4.3 Intellectual Disability (ID)

Compelling evidence demonstrated that major ID-related proteins interact with proteins enriched at synaptic compartments. These synaptic proteins play an important role in synaptic function, plasticity, and dendritic spine morphogenesis (Humeau et al. 2009; Vaillend et al. 2008). For instance, different ID proteins such as SynGAP, member of the guanine nucleotide exchange factor of proteins (IQSEC2), X-linked interleukin-1 receptor accessory protein-like 1 (IL1RAPL1), and collybistin (Cb) are found to be directly associated with postsynaptic density protein 95 (PSD-95) and gephyrin which are major proteins involved in the excitatory and inhibitory synapses, respectively (Pavlowsky et al. 2012). The major synaptic signaling pathways involved in ID are the RhoGTPase, Ras and JNK signaling, Rab, and Arf (Pavlowsky et al. 2012). Rho-class small GTPases are involved in important cellular processes such as neurogenesis, axon guidance, and synaptic plasticity. Guanine nucleotide exchange factors (GEFs) and GTPase-activating proteins regulate the RhoGTPases via protein:protein interactions. Dysregulated RhoA, CDC42, and Rac1/Rac3 have been associated with ID (Martinelli et al. 2018; Reijnders et al. 2017; Hiraide et al. 2019; Martin Lorenzo et al. 2020). Moreover, the ankyrin repeat domain 11 (*ANKRD11*) gene, which is found to be associated with ID, autism, and KBG syndrome, is required for cortical dendrite growth and arborization during brain development. Recently, *ANKRD11* was found to regulate dendrite differentiation through the BDNF/tropomyosin receptor kinase B (TrkB) pathway in a mouse model (Ka and Kim 2018). The study emphasized the significance of *ANKRD11* in regulating the expression of BDNF, TrkB, and neurite growth-related genes which are important for normal brain development (Ka and Kim 2018).

Furthermore, the ubiquitin system acts as a regulatory hub in the middle of the three crucial pathways (Wnt, mTOR, and TGFβ) that are affected in ID and ASD. The ubiquitin ligases (UBE3A, CUL4B, and HUWE1) and the deubiquitylating enzymes (USP7 and USP9X) are involved in the regulation of the members of the three crucial signaling pathways involved in the etiology of ID and ASD (Kasherman et al. 2020). Mutations in signaling pathways controlled by the ubiquitin system and in the ubiquitin-modifying proteins have been reported in ID and ASD. Several of these ubiquitin-modifying proteins were found to interact with the components of the Wnt, TGFβ, and mTOR signaling pathways (Kasherman et al. 2020). During neural development, the signaling pathways are temporally and spatially regulated and any impairment in these pathways could interfere with the processes associated with the normal brain development (Kasherman et al. 2020). Moreover, important neural processes such as axonal development, synaptic pruning, and dendritic maturation are regulated by the ubiquitin system (Ding and Shen 2008; Riccomagno and Kolodkin 2015).

2.4.4 Dyslexia

In individuals with DD, there are always complaints of the appearance of blurred letters and letters moving around during the reading process. The above complaints were found to emerge from abnormalities in the magnocellular system that is an important part of the visual system. The magnocellular system allows rapid perception of movement and changes in brightness and is specialized for the processing of temporal information (Stein and Walsh 1997). The magnocellular stream also known as the m-stream is found to be important for intact reading and in guiding visual attention. Developmental dyslexics were found to have impaired contrast sensitivity to low-luminance stimuli at low spatial frequencies (Ben-Yehudah et al. 2001). The relationship between dyslexia and deficits in the magnocellular pathway was demonstrated by many studies. Demb et al. measured the discrimination and contrast-detection thresholds under conditions for which M pathway integrity is crucial for psychological performance to test the hypothesis that dyslexia is associated with a deficit in the magnocellular (M) pathway. The results demonstrated higher psychophysical thresholds in speed discrimination and contrast-detection tasks in dyslexic subject controls (Demb et al. 1998). Many studies showed the presence of impaired visual motion sensitivity at high illumination levels and contrasts, and the most convincing evidence was demonstrated by an earlier post-mortem study that showed disordered magnocellular layers of the lateral geniculate nucleus (LGN) in five dyslexic brains (Galaburda et al. 1985). The magnocellular theory in dyslexia was further confirmed by a study by Giraldo-Chica et al. that reported reduced left LGN volume in dyslexics compared to the controls (Giraldo-Chica et al. 2015). Interestingly, another study found reduced middle temporal visual (V5/MT) activity in dyslexics compared to age-matched controls. However, when dyslexics were matched with younger controls on reading ability, no differences in the V5/MT activity were observed, suggesting that the magnocellular visual deficits are not causal but might be consequential to dyslexia (Olulade et al. 2013). It is also believed that magnocellular system deficits not only are confined to vision but also affect motion processing in neurological disorders such as schizophrenia (Kim et al. 2006). Another study identified a deficit in the magnocellular dorsal temporal oscillation in dyslexic children and suggested that the deficit in the magnocellular dorsal pathway could impair sub-lexical mechanisms that are important for reading processes (Gori et al. 2014). The implication of the estrogen signaling pathway in dyslexia was also demonstrated earlier.

DYX1C1, one of the susceptibility genes implicated in dyslexia, is found to interact with estrogen receptors (ER) (ERα and ERβ) in the presence of 17-beta-estradiol (Massinen et al. 2009). Furthermore, the overexpression of *DYX1C1* resulted in the decrease of endogenous ERα or exogenous ERβ. The above results were suggestive that the dyslexia susceptibility gene *DYX1C1* is involved in the regulation of ERs and might play an important role in estrogen-signaling effects in the brain (Massinen et al. 2009).

2.4.5 Tourette's Syndrome (TS)

In addition to the most implicated pathways, such as the serotonergic and dopaminergic pathways, there are also other pathways implicated in TS such as the glutamate, GABA, cholinergic, opioid, and endocannabinoid-associated pathways. The dopaminergic pathways include the nigrostriatal, mesolimbic, and mesocortical systems. Dopamine has an established role in various movement and learning disorders, temporal processing, cognitive function, reward processing, and sensorimotor integration (Bromberg-Martin et al. 2010). There is extensive evidence supporting the involvement of the dopaminergic genes in TS (Díaz-Anzaldúa et al. 2004; Müller-Vahl et al. 2017; Tarnok et al. 2007; He et al. 2015). Moreover, recognizing that tic has been identified as a habitual behavioral disorder, studies have shown that increased dopaminergic activity leads to the progression of habitual behaviors (de Wit et al. 2012; Singer 2016) (Maia and Conceição 2017; Hienert et al. 2018). There is an increase in both phasic and tonic DA in TS, which produces increased propensities for tic learning and expression, respectively.

Early studies have also supported that TS is related to the supersensitivity of dopamine receptors (Singer et al. 1982). A postmortem study showed that the dopamine uptake carrier sites were significantly increased in the caudate and putamen of TS patients, suggesting an enhanced dopamine innervation within the striatum in TS (Singer et al. 1991). Another study reported increased presynaptic dopaminergic activity in children with TS using positron-emission tomography (PET) (Ernst et al. 1999). The study reported a higher accumulation of PET radiotracer [18F]fluorodopa (FDOPA) in children with TS compared to healthy controls (Ernst et al. 1999). Another PET study reported that regions of dopamine release were more widespread and extended to the anterior cingulate and medial frontal gyri regions in TS subjects compared to healthy controls (Steeves et al. 2010). Compared to healthy controls, dopamine uptake by the platelet storage granules in TS patients is significantly lower (Rabey et al. 1995). The above studies reflect the dysfunction of the dopaminergic pathway in TS.

Evidence for the role of serotonin in TS was shown in several studies. Early studies showed altered serotonin metabolism in the blood and CSF of TS patients (Butler et al. 1979; Comings 1990). A negative correlation was also reported between overall tic severity and [123I]-β2-beta-carbomethoxy-3-beta (4-iodophenyl) tropane (CIT) binding to the serotonin transporter in the midbrain, thus showing the implication of serotonergic system in TS (Heinz et al. 1998). Moreover, PET imaging with [18F]-altanserin showed increased binding of the serotonin receptor 5-HT2A in multiple brain regions (Haugbøl et al. 2007). Another PET study reported a decreased uptake of alpha-[(11)C]methyl-L-tryptophan (AMT) in the dorsolateral prefrontal cortical regions and increased uptake of AMT in the caudate nucleus and thalamus of TS patients, thus providing evidence for the involvement of the serotonergic system in the pathophysiology of TS (Behen et al. 2007).

Glutamate is an excitatory neurotransmitter and is found to play important roles in corticobasal-ganglia-thalamocortical circuits (CBGTC). The evidence of glutamatergic system dysfunction was supported by several studies. For instance, a postmortem study reported reduced glutamate in the globus pallidus interna, globus pallidus externa, and substantia nigra pars reticulata in TS patients (Anderson et al. 1992). Moreover, reduced striatal and thalamic glutamate and glutamine concentrations were observed in TS patients compared to healthy controls (Kanaan et al. 2017). Also, the role of the glutamatergic system in tic behaviors was supported by studies employing optogenetic and chemogenetic methodologies (Burton 2017). Several studies also supported the use of glutamatergic modulatory therapy for TS (Singer et al. 2010). Contrastingly, a magnetic resonance spectroscopy (MRS) study found no evidence of glutamatergic neuropathology within the frontostriatal circuits in TS subjects (Naaijen et al. 2017).

Gamma aminobutyric acid (GABA) is the primary inhibitory neurotransmitter which is mainly involved in reducing neuronal excitability throughout the nervous system. The GABAergic system regulates anxiety, vigilance, epileptogenic activity, and memory functions (Rudolph 2008). The dysfunctional signaling of dopamine and GABA together was found to contribute to impairment in the functioning of cortical-striatal-thalamic-cortical (CSTC) circuits in TS (Jackson et al. 2015). Imaging and postmortem studies showed alteration in the GABAergic signaling in TS. Early studies in TS-affected individuals showed altered parvalbumin-positive neurons in the basal ganglia and decreased number of parvalbumin and cholinergic interneurons in the striatum (Kalanithi et al. 2005; Kataoka et al. 2010). A PET study using [11C]-flumazenil (a selective $GABA_A$ receptor antagonist) reported decreased binding of GABA receptors bilaterally in the globus pallidus, ventral striatum, amygdala, thalamus, and right insula of TS patients (Lerner et al. 2012). Moreover, a GABA-mediated inhibition process known as the short-interval intracortical inhibition measured by using paired-pulse transcranial magnetic stimulation (TMS) was reported to be reduced in TS individuals (Orth 2009). Furthermore, an MRS study showed that increased GABA concentrations in the supplementary motor area (SMA) in TS individuals inversely correlated with cortical excitability in the primary motor cortex. The results suggested that increased GABA contributed to enhanced control over motor excitability which could lead to the suppression of tics in TS (Draper et al. 2014). Animal studies showed that disruption of striatal GABA by GABA antagonists such as bicuculline and picrotoxin evoked tics and generated tic-like behaviors (Pogorelov et al. 2015; Bronfeld et al. 2013).

On the other hand, striatal cholinergic interneurons that are implicated in TS are involved in motor control and associative plasticity and are found to drive GABA release from dopamine terminals (Nelson et al. 2014). The evidence supporting the involvement of choline in TS was from a postmortem study that found decreased cholinergic interneurons in the associative and sensorimotor regions of the striatum in TS patients (Kataoka et al. 2010). Another study reported that targeted ablation of striatal cholinergic interneurons produced behavioral manifestations of TS in an animal model (Xu et al. 2015). Contrastingly, few early studies reported no association of the cholinergic system with TS (Singer et al. 1984, 1990).

Dysregulation of the histaminergic modulatory system was also implicated in various neuropsychiatric diseases. Histamine is a monoamine neurotransmitter and is mainly involved in inflammatory responses. The dual functionality of H3R as an auto- and hetero-receptor enables them to modulate histaminergic as well as other neurotransmitter systems (Nieto-Alamilla et al. 2016). Many promising studies showed the role of histamine 3 receptor (H3R) antagonists in various neurological disorders such as ADHD, schizophrenia, and narcolepsy (Weisler et al. 2012; Mahmood et al. 2016; Harwell and Fasinu 2020). In addition, the activation of histamine 3 (H3) receptors in the dorsal striatum was found to trigger stereotypies in a mouse model of tic disorders (Rapanelli et al. 2017).

Few studies reported the involvement of the opioid and endocannabinoid system in the pathobiology of TS. The primary evidence for the involvement of opioid receptors was from a postmortem study that reported decreased levels of dynorphin, an opioid peptide in the striatal fibers of a TS patient (Haber et al. 1986). Similarly, another study reported increased concentrations of dynorphin A (Olusanya et al. 2018; Morris-Rosendahl and Crocq 2020; Parenti et al. 2020; van Loo and Martens 2007; Mitchell 2014; D'Gama and Walsh 2018; Lynch 2010; Wilfert et al. 2017) in the cerebrospinal fluid (CSF) of TS patients (Leckman et al. 1988). URB597 is an indirect endocannabinoid agonist mitigated serotonin receptor agonist (DOI)-induced head twitches in four different mouse strains (Ceci et al. 2015). Mathews et al. also reported cannabis use during pregnancy as a critical risk factor for chronic tic disorders/TS (Mathews et al. 2014). Furthermore, a recent study identified three significant gene sets implicating lymphocytic, ligand-gated ion channel signaling, cell adhesion, and transsynaptic signaling processes by using set-based association method (SBA) analysis (Tsetsos et al. 2021).

2.5 Conclusions

Despite the broad and complex heterogeneity of the NDDs, the mutational spectrum of NDDs appears to converge on highly interconnected molecular pathways that encode genes and proteins involved in chromatin remodeling, synaptic function, and transcription mechanisms. Genetic evidence showed that impairment in synaptic functioning mechanisms was central to various NDDs. In the past few decades, many genetic and protein targets have emerged, which might serve as therapeutic targets for the treatment of NDDs. Recent advancements in technologies that are used to generate and combine large datasets of genetic information have led to the successful identification of gene mutations underlying NDDs, opening a gateway to personalized medicine and the implementation of systematic genetic counseling. Considerable evidence suggested that the most widely used technologies for the identification of NDD causal mutations such as WES and WGS could help improve the diagnosis and treatment of genetic diseases that would ultimately improve the patient health outcomes. The establishment of efficient and reliable translational models could help in the identification of potential therapeutic targets and unravel the molecular mechanisms and the mutational effects at both molecular and phenotypic levels.

Acknowledgment This work was supported by Qatar Foundation, National Priorities Research Program (NPRP10-0202-170320), and the Sidra Medicine Precision Program (SDR #200048).

References

Abelson JF et al (2005) Sequence variants in SLITRK1 are associated with Tourette's syndrome. Science 310(5746):317–320

Accogli A et al (2019) De novo pathogenic variants in N-cadherin cause a syndromic neurodevelopmental disorder with corpus collosum, axon, cardiac, ocular, and genital defects. Am J Hum Genet 105(4):854–868

Adamczyk A et al (2011) Genetic and functional studies of a missense variant in a glutamate transporter, SLC1A3, in Tourette syndrome. Psychiatr Genet 21(2):90–97

Aebi M et al (2016) Gene-set and multivariate genome-wide association analysis of oppositional defiant behavior subtypes in attention-deficit/hyperactivity disorder. Am J Med Genet B Neuropsychiatr Genet 171(5):573–588

Akutagava-Martins GC et al (2014) Glutamatergic copy number variants and their role in attention-deficit/hyperactivity disorder. Am J Med Genet B Neuropsychiatr Genet 165b(6):502–509

Anazi S et al (2017) Clinical genomics expands the morbid genome of intellectual disability and offers a high diagnostic yield. Mol Psychiatry 22(4):615–624

Anderson GM et al (1992) Postmortem analysis of subcortical monoamines and amino acids in Tourette syndrome. Adv Neurol 58:123–133

Andrews W et al (2006) Robo1 regulates the development of major axon tracts and interneuron migration in the forebrain. Development 133(11):2243–2252

Artuso R et al (2011) Investigation of modifier genes within copy number variations in Rett syndrome. J Hum Genet 56(7):508–515

Bacchelli E et al (2020) An integrated analysis of rare CNV and exome variation in Autism Spectrum Disorder using the Infinium PsychArray. Sci Rep 10(1):3198

Bae SM, Hong JY (2018) The Wnt signaling pathway and related therapeutic drugs in autism spectrum disorder. Clin Psychopharmacol Neurosci 16(2):129–135

Banerjee A et al (2016) Overexpression of Homer1a in the basal and lateral amygdala impairs fear conditioning and induces an autism-like social impairment. Mol Autism 7:16

Bauman ML, Kemper TL (2005) Neuroanatomic observations of the brain in autism: a review and future directions. Int J Dev Neurosci 23(2–3):183–187

Becker J et al (2014) Genetic analysis of dyslexia candidate genes in the European cross-linguistic NeuroDys cohort. Eur J Hum Genet 22(5):675–680

Behen M et al (2007) Abnormal brain tryptophan metabolism and clinical correlates in Tourette syndrome. Mov Disord 22(15):2256–2262

Belloso JM et al (2007) Disruption of the CNTNAP2 gene in a t(7;15) translocation family without symptoms of Gilles de la Tourette syndrome. Eur J Hum Genet 15(6):711–713

Ben-Yehudah G et al (2001) Impaired temporal contrast sensitivity in dyslexics is specific to retain-and-compare paradigms. Brain 124(Pt 7):1381–1395

Bey AL et al (2018) Brain region-specific disruption of Shank3 in mice reveals a dissociation for cortical and striatal circuits in autism-related behaviors. Transl Psychiatry 8(1):94

Bielen H, Houart C (2014) The Wnt cries many: Wnt regulation of neurogenesis through tissue patterning, proliferation, and asymmetric cell division. Dev Neurobiol 74(8):772–780

Bonvicini C, Faraone SV, Scassellati C (2016) Attention-deficit hyperactivity disorder in adults: a systematic review and meta-analysis of genetic, pharmacogenetic and biochemical studies. Mol Psychiatry 21(7):872–884

Bourgeron T (2015) From the genetic architecture to synaptic plasticity in autism spectrum disorder. Nat Rev Neurosci 16(9):551–563

Bromberg-Martin ES, Matsumoto M, Hikosaka O (2010) Dopamine in motivational control: rewarding, aversive, and alerting. Neuron 68(5):815–834

Bronfeld M et al (2013) Motor tics evoked by striatal disinhibition in the rat. Front Syst Neurosci 7: 50

Brown AB et al (2011) Relationship of DAT1 and adult ADHD to task-positive and task-negative working memory networks. Psychiatry Res 193(1):7–16

Burton FH (2017) Back to the future: circuit-testing TS & OCD. J Neurosci Methods 292:2–11

Butler IJ et al (1979) Biogenic amine metabolism in Tourette syndrome. Ann Neurol 6(1):37–39

Cavallini MC et al (2000) An association study between 5-HTTLPR polymorphism, COMT polymorphism, and Tourette's syndrome. Psychiatry Res 97(2–3):93–100

Ceci C et al (2015) Interaction between the endocannabinoid and serotonergic system in the exhibition of head twitch response in four mouse strains. Neurotox Res 27(3):275–283

Centanni TM et al (2016) Knockdown of dyslexia-gene Dcdc2 interferes with speech sound discrimination in continuous streams. J Neurosci 36(17):4895–4906

Chang J-Y et al (2019) Mechanisms of Ca2+/calmodulin-dependent kinase II activation in single dendritic spines. Nat Commun 10(1):2784

Che A et al (2016) Mutation of the dyslexia-associated gene Dcdc2 enhances glutamatergic synaptic transmission between layer 4 neurons in mouse neocortex. Cereb Cortex 26(9): 3705–3718

Chien WH et al (2013) Increased gene expression of FOXP1 in patients with autism spectrum disorders. Mol Autism 4(1):23

Chou IC et al (2013) Association analysis between Tourette's syndrome and two dopamine genes (DAT1, DBH) in Taiwanese children. Biomedicine 3(2):88–91

Chow ML et al (2012) Age-dependent brain gene expression and copy number anomalies in autism suggest distinct pathological processes at young versus mature ages. PLoS Genet 8(3): e1002592–e1002592

Clifton NE et al (2020) FMRP and CYFIP1 at the synapse and their role in psychiatric vulnerability. Complex Psychiatry 6(1–2):5–19

Comings DE (1990) Blood serotonin and tryptophan in Tourette syndrome. Am J Med Genet 36(4): 418–430

Comings DE et al (1991) The dopamine D2 receptor locus as a modifying gene in neuropsychiatric disorders. JAMA 266(13):1793–1800
Comings DE et al (1996) Polygenic inheritance of Tourette syndrome, stuttering, attention deficit hyperactivity, conduct, and oppositional defiant disorder: the additive and subtractive effect of the three dopaminergic genes—DRD2, D beta H, and DAT1. Am J Med Genet 67(3):264–288
Conciatori M et al (2004) Association between the HOXA1 A218G polymorphism and increased head circumference in patients with autism. Biol Psychiatry 55(4):413–419
Costales JL, Kolevzon A (2015) Phelan-McDermid syndrome and SHANK3: implications for treatment. Neurotherapeutics 12(3):620–630
Courchet V et al (2018) Haploinsufficiency of autism spectrum disorder candidate gene NUAK1 impairs cortical development and behavior in mice. Nat Commun 9(1):4289
Crane J et al (2011) Family-based genetic association study of DLGAP3 in Tourette Syndrome. Am J Med Genet B Neuropsychiatr Genet 156b(1):108–114
D'Andrea I et al (2015) Lack of kinase-independent activity of PI3Kγ in locus coeruleus induces ADHD symptoms through increased CREB signaling. EMBO Mol Med 7(7):904–917
D'Gama AM, Walsh CA (2018) Somatic mosaicism and neurodevelopmental disease. Nat Neurosci 21(11):1504–1514
Dachtler J et al (2014) Deletion of α-neurexin II results in autism-related behaviors in mice. Transl Psychiatry 4(11):e484
Dahdouh F et al (2009) Further evidence for DYX1C1 as a susceptibility factor for dyslexia. Psychiatr Genet 19(2):59–63
Das BC et al (2014) Retinoic acid signaling pathways in development and diseases. Bioorg Med Chem 22(2):673–683
de la Torre-Ubieta L et al (2016) Advancing the understanding of autism disease mechanisms through genetics. Nat Med 22(4):345–361
De Rubeis S et al (2014) Synaptic, transcriptional and chromatin genes disrupted in autism. Nature 515(7526):209–215
de Wit S et al (2012) Reliance on habits at the expense of goal-directed control following dopamine precursor depletion. Psychopharmacology 219(2):621–631
Deciphering Developmental Disorders Study (2015) Large-scale discovery of novel genetic causes of developmental disorders. Nature 519(7542):223–228
Dehning S et al (2010) A genetic variant of HTR2C may play a role in the manifestation of Tourette syndrome. Psychiatr Genet 20(1):35–38
Delattre V et al (2013) Nlgn4 knockout induces network hypo-excitability in juvenile mouse somatosensory cortex in vitro. Sci Rep 3:2897
Demb JB et al (1998) Psychophysical evidence for a magnocellular pathway deficit in dyslexia. Vis Res 38(11):1555–1559
Demontis D et al (2016) Whole-exome sequencing reveals increased burden of rare functional and disruptive variants in candidate risk genes in individuals with persistent attention-deficit/hyperactivity disorder. J Am Acad Child Adolesc Psychiatry 55(6):521–523
Demontis D et al (2019) Discovery of the first genome-wide significant risk loci for attention deficit/hyperactivity disorder. Nat Genet 51(1):63–75
Devlin B et al (2002) No evidence for linkage of liability to autism to HOXA1 in a sample from the CPEA network. Am J Med Genet 114(6):667–672
Díaz-Anzaldúa A et al (2004) Tourette syndrome and dopaminergic genes: a family-based association study in the French Canadian founder population. Mol Psychiatry 9(3):272–277
Ding M, Shen K (2008) The role of the ubiquitin proteasome system in synapse remodeling and neurodegenerative diseases. Bioessays 30(11–12):1075–1083
Dong F et al (2016) Deletion of CTNNB1 in inhibitory circuitry contributes to autism-associated behavioral defects. Hum Mol Genet 25(13):2738–2751
Draper A et al (2014) Increased GABA contributes to enhanced control over motor excitability in Tourette syndrome. Curr Biol 24(19):2343–2347

Durand CM et al (2007) Mutations in the gene encoding the synaptic scaffolding protein SHANK3 are associated with autism spectrum disorders. Nat Genet 39(1):25–27

Elia J et al (2011) Genome-wide copy number variation study associates metabotropic glutamate receptor gene networks with attention deficit hyperactivity disorder. Nat Genet 44(1):78–84

Enard W et al (2009) A humanized version of Foxp2 affects cortico-basal ganglia circuits in mice. Cell 137(5):961–971

Ernst M et al (1999) High presynaptic dopaminergic activity in children with Tourette's disorder. J Am Acad Child Adolesc Psychiatry 38(1):86–94

Esnafoglu E, Ayyıldız SN (2017) Decreased levels of serum fibroblast growth factor-2 in children with autism spectrum disorder. Psychiatry Res 257:79–83

Farwell KD et al (2015) Enhanced utility of family-centered diagnostic exome sequencing with inheritance model–based analysis: results from 500 unselected families with undiagnosed genetic conditions. Genet Med 17(7):578–586

Féron F et al (2016) Olfactory stem cells reveal MOCOS as a new player in autism spectrum disorders. Mol Psychiatry 21(9):1215–1224

Feyder M et al (2010) Association of mouse Dlg4 (PSD-95) gene deletion and human DLG4 gene variation with phenotypes relevant to autism spectrum disorders and Williams' syndrome. Am J Psychiatry 167(12):1508–1517

Field M et al (2007) Mutations in the BRWD3 gene cause X-linked mental retardation associated with macrocephaly. Am J Hum Genet 81(2):367–374

Fisher SE, DeFries JC (2002) Developmental dyslexia: genetic dissection of a complex cognitive trait. Nat Rev Neurosci 3(10):767–780

Francks C et al (2004) A 77-kilobase region of chromosome 6p22.2 is associated with dyslexia in families from the United Kingdom and from the United States. Am J Hum Genet 75(6): 1046–1058

Franke B et al (2012) The genetics of attention deficit/hyperactivity disorder in adults, a review. Mol Psychiatry 17(10):960–987

Fujisawa T, Filippakopoulos P (2017) Functions of bromodomain-containing proteins and their roles in homeostasis and cancer. Nat Rev Mol Cell Biol 18(4):246–262

Fujita E et al (2008) Ultrasonic vocalization impairment of Foxp2 (R552H) knock-in mice related to speech-language disorder and abnormality of Purkinje cells. Proc Natl Acad Sci U S A 105(8): 3117–3122

Fukushi D, Mizuno S, Yamada K, Kimura R, Yamada Y, Kumagai T, Wakamatsu N (2012) Aneuploidy and intellectual disability. In: Storchova Z (ed) Aneuploidy in health and disease. ISBN: 978-953-51-0608-1

Gade R et al (1998) Correlation of length of VNTR alleles at the X-linked MAOA gene and phenotypic effect in Tourette syndrome and drug abuse. Mol Psychiatry 3(1):50–60

Galaburda AM et al (1985) Developmental dyslexia: four consecutive patients with cortical anomalies. Ann Neurol 18(2):222–233

Garcia-Martínez I et al (2017) Gene-wide association study reveals RNF122 ubiquitin ligase as a novel susceptibility gene for attention deficit hyperactivity disorder. Sci Rep 7(1):5407

Gargus JJ, Schmunk G (2014) Dysregulation of neurogenic calcium signaling and autism. In: Patel VB, Preedy VR, Martin CR (eds) Comprehensive guide to autism. Springer, New York, pp 1285–1312

Gazestani VH et al (2018) Transcriptional organization of autism spectrum disorder and its connection to ASD risk genes and phenotypic variation. bioRxiv 435917

Gdalyahu A et al (2015) The autism related protein contactin-associated protein-like 2 (CNTNAP2) stabilizes new spines: an in vivo mouse study. PLoS One 10(5):e0125633

Gelernter J et al (1994) D2 dopamine receptor alleles do not influence severity of Tourette's syndrome. Results from four large kindreds. Arch Neurol 51(4):397–400

Giovedí S et al (2014) Involvement of synaptic genes in the pathogenesis of autism spectrum disorders: the case of synapsins. Front Pediatr 2:94

Giraldo-Chica M, Hegarty JP 2nd, Schneider KA (2015) Morphological differences in the lateral geniculate nucleus associated with dyslexia. Neuroimage Clin 7:830–836
Gizer IR, Ficks C, Waldman ID (2009) Candidate gene studies of ADHD: a meta-analytic review. Hum Genet 126(1):51–90
Gori S et al (2014) Magnocellular-dorsal pathway and sub-lexical route in developmental dyslexia. Front Hum Neurosci 8:460
Grayton HM et al (2013) Altered social behaviours in neurexin 1α knockout mice resemble core symptoms in neurodevelopmental disorders. PLoS One 8(6):e67114
Griswold AJ et al (2012) Evaluation of copy number variations reveals novel candidate genes in autism spectrum disorder-associated pathways. Hum Mol Genet 21(15):3513–3523
Guang S et al (2018) Synaptopathology involved in autism spectrum disorder. Front Cell Neurosci 12:470–470
Guilmatre A et al (2014) The emerging role of SHANK genes in neuropsychiatric disorders. Dev Neurobiol 74(2):113–122
Guo Y et al (2012) Analysis of the BTBD9 and HTR2C variants in Chinese Han patients with Tourette syndrome. Psychiatr Genet 22(6):300–303
Gyuris A et al (2009) The chromatin-targeting protein Brd2 is required for neural tube closure and embryogenesis. Biochim Biophys Acta 1789(5):413–421
Haber SN et al (1986) Gilles de la Tourette's syndrome. A postmortem neuropathological and immunohistochemical study. J Neurol Sci 75(2):225–241
Harripaul R et al (2018) Mapping autosomal recessive intellectual disability: combined microarray and exome sequencing identifies 26 novel candidate genes in 192 consanguineous families. Mol Psychiatry 23(4):973–984
Harwell V, Fasinu PS (2020) Pitolisant and other histamine-3 receptor antagonists-an update on therapeutic potentials and clinical prospects. Medicines (Basel) 7(9):55
Haugbøl S et al (2007) Cerebral 5-HT2A receptor binding is increased in patients with Tourette's syndrome. Int J Neuropsychopharmacol 10(2):245–252
Hawi Z et al (2017) Rare DNA variants in the brain-derived neurotrophic factor gene increase risk for attention-deficit hyperactivity disorder: a next-generation sequencing study. Mol Psychiatry 22(4):580–584
Hayman V, Fernandez TV (2018) Genetic insights into ADHD biology. Front Psychiatry 9:251–251
He F et al (2015) Association between Tourette syndrome and the dopamine D3 receptor gene rs6280. Chin Med J 128(5):654–658
Heinz A et al (1998) Tourette's syndrome: [I-123]beta-CIT SPECT correlates of vocal tic severity. Neurology 51(4):1069–1074
Herzberg I et al (2010) Association of DRD2 variants and Gilles de la Tourette syndrome in a family-based sample from a South American population isolate. Psychiatr Genet 20(4):179–183
Hienert M et al (2018) Striatal dopaminergic alterations in Tourette's syndrome: a meta-analysis based on 16 PET and SPECT neuroimaging studies. Transl Psychiatry 8(1):143
Hildonen M et al (2021) Elevated expression of SLC6A4 encoding the serotonin transporter (SERT) in Gilles de la Tourette syndrome. Genes 12(1):86
Hiraide T et al (2019) A de novo variant in RAC3 causes severe global developmental delay and a middle interhemispheric variant of holoprosencephaly. J Hum Genet 64(11):1127–1132
Horder J et al (2018) Glutamate and GABA in autism spectrum disorder—a translational magnetic resonance spectroscopy study in man and rodent models. Transl Psychiatry 8(1):106
Hosie S et al (2018) Altered amygdala excitation and CB1 receptor modulation of aggressive behavior in the neuroligin-3(R451C) mouse model of autism. Front Cell Neurosci 12:234
Hosseini M et al (2019) GPR126: a novel candidate gene implicated in autosomal recessive intellectual disability. Am J Med Genet A 179(1):13–19
Hu VW et al (2006) Gene expression profiling of lymphoblastoid cell lines from monozygotic twins discordant in severity of autism reveals differential regulation of neurologically relevant genes. BMC Genomics 7:118–118

Hu H et al (2019) Genetics of intellectual disability in consanguineous families. Mol Psychiatry 24(7):1027–1039

Huang AY et al (2017) Rare copy number variants in NRXN1 and CNTN6 increase risk for Tourette syndrome. Neuron 94(6):1101–1111.e7

Humeau Y et al (2009) X-linked mental retardation: focus on synaptic function and plasticity. J Neurochem 109(1):1–14

Iossifov I, Ronemus M, Levy D, Wang Z, Hakker I, Rosenbaum J, Yamrom B, Lee YH, Narzisi G, Leotta A, Kendall J, Grabowska E, Ma B, Marks S, Rodgers L, Stepansky A, Troge J, Andrews P, Bekritsky M, Pradhan K, Ghiban E, Kramer M, Parla J, Demeter R, Fulton LL, Fulton RS, Magrini VJ, Ye K, Darnell JC, Darnell RB, Mardis ER, Wilson RK, Schatz MC, McCombie WR, Wigler M (2012) De novo gene disruptions in children on the autistic spectrum. Neuron 74 (2):285–299. https://doi.org/10.1016/j.neuron.2012.04.009. PMID: 22542183; PMCID: PMC3619976

Jackson GM et al (2015) Inhibition, disinhibition, and the control of action in Tourette syndrome. Trends Cogn Sci 19(11):655–665

Jamain S et al (2003) Mutations of the X-linked genes encoding neuroligins NLGN3 and NLGN4 are associated with autism. Nat Genet 34(1):27–29

Jarick I et al (2014) Genome-wide analysis of rare copy number variations reveals PARK2 as a candidate gene for attention-deficit/hyperactivity disorder. Mol Psychiatry 19(1):115–121

Jiang YH et al (2004) A mixed epigenetic/genetic model for oligogenic inheritance of autism with a limited role for UBE3A. Am J Med Genet A 131(1):1–10

Ka M, Kim W-Y (2018) ANKRD11 associated with intellectual disability and autism regulates dendrite differentiation via the BDNF/TrkB signaling pathway. Neurobiol Dis 111:138–152

Ka M et al (2016) Essential roles for ARID1B in dendritic arborization and spine morphology of developing pyramidal neurons. J Neurosci 36(9):2723–2742

Kahrizi K et al (2019) Effect of inbreeding on intellectual disability revisited by trio sequencing. Clin Genet 95(1):151–159

Kalanithi PS et al (2005) Altered parvalbumin-positive neuron distribution in basal ganglia of individuals with Tourette syndrome. Proc Natl Acad Sci U S A 102(37):13307–13312

Kanaan AS et al (2017) Pathological glutamatergic neurotransmission in Gilles de la Tourette syndrome. Brain 140(1):218–234

Kang H et al (2016) Slitrk missense mutations associated with neuropsychiatric disorders distinctively impair slitrk trafficking and synapse formation. Front Mol Neurosci 9:104

Kasherman MA et al (2020) The ubiquitin system: a regulatory hub for intellectual disability and autism spectrum disorder. Mol Neurobiol 57(5):2179–2193

Kataoka Y et al (2010) Decreased number of parvalbumin and cholinergic interneurons in the striatum of individuals with Tourette syndrome. J Comp Neurol 518(3):277–291

Katzman MA et al (2017) Adult ADHD and comorbid disorders: clinical implications of a dimensional approach. BMC Psychiatry 17(1):302–302

Keen-Kim D et al (2006) Overrepresentation of rare variants in a specific ethnic group may confuse interpretation of association analyses. Hum Mol Genet 15(22):3324–3328

Kelleher RJ 3rd et al (2012) High-throughput sequencing of mGluR signaling pathway genes reveals enrichment of rare variants in autism. PLoS One 7(4):e35003

Kere J (2014) The molecular genetics and neurobiology of developmental dyslexia as model of a complex phenotype. Biochem Biophys Res Commun 452(2):236–243

Khalil R et al (2018) PSMD12 haploinsufficiency in a neurodevelopmental disorder with autistic features. Am J Med Genet B Neuropsychiatr Genet 177(8):736–745

Kidd KK, Prusoff BA, Cohen DJ (1980) Familial pattern of Gilles de la Tourette syndrome. Arch Gen Psychiatry 37(12):1336–1339

Kim D et al (2006) Magnocellular contributions to impaired motion processing in schizophrenia. Schizophr Res 82(1):1–8

Kim H-G et al (2008) Disruption of neurexin 1 associated with autism spectrum disorder. Am J Hum Genet 82(1):199–207

Kim DS et al (2017) Sequencing of sporadic Attention-Deficit Hyperactivity Disorder (ADHD) identifies novel and potentially pathogenic de novo variants and excludes overlap with genes

associated with autism spectrum disorder. Am J Med Genet B Neuropsychiatr Genet 174(4): 381–389
Kin Ting Kam R et al (2012) Retinoic acid synthesis and functions in early embryonic development. Cell Biosci 2(1):11
Kong R et al (2016) Genetic variant in DIP2A gene is associated with developmental dyslexia in Chinese population. Am J Med Genet B Neuropsychiatr Genet 171b(2):203–208
Krumm N et al (2014) A de novo convergence of autism genetics and molecular neuroscience. Trends Neurosci 37(2):95–105
Kuc K et al (2020) The SLC6A3 gene polymorphism is related to the development of attentional functions but not to ADHD. Sci Rep 10(1):6176
Kuechler A et al (2015) De novo mutations in beta-catenin (CTNNB1) appear to be a frequent cause of intellectual disability: expanding the mutational and clinical spectrum. Hum Genet 134(1): 97–109
Kumar S et al (2019) Impaired neurodevelopmental pathways in autism spectrum disorder: a review of signaling mechanisms and crosstalk. J Neurodev Disord 11(1):10
Kurki MI et al (2019) Contribution of rare and common variants to intellectual disability in a sub-isolate of Northern Finland. Nat Commun 10(1):410
Kuwano Y et al (2011) Autism-associated gene expression in peripheral leucocytes commonly observed between subjects with autism and healthy women having autistic children. PLoS One 6(9):e24723–e24723
Lai CS et al (2001) A forkhead-domain gene is mutated in a severe speech and language disorder. Nature 413(6855):519–523
Lai CS et al (2003) FOXP2 expression during brain development coincides with adult sites of pathology in a severe speech and language disorder. Brain 126(Pt 11):2455–2462
Lai X et al (2018) Vitamin A deficiency induces autistic-like behaviors in rats by regulating the RARβ-CD38-oxytocin axis in the hypothalamus. Mol Nutr Food Res 62(5)
Lalli MA et al (2016) Haploinsufficiency of BAZ1B contributes to Williams syndrome through transcriptional dysregulation of neurodevelopmental pathways. Hum Mol Genet 25(7): 1294–1306
Lasalle JM (2013) Autism genes keep turning up chromatin. OA Autism 1(2):14–14
Lawson-Yuen A et al (2008) Familial deletion within NLGN4 associated with autism and Tourette syndrome. Eur J Hum Genet 16(5):614–618
Leckman JF et al (1988) Elevated CSF dynorphin A [1-8] in Tourette's syndrome. Life Sci 43(24): 2015–2023
Lee CC et al (2005) Dopamine receptor D2 gene polymorphisms are associated in Taiwanese children with Tourette syndrome. Pediatr Neurol 33(4):272–276
Lee S et al (2018) Shank2 deletion in parvalbumin neurons leads to moderate hyperactivity, enhanced self-grooming and suppressed seizure susceptibility in mice. Front Mol Neurosci 11:209
Lerner A et al (2012) Widespread abnormality of the γ-aminobutyric acid-ergic system in Tourette syndrome. Brain 135(Pt 6):1926–1936
Lesch KP et al (2008) Molecular genetics of adult ADHD: converging evidence from genome-wide association and extended pedigree linkage studies. J Neural Transm (Vienna) 115(11): 1573–1585
Lesch KP et al (2011) Genome-wide copy number variation analysis in attention-deficit/ hyperactivity disorder: association with neuropeptide Y gene dosage in an extended pedigree. Mol Psychiatry 16(5):491–503
Li J, Zhao G, Gao X (2013) Development of neurodevelopmental disorders: a regulatory mechanism involving bromodomain-containing proteins. J Neurodev Disord 5(1):4
Liao X, Li Y (2020) Genetic associations between voltage-gated calcium channels and autism spectrum disorder: a systematic review. Mol Brain 13(1):96
Liu SG et al (2011) An association analysis between 5-HTTLPR polymorphism and obsessive-compulsive disorder, Tourette syndrome in a Chinese Han population. CNS Neurosci Ther 17(6):793–795

Liu S et al (2015) Family-based association study between monoamine oxidase A (MAOA) gene promoter VNTR polymorphism and Tourette's syndrome in Chinese Han population. Neurocase 21(1):106–108
Ludwig KU et al (2010) Variation in GRIN2B contributes to weak performance in verbal short-term memory in children with dyslexia. Am J Med Genet B Neuropsychiatr Genet 153b(2):503–511
Lynch M (2010) Rate, molecular spectrum, and consequences of human mutation. Proc Natl Acad Sci U S A 107(3):961–968
Mahmood D et al (2016) Histamine H3 receptor antagonists display antischizophrenic activities in rats treated with MK-801. J Basic Clin Physiol Pharmacol 27(5):463–471
Maia TV, Conceição VA (2017) The roles of phasic and tonic dopamine in tic learning and expression. Biol Psychiatry 82(6):401–412
Manning M et al (2010) Array-based technology and recommendations for utilization in medical genetics practice for detection of chromosomal abnormalities. Genet Med 12(11):742–745
Marangi G et al (2013) TRAPPC9-related autosomal recessive intellectual disability: report of a new mutation and clinical phenotype. Eur J Hum Genet 21(2):229–232
Marino C et al (2007) Association of short-term memory with a variant within DYX1C1 in developmental dyslexia. Genes Brain Behav 6(7):640–646
Martin CL, Ledbetter DH (2017) Chromosomal microarray testing for children with unexplained neurodevelopmental disorders. JAMA 317(24):2545–2546
Martin Lorenzo S et al (2020) Targeting the RHOA pathway improves learning and memory in Kctd13 and 16p11.2 deletion mouse models. bioRxiv 2020.05.22.110098
Martinelli S et al (2018) Functional dysregulation of CDC42 causes diverse developmental phenotypes. Am J Hum Genet 102(2):309–320
Mascheretti S et al (2015) GRIN2B mediates susceptibility to intelligence quotient and cognitive impairments in developmental dyslexia. Psychiatr Genet 25(1):9–20
Massinen S et al (2009) Functional interaction of DYX1C1 with estrogen receptors suggests involvement of hormonal pathways in dyslexia. Hum Mol Genet 18(15):2802–2812
Mathews CA, Grados MA (2011) Familiality of Tourette syndrome, obsessive-compulsive disorder, and attention-deficit/hyperactivity disorder: heritability analysis in a large sib-pair sample. J Am Acad Child Adolesc Psychiatry 50(1):46–54
Mathews CA et al (2014) Association between pre- and perinatal exposures and Tourette syndrome or chronic tic disorder in the ALSPAC cohort. Br J Psychiatry 204(1):40–45
Matsson H et al (2015) Polymorphisms in DCDC2 and S100B associate with developmental dyslexia. J Hum Genet 60(7):399–401
McKenzie K et al (2016) Systematic review of the prevalence and incidence of intellectual disabilities: current trends and issues. Curr Dev Disord Rep 3(2):104–115
McSherry M et al (2018) Identification of candidate gene FAM183A and novel pathogenic variants in known genes: high genetic heterogeneity for autosomal recessive intellectual disability. PLoS One 13(11):e0208324
Mick E et al (2008) Genome-wide association study of response to methylphenidate in 187 children with attention-deficit/hyperactivity disorder. Am J Med Genet B Neuropsychiatr Genet 147b(8): 1412–1418
Mignot C et al (2016) Genetic and neurodevelopmental spectrum of SYNGAP1-associated intellectual disability and epilepsy. J Med Genet 53(8):511–522
Milani D et al (2015) Rubinstein-Taybi syndrome: clinical features, genetic basis, diagnosis, and management. Ital J Pediatr 41:4
Mishra N et al (2016) Chromosome 12p deletion spanning the GRIN2B gene presenting with a neurodevelopmental phenotype. A case report and review of literature. Child Neurol Open 3: 2329048x16629980
Mitchell KJ (2014) The genetic architecture of neurodevelopmental disorders. bioRxiv 009449
Mitra I et al (2017) Reverse pathway genetic approach identifies epistasis in autism spectrum disorders. PLoS Genet 13(1):e1006516–e1006516

Modi ME et al (2018) Hyperactivity and hypermotivation associated with increased striatal mGluR1 signaling in a Shank2 rat model of autism. Front Mol Neurosci 11:107
Mooney MA et al (2016) Pathway analysis in attention deficit hyperactivity disorder: an ensemble approach. Am J Med Genet B Neuropsychiatr Genet 171(6):815–826
Moreno-Ramos OA et al (2015) Whole-exome sequencing in a South American cohort links ALDH1A3, FOXN1 and retinoic acid regulation pathways to autism spectrum disorders. PLoS One 10(9):e0135927–e0135927
Morris-Rosendahl DJ, Crocq M-A (2020) Neurodevelopmental disorders-the history and future of a diagnostic concept. Dialogues Clin Neurosci 22(1):65–72
Mössner R et al (2007) Role of the novel tryptophan hydroxylase-2 gene in Tourette syndrome. Mol Psychiatry 12(7):617–619
Moya PR et al (2013) Common and rare alleles of the serotonin transporter gene, SLC6A4, associated with Tourette's disorder. Mov Disord 28(9):1263–1270
Müller-Vahl KR et al (2017) Gilles de la Tourette syndrome is associated with hypermethylation of the dopamine D2 receptor gene. J Psychiatr Res 86:1–8
Muscarella LA et al (2010) Candidate gene study of HOXB1 in autism spectrum disorder. Mol Autism 1(1):9–9
Naaijen J et al (2017) Fronto-striatal glutamate in children with Tourette's disorder and attention-deficit/hyperactivity disorder. Neuroimage Clin 13:16–23
Nag A et al (2013) CNV analysis in Tourette syndrome implicates large genomic rearrangements in COL8A1 and NRXN1. PLoS One 8(3):e59061
Nayar K et al (2019) Language processing skills linked to FMR1 variation: a study of gaze-language coordination during rapid automatized naming among women with the FMR1 premutation. PLoS One 14(7):e0219924
Nelson AB et al (2014) Striatal cholinergic interneurons Drive GABA release from dopamine terminals. Neuron 82(1):63–70
Neri G et al (2018) X-linked intellectual disability update 2017. Am J Med Genet A 176(6):1375–1388
Nguyen A et al (2010) Global methylation profiling of lymphoblastoid cell lines reveals epigenetic contributions to autism spectrum disorders and a novel autism candidate gene, RORA, whose protein product is reduced in autistic brain. FASEB J 24(8):3036–3051
Nguyen RL et al (2018) Intracellular calcium dysregulation in autism spectrum disorder: an analysis of converging organelle signaling pathways. Biochim Biophys Acta, Mol Cell Res 1865(11, Part B):1718–1732
Niemi MEK et al (2018) Common genetic variants contribute to risk of rare severe neurodevelopmental disorders. Nature 562(7726):268–271
Nieto-Alamilla G et al (2016) The histamine H3 receptor: structure, pharmacology, and function. Mol Pharmacol 90(5):649–673
Nisar S et al (2019) Association of genes with phenotype in autism spectrum disorder. Aging 11(22):10742–10770
Norris RHC et al (2019) Mutations in neuroligin-3 in male mice impact behavioral flexibility but not relational memory in a touchscreen test of visual transitive inference. Mol Autism 10(1):42
Nöthen MM et al (1994) Association analysis of the dopamine D2 receptor gene in Tourette's syndrome using the haplotype relative risk method. Am J Med Genet 54(3):249–252
O'Roak BJ et al (2011) Exome sequencing in sporadic autism spectrum disorders identifies severe de novo mutations. Nat Genet 43(6):585–589
O'Roak BJ et al (2012a) Sporadic autism exomes reveal a highly interconnected protein network of de novo mutations. Nature 485(7397):246–250
O'Roak BJ et al (2012b) Multiplex targeted sequencing identifies recurrently mutated genes in autism spectrum disorders. Science (New York, NY) 338(6114):1619–1622
Olulade OA, Napoliello EM, Eden GF (2013) Abnormal visual motion processing is not a cause of dyslexia. Neuron 79(1):180–190

Olusanya BO et al (2018) Developmental disabilities among children younger than 5 years in 195 countries and territories, 1990-2016: a systematic analysis for the Global Burden of Disease Study 2016. Lancet Glob Health 6(10):e1100–e1121

Orth M (2009) Transcranial magnetic stimulation in Gilles de la Tourette syndrome. J Psychosom Res 67(6):591–598

Ozbay F et al (2006) Analysis of the dopamine beta hydroxylase gene in Gilles de la Tourette syndrome. Am J Med Genet B Neuropsychiatr Genet 141b(6):673–677

Pagnamenta AT et al (2010) Characterization of a family with rare deletions in CNTNAP5 and DOCK4 suggests novel risk loci for autism and dyslexia. Biol Psychiatry 68(4):320–328

Palladino VS et al (2020) Energy metabolism disturbances in cell models of PARK2 CNV carriers with ADHD. J Clin Med 9(12):4092

Parenti I et al (2020) Neurodevelopmental disorders: from genetics to functional pathways. Trends Neurosci 43(8):608–621

Pauls DL et al (1981) Familial pattern and transmission of Gilles de la Tourette syndrome and multiple tics. Arch Gen Psychiatry 38(10):1091–1093

Pavlowsky A, Chelly J, Billuart P (2012) Emerging major synaptic signaling pathways involved in intellectual disability. Mol Psychiatry 17(7):682–693

Peñagarikano O et al (2011) Absence of CNTNAP2 leads to epilepsy, neuronal migration abnormalities, and core autism-related deficits. Cell 147(1):235–246

Pengelly RJ et al (2016) Mutations specific to the Rac-GEF domain of TRIO cause intellectual disability and microcephaly. J Med Genet 53(11):735–742

Persico AM et al (2001) Reelin gene alleles and haplotypes as a factor predisposing to autistic disorder. Mol Psychiatry 6(2):150–159

Peter B et al (2011) Replication of CNTNAP2 association with nonword repetition and support for FOXP2 association with timed reading and motor activities in a dyslexia family sample. J Neurodev Disord 3(1):39–49

Pinto D et al (2010) Functional impact of global rare copy number variation in autism spectrum disorders. Nature 466(7304):368–372

Pinto D et al (2014) Convergence of genes and cellular pathways dysregulated in autism spectrum disorders. Am J Hum Genet 94(5):677–694

Pogorelov V et al (2015) Corticostriatal interactions in the generation of tic-like behaviors after local striatal disinhibition. Exp Neurol 265:122–128

Poot M et al (2010) Disruption of CNTNAP2 and additional structural genome changes in a boy with speech delay and autism spectrum disorder. Neurogenetics 11(1):81–89

Prchalova D, Havlovicova M, Sterbova K, Stranecky V, Hancarova M, Sedlacek Z (2017) Analysis of 31-year-old patient with SYNGAP1 gene defect points to importance of variants in broader splice regions and reveals developmental trajectory of SYNGAP1-associated phenotype: case report. BMC Med Genet 18(1):62. https://doi.org/10.1186/s12881-017-0425-4

Rabbani B, Tekin M, Mahdieh N (2014) The promise of whole-exome sequencing in medical genetics. J Hum Genet 59(1):5–15

Rabey JM et al (1995) Decreased dopamine uptake into platelet storage granules in Gilles de la Tourette disease. Biol Psychiatry 38(2):112–115

Rahi S, Mehan S (2020) Understanding abnormal SMO-SHH signaling in autism spectrum disorder: potential drug target and therapeutic goals. Cell Mol Neurobiol. https://doi.org/10.1007/s10571-020-01010-1. Epub ahead of print. PMID: 33206287

Rapanelli M et al (2017) Histamine H3R receptor activation in the dorsal striatum triggers stereotypies in a mouse model of tic disorders. Transl Psychiatry 7(1):e1013

Rasmussen AH, Rasmussen HB, Silahtaroglu A (2017) The DLGAP family: neuronal expression, function and role in brain disorders. Mol Brain 10(1):43

Reijnders MRF et al (2017) RAC1 missense mutations in developmental disorders with diverse phenotypes. Am J Hum Genet 101(3):466–477

Reinson K et al (2016) Biallelic CACNA1A mutations cause early onset epileptic encephalopathy with progressive cerebral, cerebellar, and optic nerve atrophy. Am J Med Genet A 170(8): 2173–2176

Rendall AR, Truong DT, Fitch RH (2016) Learning delays in a mouse model of Autism Spectrum Disorder. Behav Brain Res 303:201–207

Reuter MS et al (2017) Diagnostic yield and novel candidate genes by exome sequencing in 152 consanguineous families with neurodevelopmental disorders. JAMA Psychiatry 74(3): 293–299

Ribasés M et al (2012) An association study of sequence variants in the forkhead box P2 (FOXP2) gene and adulthood attention-deficit/hyperactivity disorder in two European samples. Psychiatr Genet 22(4):155–160

Riccomagno MM, Kolodkin AL (2015) Sculpting neural circuits by axon and dendrite pruning. Annu Rev Cell Dev Biol 31:779–805

Richter M et al (2019) Altered TAOK2 activity causes autism-related neurodevelopmental and cognitive abnormalities through RhoA signaling. Mol Psychiatry 24(9):1329–1350

Rodenas-Cuadrado P, Ho J, Vernes SC (2014) Shining a light on CNTNAP2: complex functions to complex disorders. Eur J Hum Genet 22(2):171–178

Roeske D et al (2011) First genome-wide association scan on neurophysiological endophenotypes points to trans-regulation effects on SLC2A3 in dyslexic children. Mol Psychiatry 16(1):97–107

Romano V et al (2003) Lack of association of HOXA1 and HOXB1 mutations and autism in Sicilian (Italian) patients. Mol Psychiatry 8(8):716–717

Ropers HH (2008) Genetics of intellectual disability. Curr Opin Genet Dev 18(3):241–250

Rosso SB, Inestrosa NC (2013) WNT signaling in neuronal maturation and synaptogenesis. Front Cell Neurosci 7:103

Rudolph U (2008) GABAergic system. In: Offermanns S, Rosenthal W (eds) Encyclopedia of molecular pharmacology. Springer, Berlin, Heidelberg, pp 515–519

Sánchez-Mora C et al (2015) Case-control genome-wide association study of persistent attention-deficit hyperactivity disorder identifies FBXO33 as a novel susceptibility gene for the disorder. Neuropsychopharmacology 40(4):915–926

Sanders SJ et al (2012) De novo mutations revealed by whole-exome sequencing are strongly associated with autism. Nature 485(7397):237–241

Santos-Cortez RLP et al (2018) Novel candidate genes and variants underlying autosomal recessive neurodevelopmental disorders with intellectual disability. Hum Genet 137(9):735–752

Sarachana T, Hu VW (2013) Genome-wide identification of transcriptional targets of RORA reveals direct regulation of multiple genes associated with autism spectrum disorder. Mol Autism 4(1):14

Savova V et al (2017) Risk alleles of genes with monoallelic expression are enriched in gain-of-function variants and depleted in loss-of-function variants for neurodevelopmental disorders. Mol Psychiatry 22(12):1785–1794

Scerri TS et al (2004) Putative functional alleles of DYX1C1 are not associated with dyslexia susceptibility in a large sample of sibling pairs from the UK. J Med Genet 41(11):853–857

Scerri TS et al (2011) DCDC2, KIAA0319 and CMIP are associated with reading-related traits. Biol Psychiatry 70(3):237–245

Schmunk G et al (2017) High-throughput screen detects calcium signaling dysfunction in typical sporadic autism spectrum disorder. Sci Rep 7:40740–40740

Schulte-Körne G, Warnke A, Remschmidt H (2006) [Genetics of dyslexia]. Z Kinder Jugendpsychiatr Psychother 34(6):435–444

Scott KE et al (2018) Altered auditory processing, filtering, and reactivity in the Cntnap2 knock-out rat model for neurodevelopmental disorders. J Neurosci 38(40):8588–8604

Segura M et al (2015) Neurotrophin blood-based gene expression and social cognition analysis in patients with autism spectrum disorder. Neurogenetics 16(2):123–131

Serajee FJ et al (2003) Association of INPP1, PIK3CG, and TSC2 gene variants with autistic disorder: implications for phosphatidylinositol signalling in autism. J Med Genet 40(11):e119–e119

Shaywitz SE (1998) Dyslexia. N Engl J Med 338(5):307–312

Shu W et al (2005) Altered ultrasonic vocalization in mice with a disruption in the Foxp2 gene. Proc Natl Acad Sci U S A 102(27):9643–9648

Simon V et al (2009) Prevalence and correlates of adult attention-deficit hyperactivity disorder: meta-analysis. Br J Psychiatry 194(3):204–211

Singer HS (2016) Habitual and goal-directed behaviours and Tourette syndrome. Brain 139(Pt 2): 312–316

Singer HS et al (1982) Clinical symptomatology, CSF neurotransmitter metabolites, and serum haloperidol levels in Tourette syndrome. Adv Neurol 35:177–183

Singer HS, Oshida L, Coyle JT (1984) CSF cholinesterase activity in Gilles de la Tourette's syndrome. Arch Neurol 41(7):756–757

Singer HS et al (1990) Tourette's syndrome: a neurochemical analysis of postmortem cortical brain tissue. Ann Neurol 27(4):443–446

Singer HS, Hahn IH, Moran TH (1991) Abnormal dopamine uptake sites in postmortem striatum from patients with Tourette's syndrome. Ann Neurol 30(4):558–562

Singer HS, Morris C, Grados M (2010) Glutamatergic modulatory therapy for Tourette syndrome. Med Hypotheses 74(5):862–867

Skeide MA et al (2016) NRSN1 associated grey matter volume of the visual word form area reveals dyslexia before school. Brain 139(Pt 10):2792–2803

Smith KT, Nicholls RD, Reines D (2006) The gene encoding the fragile X RNA-binding protein is controlled by nuclear respiratory factor 2 and the CREB family of transcription factors. Nucleic Acids Res 34(4):1205–1215

Spellmann I et al (2011) Homer-1 polymorphisms are associated with psychopathology and response to treatment in schizophrenic patients. J Psychiatr Res 45(2):234–241

Steeves TD et al (2010) Extrastriatal dopaminergic dysfunction in tourette syndrome. Ann Neurol 67(2):170–181

Stein J, Walsh V (1997) To see but not to read; the magnocellular theory of dyslexia. Trends Neurosci 20(4):147–152

Stephenson JR et al (2017) A novel human CAMK2A mutation disrupts dendritic morphology and synaptic transmission, and causes ASD-related behaviors. J Neurosci 37(8):2216–2233

Stern D et al (2017) Association of the missense variant p.Arg203Trp in PACS1 as a cause of intellectual disability and seizures. Clin Genet 92(2):221–223

Strømme P et al (2002) Infantile spasms, dystonia, and other X-linked phenotypes caused by mutations in Aristaless related homeobox gene, ARX. Brain Dev 24(5):266–268

Subramanian M et al (2015) Characterizing autism spectrum disorders by key biochemical pathways. Front Neurosci 9:313

Taipale M et al (2003) A candidate gene for developmental dyslexia encodes a nuclear tetratricopeptide repeat domain protein dynamically regulated in brain. Proc Natl Acad Sci U S A 100(20):11553–11558

Talebizadeh Z et al (2002) No association between HOXA1 and HOXB1 genes and autism spectrum disorders (ASD). J Med Genet 39(11):e70

Tarnok Z et al (2007) Dopaminergic candidate genes in Tourette syndrome: association between tic severity and 3′ UTR polymorphism of the dopamine transporter gene. Am J Med Genet B Neuropsychiatr Genet 144b(7):900–905

Tora D et al (2017) Cellular functions of the autism risk factor PTCHD1 in mice. J Neurosci 37(49): 11993–12005

Truong DT et al (2014) Mutation of Dcdc2 in mice leads to impairments in auditory processing and memory ability. Genes Brain Behav 13(8):802–811

Truong DT et al (2015) Auditory processing and morphological anomalies in medial geniculate nucleus of Cntnap2 mutant mice. Behav Neurosci 129(6):731–743

Tsetsos F et al (2021) Synaptic processes and immune-related pathways implicated in Tourette syndrome. Transl Psychiatry 11(1):56

Uehara T et al (2019) CNOT2 haploinsufficiency causes a neurodevelopmental disorder with characteristic facial features. Am J Med Genet A 179(12):2506–2509

Vaccarino FM et al (2009) Regulation of cerebral cortical size and neuron number by fibroblast growth factors: implications for autism. J Autism Dev Disord 39(3):511–520

Vaillend C, Poirier R, Laroche S (2008) Genes, plasticity and mental retardation. Behav Brain Res 192(1):88–105

van Loo KMJ, Martens GJM (2007) Genetic and environmental factors in complex neurodevelopmental disorders. Curr Genomics 8(7):429–444

Verkerk AJ et al (2003) CNTNAP2 is disrupted in a family with Gilles de la Tourette syndrome and obsessive-compulsive disorder. Genomics 82(1):1–9

Vernes SC et al (2011) Foxp2 regulates gene networks implicated in neurite outgrowth in the developing brain. PLoS Genet 7(7):e1002145

Verpelli C et al (2012) Scaffold proteins at the postsynaptic density. Adv Exp Med Biol 970:29–61

Voyiaziakis E et al (2011) Association of SLC6A4 variants with obsessive-compulsive disorder in a large multicenter US family study. Mol Psychiatry 16(1):108–120

Waltes R et al (2014) Common variants in genes of the postsynaptic FMRP signalling pathway are risk factors for autism spectrum disorders. Hum Genet 133(6):781–792

Walther DJ et al (2003) Synthesis of serotonin by a second tryptophan hydroxylase isoform. Science 299(5603):76

Wang L-J et al (2020) Gray matter volume and microRNA levels in patients with attention-deficit/hyperactivity disorder. Eur Arch Psychiatry Clin Neurosci 270(8):1037–1045

Weinshenker D (2007) Dopamine beta-hydroxylase. In: Enna SJ, Bylund DB (eds) xPharm: the comprehensive pharmacology reference. Elsevier, New York, pp 1–15

Weisler RH et al (2012) Randomized clinical study of a histamine H3 receptor antagonist for the treatment of adults with attention-deficit hyperactivity disorder. CNS Drugs 26(5):421–434

Weißflog L et al (2013) KCNIP4 as a candidate gene for personality disorders and adult ADHD. Eur Neuropsychopharmacol 23(6):436–447

Wengert ER et al (2019) Biallelic inherited SCN8A variants, a rare cause of SCN8A-related developmental and epileptic encephalopathy. Epilepsia 60(11):2277–2285

Wieczorek D (2018) Autosomal dominant intellectual disability. Med Genet 30(3):318–322

Wilfert AB et al (2017) Recurrent de novo mutations in neurodevelopmental disorders: properties and clinical implications. Genome Med 9(1):101–101

Willcutt EG (2012) The prevalence of DSM-IV attention-deficit/hyperactivity disorder: a meta-analytic review. Neurotherapeutics 9(3):490–499

Williams NM et al (2012) Genome-wide analysis of copy number variants in attention deficit hyperactivity disorder: the role of rare variants and duplications at 15q13.3. Am J Psychiatry 169(2):195–204

Williams SM et al (2019) An integrative analysis of non-coding regulatory DNA variations associated with autism spectrum disorder. Mol Psychiatry 24(11):1707–1719

Winczewska-Wiktor A et al (2016) A de novo CTNNB1 nonsense mutation associated with syndromic atypical hyperekplexia, microcephaly and intellectual disability: a case report. BMC Neurol 16:35

Wiśniowiecka-Kowalnik B, Nowakowska BA (2019) Genetics and epigenetics of autism spectrum disorder—current evidence in the field. J Appl Genet 60(1):37–47

Woodbury-Smith M, Scherer SW (2018) Progress in the genetics of autism spectrum disorder. Dev Med Child Neurol 60(5):445–451

Wu N, Wang Y, Jia JY, Pan YH, Yuan XB (2021) Association of CDH11 with autism spectrum disorder revealed by matched-gene co-expression analysis and mouse behavioral studies. Neurosci Bull. https://doi.org/10.1007/s12264-021-00770-0. Epub ahead of print. PMID: 34523068

Xu M et al (2015) Targeted ablation of cholinergic interneurons in the dorsolateral striatum produces behavioral manifestations of Tourette syndrome. Proc Natl Acad Sci U S A 112(3): 893–898

Xu X et al (2018) Excessive UBE3A dosage impairs retinoic acid signaling and synaptic plasticity in autism spectrum disorders. Cell Res 28(1):48–68

Yang B et al (2007) A meta-analysis of association studies between the 10-repeat allele of a VNTR polymorphism in the 3′-UTR of dopamine transporter gene and attention deficit hyperactivity disorder. Am J Med Genet B Neuropsychiatr Genet 144b(4):541–550

Yasin H, Zahir FR (2020) Chromodomain helicase DNA-binding proteins and neurodevelopmental disorders. J Transl Genet Genomics 4(4):307–319

Yoon DY et al (2007) Dopaminergic polymorphisms in Tourette syndrome: association with the DAT gene (SLC6A3). Am J Med Genet B Neuropsychiatr Genet 144b(5):605–610

Yuen RKC et al (2015) Whole-genome sequencing of quartet families with autism spectrum disorder. Nat Med 21(2):185–191

Zarrei M et al (2019) A large data resource of genomic copy number variation across neurodevelopmental disorders. npj Genomic Med 4(1):26

Zayats T et al (2016) Exome chip analyses in adult attention deficit hyperactivity disorder. Transl Psychiatry 6(10):e923

Zheng F et al (2016) Mutation of the CH1 domain in the histone acetyltransferase CREBBP results in autism-relevant behaviors in mice. PLoS One 11(1):e0146366

Neurodevelopmental Disorders: Epigenetic Implications and Potential Analysis Methods

3

Rwik Sen

Abstract

Neurodevelopment is a complex process that begins when the embryo starts to develop. The process is governed by various factors under tight temporal and spatial orchestration. Neurodevelopmental disorders arise from abnormalities in one or more of their regulatory factors in terms of mutation, absence, overexpression, or expression in the wrong location at the wrong time. Although several mutations have been identified in many neurodevelopmental disorders, often, there is no proper correlation between the genotype and phenotypic severity of the disease. Patients with similar mutations show considerable variation in disease phenotypes. Hence, in addition to studying genetic mutations in neurodevelopmental disorders, understanding the regulation of the genes has gained focus. In this direction, several studies unraveled that the regulation of gene expression by epigenetic factors plays a crucial role in neurodevelopment, and associated biological and psychiatric disorders. Therefore, this chapter focuses on the epigenetic regulation of neurodevelopmental disorders and a few methods to study the landscape and protein interactions of the epigenome. The state-of-the-art methods are reliable, robust, and high-throughput tools for obtaining genome-wide information on the epigenetics of neurodevelopmental disorders at very high resolutions.

Keywords

Epigenetics · Chromatin · Transcription factor · Next generation sequencing · High-throughput · Single-cell profiling · ChIP-Seq · Protein interactome · CUT&Tag · ATAC-seq

R. Sen (✉)
Active Motif Inc., Carlsbad, CA, USA
e-mail: rsen@activemotif.com

M. W. Qoronfleh et al. (eds.), *Proteins Associated with Neurodevelopmental Disorders*, Nutritional Neurosciences, https://doi.org/10.1007/978-981-15-9781-7_3

3.1 Introduction

Neurodevelopmental disorders comprise a spectrum of pathologies with varying prevalence based on the disease and geographical location. DSM-5 or the Diagnostic and Statistical Manual of Mental Disorders, Fifth Edition, contains the official definition and classifications of neurodevelopmental disorders (Morris-Rosendahl and Crocq 2020). Several factors contribute to the etiology of the diseases, among which epigenetics plays an important role. Epigenetics comprises a heritable phenomenon that regulates gene expression without altering the underlying DNA sequences. DNA exists in compact packaging inside a eukaryotic nucleus where it wraps around octamers comprised of core histone proteins, namely H2A, H2B, H3, and H4. Histone octamer is formed when a tetramer of two copies of H3 and H4 forms a complex with two dimers of H2A-H2B.

Epigenetic mechanisms regulate gene expression by tightening or loosening the DNA packaging around histone octamers. If the packaging is tight, then the gene is less accessible to transcription factors and not expressed. On the other hand, if the packaging is loosened, then transcription factors can access the DNA for gene activation. The packaging of 146 base pairs of DNA around a histone octamer is called nucleosome core particle (NCP) (Luger et al. 1997). Another histone known as H1 associates with the NCPs, near the entry and exit locations of DNA for protecting free linker DNA between NCPs (Brockers and Schneider 2019). The complete nucleosomes constitute higher order chromatin structure, and are known as fundamental units of chromatin (Allan et al. 1981). Chromatin structure governs gene expression. Hence, transcription is regulated by factors that alter chromatin dynamics, which include histone-modifying enzymes, ATP-dependent chromatin remodelers, noncoding RNA, and prions.

Studies have shown that several epigenetic factors are implicated in various neurodevelopmental disorders. Hence, it is beneficial to focus on some important and popular methodology to study epigenetics, which can reveal further details about neurodevelopmental disease mechanisms. Advancements in high-throughput methods to study epigenetics will enable accurate modeling of neurodevelopmental disorders and identify novel targets for therapeutic developments. In this direction, an overview of the epigenetics of neurodevelopmental disorders and some significant methods to study epigenetics have been presented in this chapter.

3.2 Epigenetics of Neurodevelopmental Disorders

Epigenetic abnormalities have been implicated in neurodevelopmental disorders like mental retardation/intellectual disability (MR/ID) syndromes and autism spectrum disorders (ASDs), neurofunctional diseases like schizophrenia (SCZ) and bipolar disorder (BD), and neurodegeneration like Alzheimer's, Parkinson, and Huntington's disease (Zahir and Brown 2011; Miyake et al. 2012; Urdinguio et al. 2009). Epigenetic and transcriptional regulators belong to the major category of genes that are found to be pathogenic for MR/ID (Chelly et al. 2006).

Epigenetic processes like DNA methylation changes are known to impact long-term changes in developmental programs, which make people prone to major depressive disorders (MDD) and SCZ (Hoffmann et al. 2017). A predominant risk factor for SCZ/MDD is early life adversity which sets off persistent DNA methylation alterations throughout the genome that affect genes regulating early and mature brain function, neural proliferation, differentiation, and synaptic plasticity (Hoffmann et al. 2017). Genetic changes that regulate dynamic DNA methylation during early development are predicted to impact future epigenomic variations in SCZ (Hoffmann et al. 2017).

In addition to epigenetic processes, mutations in several epigenetic factors are implicated in various neurodevelopmental disorders (Zahir and Brown 2011). Mutation in DNA methyltransferase 3 beta (DNMT3B) is implicated in immunodeficiency, centromeric region instability, facial anomalies (ICF) syndrome, which also shows psychomotor retardation (Lander et al. 2001; Kramer and van Bokhoven 2009). Methyl-CpG-binding protein 2 (MECP2) is mutated in ASD (Esteller 2007), SCZ (Trivier et al. 1996), and MR (Amir et al. 2000). Mutation in ATP-dependent chromodomain-helicase-DNA-binding protein 7 (CHD7) is implicated in CHARGE syndrome with cranial nerve abnormalities (Buiting 2010).

Abnormalities with enzymes that epigenetically modify histones and RNA are also associated with neurodevelopmental disorders. Histone acetyltransferase (HAT) domain-containing CREB-binding protein (CREBBP) is mutated in Rubinstein-Taybi syndrome with physical abnormalities and moderate-to-severe intellectual disability (Graff and Mansuy 2009). Mutation in histone deacetylase 4 (HDAC4) is implicated in MR and SCZ (Inlow and Restifo 2004). Abnormality in nucleosome remodeling and deacetylase (NuRD) complex for chromatin remodeling is associated with SCZ and BD (Hoffmann and Spengler 2019). H3K36 methylation regulates neurological development and disease, including brain formation (Zaghi et al. 2019; Kim et al. 2017). RNA modification N^6-methyladenosine (m6A) is implicated in neurogenesis (Rockwell and Hongay 2019).

Overall, epigenetic mechanisms regulate neurodevelopment (Cariaga-Martinez et al. 2018; Gabriele et al. 2018; Mossink et al. 2021; Li et al. 2013; Timpano and Picketts 2020; Fallah et al. 2020; Zamora-Moratalla et al. 2021), and their abnormalities lead to neurodevelopmental disorders, including a variety of frequent as well as rare diseases. Hence, therapies against neurodevelopmental disorders can be developed if the associated epigenetic regulations are understood better. In this direction, some state-of-the-art methods to study epigenetic phenomena have been described in the next section.

3.3 ChIP-seq or *Ch*romatin *I*mmuno*p*recipitation and *Seq*uencing

3.3.1 ChIP-seq Background

In the epigenetics field, ChIP is a very popular, long-standing, and evolving technique which is considered a gold standard for studying interactions between proteins and DNA/chromatin. The earliest reports of ChIP date back to circa 1984 when John T. Lis and David Gilmour studied the association of RNA polymerase II with genes of interest in vivo, in bacterial (Gilmour and Lis 1984) and *Drosophila (*Gilmour and Lis 1985*)* cells. Thereafter, several studies contributed to the development and evolution of the technique (Varshavsky 2008; Solomon et al. 1988; Orlando 2000; Hebbes et al. 1988; O'Neill and Turner 2003). The advent of next-generation sequencing (NGS) technology led to the development of ChIP-seq, where the association of proteins of interest can be detected genome-wide. The earliest publications on ChIP-seq were reported in 2007 (Park 2009; Johnson et al. 2007; Barski et al. 2007; Robertson et al. 2007; Mikkelsen et al. 2007). Usually, ChIP/ChIP-seq requires a very large number of cells as starting material which might be difficult to obtain for some sample types. However, the issue has been solved by recent advances in chromatin extraction from limited starting material (Lorentsen et al. 2018). Today, ChIP-seq is a very popular technique for discovery-based and translational research, with applications in studying diseases like cancers, developmental disorders, and neurodegenerative disorders, and drug development.

3.3.2 ChIP-seq Method Outline

ChIP involves cross-linking and extraction of chromatin from cells or tissue, followed by fragmentation of chromatin, generally into 200–400 base pairs (bp) length fragments. The fragments consist of proteins cross-linked to DNA which are immunoprecipitated on beads using antibodies against target proteins that occupy the DNA. The immunoprecipitated DNA is purified and subjected to library preparation, sequencing, and bioinformatics analysis. In the absence of sequencing, the purified DNA can be alternatively analyzed by PCR using primers against genes of interest.

Each of the above steps is very important, but fragmentation and precipitation by antibody are among the two most crucial steps of ChIP. Fragmentation is significant because proper immunoprecipitation will not occur without consistency in fragment sizes. Immunoprecipitation step using antibody is crucial because binding of the antibody to the target protein should be specific for downstream processes to succeed and eventually show accurate chromatin localization of the protein of interest. If the antibody interacts nonspecifically with other proteins, the ultimate results will contain excessive background that interferes with accuracy. Various steps of ChIP-seq are described below (Fig. 3.1).

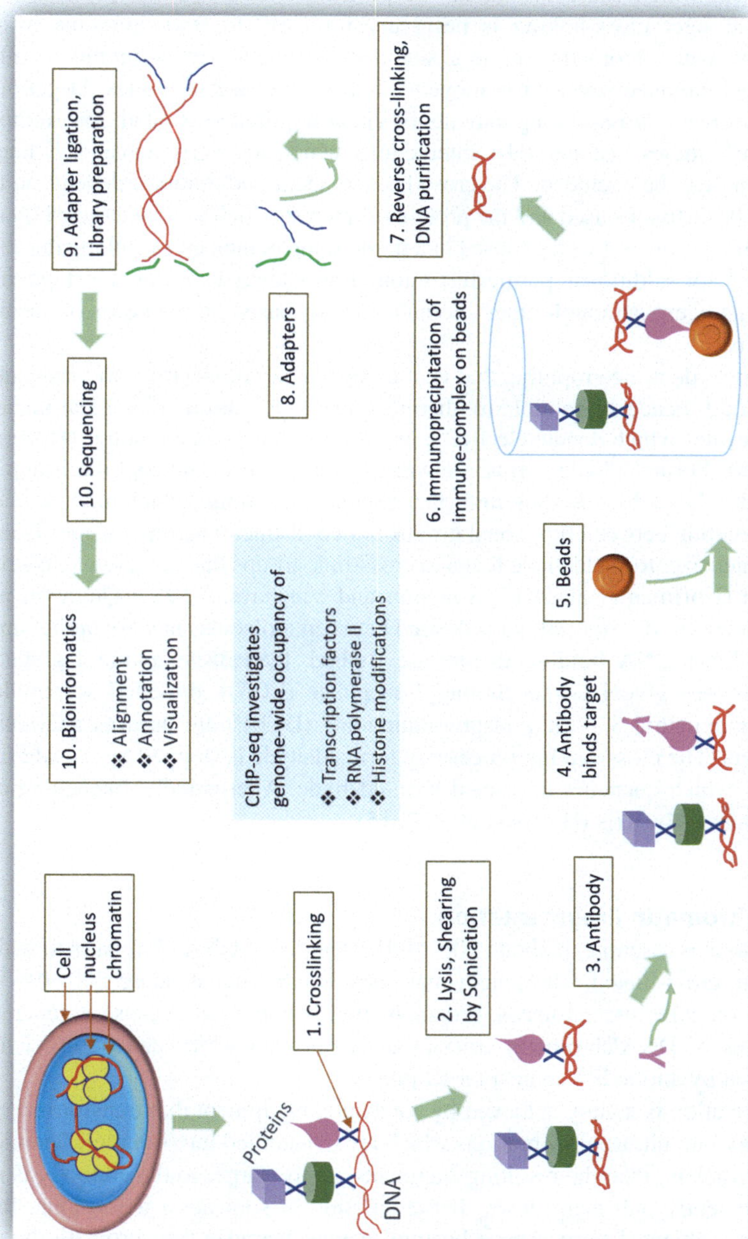

Fig. 3.1 ChIP-seq

3.3.2.1 Cross-Linking

In order to affix proteins to their precise locations on DNA/chromatin, cross-linking is performed, which causes "fixation" of proteins with DNA and proteins with proteins. Stronger cross-linking is necessary for ChIP to study proteins whose associations with chromatin are less stable or transient. Cross-linking is often performed simultaneously with homogenization in the case of tissues. Depending on the experiment, cross-linking may not be either required or wanted; for example, native ChIP studies histone-DNA interactions which are very stable, and hence cross-linking can be excluded. The cross-linking agent and duration depend on the types of cells or tissues used and the protein of interest which will be targeted by the antibody. If cells are not well exposed in a tissue sample, then cross-linking duration is increased for additional permeabilization. Formaldehyde is the most popular cross-linking agent, although other alternatives exist based on the needs of specific experiments.

Formaldehyde is electrophilic; that is, it accepts a pair of electrons to form a new covalent bond, hence susceptible to chemical attack by various biological nucleophilic molecules which donate electron pairs to form new covalent bonds (Hoffman et al. 2015). Formaldehyde forms a covalent bond with a nucleophilic group of amino acid or DNA base leading to further chemical reactions, which might include covalent bonding between functional groups of two distinct macromolecules. Due to its own small size, formaldehyde helps to cross-link groups that are a few Angstrom units apart (Hoffman et al. 2015; Solomon and Varshavsky 1985; Quievryn and Zhitkovich 2000). If cross-linking is desired between molecules that are farther apart or if non-direct DNA binding factors are studied, then disuccinimidyl glutarate (DSG), ethylene glycol bis(succinimidyl succinate (EGS)), dimethyl adipimidate (DMA), or dimethyl 3,3′-dithiobispropionimidate (DTBP) are used as alternative cross-linkers. The cross-linking process by formaldehyde is stopped by the addition of glycine, which quenches unreacted formaldehyde. Alternatively, quenching can also be achieved by Tris (Hoffman et al. 2015).

3.3.2.2 Chromatin Fragmentation

Fragmentation is essential to obtain 200–400 bp lengths of DNA/chromatin to which proteins are cross-linked. The length may vary beyond the standard 200–400 bp, depending on experimental needs. Fragmentation is important because it generates small pieces of DNA/chromatin cross-linked to protein which can be efficiently pulled down by antibodies against target proteins.

Fragmentation is mainly achieved by sonication equipment that transforms electrical energy into ultrasonic vibrations which are transmitted into samples. If sonication is insufficient, then the resulting fragments are too large, so the antibody is not able to efficiently pull them down. If the duration of sonication is too long, then samples are subjected to excessive heating, which degrades the chromatin-bound proteins to an extent where antibodies are no longer able to recognize them. Hence, sonication is an extremely critical step and needs careful attention, which consumes excessive time and labor. However, the above bottlenecks are addressed with recent advances in high-throughput sonication technology involving low setup and run times, e.g., PIXUL (Bomsztyk et al. 2019).

A sonicator generates sound waves with alternate cycles of compression and expansion whose rates are regulated by the frequency of the waves. Tiny vacuum bubbles are generated by high-intensity ultrasonic waves during cycles of low pressure. During the alternative high-pressure cycle, the bubbles become unable to absorb further energy and undergo a violent collapse resulting in a process called cavitation. Sonication is caused by the force of cavitation, leading to fragmentation of DNA/chromatin. Sonication and lysis conditions need to be optimized depending on cell and tissue types, chromatin compaction, desired fragment lengths, targeted proteins, etc. Alternatively, fragmentation can be achieved enzymatically using micrococcal nuclease or MNase.

3.3.2.3 Immunoprecipitation by Antibody

Apart from fragmentation, immunoprecipitation is the other most crucial step in ChIP. Selection of antibody is critical for achieving strong targeted binding between the antibody and its target protein under fixation conditions to detect true protein-DNA/chromatin interactions. Simultaneously, the antibody must not engage in nonspecific binding, which will generate a background signal in the output called "noise." It is difficult to distinguish true signals from noise. Monoclonal, polyclonal, and recombinant antibodies are all known to be successful in ChIP experiments.

There are several advanced technologies and platforms for generating antibodies that improve over existing versions. One such technology generates the AbFlex® recombinant antibodies whose fragment crystallizable (Fc) regions have a transpeptidase sortase recognition sequence for site-specific tethering of various labels and tags, thus allowing for multiplexing. The above antibodies have been used in several publications, including a study on how development and neuronal activity are impacted by brain methylome, which is regulated by the memory transcription factor EGR1 and DNA demethylase TET1 (Sun et al. 2019).

3.3.2.4 Immobilization on Beads

Subsequently, the entire complex of 'antibody bound to cross-linked protein-DNA/chromatin' is isolated using beads conjugated to protein A or G. The immune complex is immobilized on beads upon incubation because proteins A/G mainly bind to Fc region and other domains of the antibodies. Following immunoprecipitation on beads, multiple washing steps are implemented to remove nonspecific interactions. Beads can be magnetic or agarose. Magnetic beads are quick and simple for washing, automation compatible, and visible so that users do not remove the immune complex while pipetting. The alternative option of agarose beads provides increased sensitivity and reduced backgrounds, but the washing becomes complicated and lengthy due to centrifugation at every step.

3.3.2.5 Analysis of ChIP Results

Washing is followed by reverse cross-linking and further treatments to finally retrieve the DNA fragments that were pulled down by immunoprecipitation. The

DNA might be subjected to PCR using primers against gene loci of interest. Alternatively, unbiased analysis of the global epigenetic landscape might be performed by subjecting the DNA to library preparation, sequencing, and bioinformatics analysis. The library is also prepared for sonicated chromatin that is not immunoprecipitated, called input, for signal normalization and control of biases. The libraries for NGS are prepared by ligating small oligonucleotides, or adapters, to the ends of the DNA so that the sequencer can "read" the DNA. Care is needed to optimize adapter ligation and cleanup processes so that the maximum sample is retained. Further, adapter dimerization must be minimized so that adapters ligate to the sample and not among themselves. Hence, proper attention to the adapter ligation step is essential to produce superior quality data in useful quantities.

Two major parameters to consider during sequencing are read length and single-end versus paired-end sequencing, which depend on experimental needs and budget. Read length indicates how many base pairs of the DNA are being analyzed at a time by the sequencer. Since longer reads offer information about the relative positions of specific base pairs with greater reliability, they are more expensive to produce. Two critical characteristics of sequencing are depth and coverage. Sequencing depth and coverage have been extensively discussed in an excellent review by Sims et al., 2015, where theoretical or expected coverage is defined as "*the average number of times that each nucleotide is expected to be sequenced given a certain number of reads of a given length and the assumption that reads are randomly distributed across an idealized genome*" (Lander et al. 2001; Sims et al. 2014).

Regarding depth, Sims et al. stated, "*Actual empirical per-base coverage represents the exact number of times that a base in the reference is covered by a high-quality aligned read from a given sequencing experiment. Redundancy of coverage is also called the depth or the depth of coverage*" (Sims et al. 2014). In other words, the average theoretical depth of sequencing coverage is the product of read length and number of reads divided by the haploid genome length (Sims et al. 2014). Coverage might denote the average raw or aligned read depth in NGS studies, which Sims et al. refers to as "*the expected coverage based on the number and the length of high-quality reads before or after alignment to the reference*" (Sims et al. 2014).

After sequencing, the raw data containing sequencing reads and quality scores might be obtained as FASTQ files, which undergo bioinformatics analysis (Mistry et al. 2022). Sequence output or reads from the experimental genome are aligned to the reference genome of that particular species. Several free, open-source, and commercial alignment software packages are available for this purpose. A process called *peak calling* is applied to detect regions in the genome that are enriched with aligned reads. An important measure of the success of ChIP is the *f*raction of *r*eads *i*n *p*eak regions or FRiP whose value of 1% or higher is deemed acceptable by ENCODE (Encyclopedia of DNA Elements) (Landt et al. 2012).

Data containing uniquely mapped reads need to be stored in a format that can be visualized for our understanding. In this direction, a popular visualization option is provided by the University of California Santa Cruz (UCSC) Genome Browser genome.ucsc.edu, which is described by Karolchik et al. as an online "*tool for quickly displaying a requested portion of a genome at any scale*" (Karolchik et al. 2011).

For visualization using the UCSC Genome Browser, the data need to be annotated or labeled. Bioinformatic genome annotation refers to the genome-wide mapping of genes and their coding regions, along with their functional determination. If required, a method called motif enrichment analysis is implemented to identify binding motifs of transcription factors in the regulatory regions of genes. Hence, motif enrichment predicts which transcription factors regulate a specific set of genes. Another output of ChIP-seq analysis is the identification of super-enhancers, which are genomic regions with multiple enhancers where several transcription factors associate and promote transcription.

Overall, bioinformatics analysis of ChIP-seq generates a variety of high-quantity data in various formats, including Sequence Alignment Map (SAM) (Li et al. 2009) and its compressed binary version Binary Alignment Map (BAM) as annotation, Browser Extensible Data (BED) as peak calling, BigWig as visualization, etc. In addition, details on ChIP-seq analysis methods have been published (Landt et al. 2012; Nakato and Sakata 2021).

ChIP-seq originated as an expensive method compared to ChIP-PCR, but it gained wide popularity over time for several reasons. ChIP-seq generates large quantities of genome-wide data. Epigenetics studies using ChIP-seq show a trend towards getting published more in high-profile scientific journals compared to the work using ChIP-PCR. Bioinformatics resources, including open-source analysis platforms, have made bioinformatics analysis simpler. More personnel with bioinformatics skills are now available.

Further, innovation has lowered NGS expenses compared to the past and made the analysis and visualization of complex ChIP-seq results quite convenient, which can be integrated with other NGS-based assays. ChIP is now routinely performed in several labs across the world. Many institutions offer core facilities for the NGS portion of the workflow. At the same time, complete end-to-end services, i.e., from sample preparation to publication-grade figures derived from bioinformatics analysis, are provided by experts in the field of epigenetics like Active Motif.

3.3.3 Modifications to ChIP

Sometimes, an additional step called spike-in normalization is performed to uncover differences in ChIP output for robust histone marks, which might not be detected otherwise. Spike-in normalization can be performed by adding the following: (1) chromatin from a species that is not related to the experimental species and (2) antibody that only recognizes a histone modification of the unrelated species (Egan et al. 2016). The additions are made to the experimental chromatin when the experimental ChIP antibody is added. Essential considerations during spike-in normalization are the ratio of experimental chromatin to spike-in chromatin and increased downsampling, which might obscure site-specific changes. In addition, during the bioinformatics analysis, the number of reads from the unrelated species in each ChIP reaction is used to normalize the number of reads from the experimental species to reveal actual differences in ChIP signals.

Many modified versions of ChIP have been developed, and more variations continue to be designed. One such modification which maps protein-binding loci on the genome with higher resolution is called transposase-assisted ChIP (TAM-ChIP) (U.S. Patent Nos. 10,689,643 and 9,938,524). Here, the regular immunoprecipitation step is followed by incubation with a secondary antibody, carrying a Tn5 transposase enzyme and NGS adapter sequences. Subsequent steps involve chromatin immobilization on beads and activation of Tn5 by Mg^{2+}, leading to a process called tagmentation. Here, activated Tn5 cleaves DNA that is not occupied by proteins and simultaneously inserts the NGS adapter sequences on the DNA at the site of cleavage. Hence, regions of DNA that are occupied by proteins, i.e., parts of DNA that are in the immune complex, remain protected from cleavage. Both ends of such protected DNA get tagged with adapters. The subsequent steps of reverse cross-linking, protein digestion, etc. resemble regular ChIP protocol, which ultimately yield DNA flanked by adapter sequences that can undergo PCR and sequencing.

3.3.4 Application of ChIP-seq to Study Neurodevelopmental Disorders

ChIP-seq has been employed in several studies on neurodevelopmental disorders. In a study on ASD, ChIP-seq was performed using antibodies against the chromodomain helicase CHD8, strongly associated with ASD and H3K27 acetylation, a histone mark for active transcription and enhancers (Cotney et al. 2015). ChIP-seq was used to detect genes that are targeted by CHD8 during human and murine neurodevelopment in human midfetal brain and neural stem cells (NSCs) and murine embryonic cortex. Human midfetal cortex shows co-expression of ASD-risk genes, indicating their possible convergence in neurodevelopmental gene regulatory networks, which the study investigated through detection of genes targeted by CHD8 (Cotney et al. 2015).

In the above-discussed direction, ChIP-seq was performed on cortical tissue dissected from embryonic day 17.5 mouse embryos and human period 5 (16–19 PCWs or postconceptional weeks) fetal brain tissue. The human tissue was obtained from dissection of the striatum, cerebellum, primary visual cortex, and dorsal frontal cortex of two specimens (Cotney et al. 2015). Following the library preparation and sequencing, read alignment was performed against mm9 (*Mus musculus* genome assembly) and hg19 (human reference genome), for murine and human ChIP-seq results, respectively. Bioinformatics analysis was performed, including peak calling and identification of promoter and enhancer peaks. Motif enrichment was also performed to match the results to established binding sites for transcription factors. Hence, using ChIP-seq, the study led to the discovery of strong enrichment of CHD8 targets for other ASD-risk genes during human and murine neurodevelopment, which show convergence in co-expression networks related to ASD in the human midfetal cortex (Cotney et al. 2015). The study further showed that ASD-risk genes, which are direct targets of CHD8, are dysregulated upon CHD8 knockdown in NSCs. Detection of risk genes was improved when CHD8 occupancy results were integrated into risk models of ASD. Hence, the study sheds new insights into the epigenetic regulation of ASD through CHD8, whose absence likely impacts an ancient network of human brain developmental gene regulation (Cotney et al. 2015).

ChIP-seq was employed in another study on ACTL6B or BAF53B (Bell et al. 2019), a neuronal specific subunit of ATP-dependent chromatin remodeling BAF complex, which is mutated in neurodevelopmental disorders (Wenderski et al. 2020; Sokpor et al. 2017). In the present study, ChIP-seq was performed for BRG1, an essential subunit of BAF complex, followed by read alignment to hg19, peak identification, and analysis of differential binding sites. In a pathogenic mutant of *ACTL6B* that is associated with neurodevelopmental disorders, BRG1 occupancy was increased along certain genomic regions which affected the expression of some genes including microtubule-binding *TPP* and neurite-regulating *FSCN1 (*Bell et al. 2019*)*. Hence, using ChIP-seq, the study showed how a mutation in *ACTL6B* regulates the genome-wide association of an epigenetic factor to impact transcription leading to neurodevelopmental and neuronal defects (Bell et al. 2019).

ChIP-seq has also been used in a study on the epigenetic regulation of developmental learning (Kelly et al. 2018). In the present study, ChIP-seq was performed on the auditory forebrain of zebra finch birds to investigate genome-wide occupancy of RNA polymerase II and histone modifications, which showed that ChIP-seq could be employed across a wide spectrum of neurodevelopmental research.

3.4 RIME or Rapid Immunoprecipitation Mass Spectrometry of Endogenous Proteins

3.4.1 RIME Background

A portion of the ChIP protocol has been incorporated with mass spectrometry to develop a method called RIME or rapid immunoprecipitation mass spectrometry of endogenous proteins to study the interactome of chromatin-associated proteins, transcriptional cofactors, etc. (Fig. 3.2). The initial part of RIME resembles ChIP which involves formaldehyde-based cross-linking, cell lysis, and affinity purification of proteins by antibodies on beads. Interaction among proteins that co-purify with the target protein of interest is detected by mass spectrometry. RIME is a very sensitive assay because it captures transient and low-affinity interactions among endogenous proteins with low quantities of sample material.

RIME was first reported by the Carroll lab, where it was used to purify estrogen receptor (ER), a critical transcription factor in luminal breast cancer, from the limited content of primary samples (Mohammed et al. 2013). In the quest for identifying new cofactors that interact with ER, the study found that ER recruits GREB1 to chromatin for transcription activation (Mohammed et al. 2013). Interestingly, the paralog of GREB1 has neurodevelopmental roles because it regulates the formation

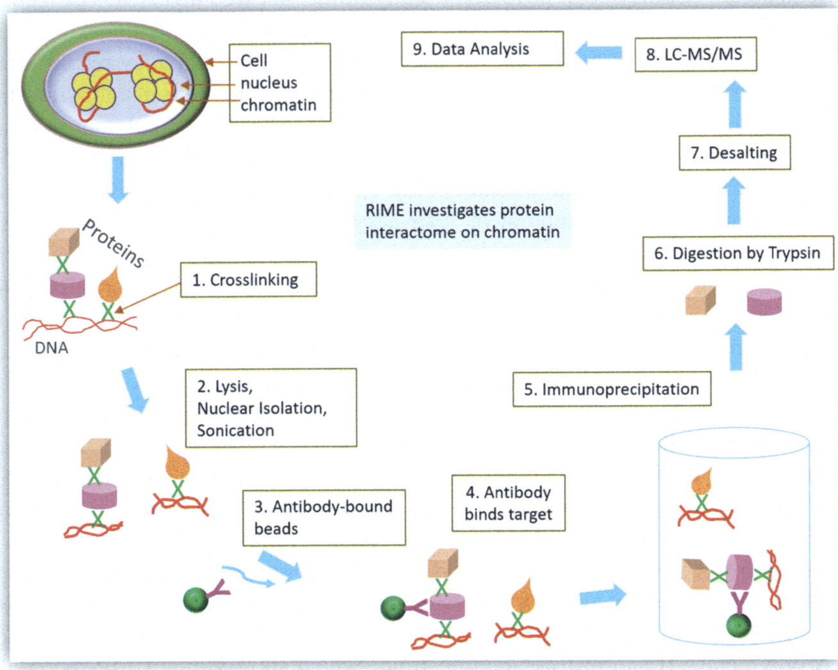

Fig. 3.2 RIME

of the inner ear, sensory epithelia innervation, cranial nerve, early craniofacial tissue, and serves as a pre-migratory neural crest regulatory molecule (Schrauwen et al. 2018). The Carroll lab published a detailed RIME protocol (Mohammed et al. 2016) and further work using RIME to reveal higher resolution interactions between progesterone receptor and ER (Mohammed et al. 2015).

3.4.2 RIME Method Outline

The RIME workflow is similar to ChIP up to the immunoprecipitation step, with one of the modifications being cell lysis followed by nuclear isolation before sonication. After immunoprecipitation, the immunoprecipitated proteins undergo digestion by trypsin for identification using mass spectrometry. The sensitivity of RIME is enhanced by excluding the process of alkylation or reduction of samples prior to digestion to limit the quantity of IgG peptides. Following digestion, the peptides are subjected to desalting by solid-phase extraction, which makes them ready for LC-MS/MS (liquid chromatography with tandem mass spectrometry) analysis. Supernatant from the digestion step contains aggregated proteins or high-molecular-weight entities which can impart back pressure on high-performance liquid chromatography (HPLC). Hence, desalting is performed to reduce such pressure. In addition, the peptide flow-through might be reloaded thrice to optimize the protein-binding process. Subsequently, proteins are separated and identified by LC-MS/MS, a method known for high specificity and sensitivity.

3.4.2.1 RIME Data Analysis

Raw mass spectra data obtained through the above steps can be processed to identify and quantify proteins using various software packages, data analysis programs, and search engines (Mohammed et al. 2016). The raw data can be visualized by some free software apps, e.g., Scaffold www.proteomesoftware.com/products. The methodology published by the Carroll lab recommends that the search should be performed using a database which is in concatenation with a *decoy* database containing the sequence order of amino acids in randomization or reversal (Mohammed et al. 2016). The purpose of the recommendation is to calculate the false-positive rate based on decoy hits, where the cutoff rate to identify peptides is 1%.

The methodology of Carroll lab further mentions how certain amino acid modifications are taken into account and how false discovery rate (FDR) parameters are used to screen high-confidence peptide spectrum matches (PSMs). PSMs refer to the total number of identified peptide sequences pertaining to a protein. Ideally, output from RIME identifies about 300–900 proteins along with their accession number, descriptions, and search statistics, where 5–10% of the output contains the expected specific interactors. The target protein should appear among the top ranks (Mohammed et al. 2016). Gene ontology analysis might be performed to detect biological pathways in which the target proteins participate. In this analysis, results are interpreted based on sets of genes that are classified or assigned to predefined sets, or *bins*, according to their roles.

3.4.3 Expansion of RIME Methodology

The Carroll lab developed a more high-throughput, sensitive, and reproducible version of RIME called quantitative multiPLEXed Rapid Immunoprecipitation Mass spectrometry of Endogenous proteins (qPLEX-RIME) (Papachristou et al. 2018). qPLEX-RIME is a multiplexed assay that conjugates RIME with chemical isobaric labeling with tandem mass tags (TMTs). One of the modifications in this workflow occurs during cross-linking where *double fixation* is performed using formaldehyde and disuccinimidyl glutarate (DSG) to address transient interactions. The method has implemented a bioinformatics protocol called qPLEXanalyzer, which encompasses data processing, visualizing, normalizing, and statistical analysis (Papachristou et al. 2018).

3.4.4 Application of RIME to Study Neurodevelopmental Disorders

Being relatively new and proteomics-based technology, RIME has been applied to a limited number of studies so far. Nonetheless, publications on RIME show its potential for studying neurodevelopmental disorders. A study using RIME shed new insights into the role of H3K27 acetylation in enhancer activity (Raisner et al. 2018). RIME was performed on cell lines using an antibody against transcriptional co-activator P300, which complexes with CBP (cAMP response element-binding protein (CREB) binding protein). The study showed that H3K27 acetylation by bromodomain of CBP/P300 is essential for enhancer activity and transcription (Raisner et al. 2018).

Although the above study did not specifically focus on neurodevelopmental disorders, its results are immensely significant to understanding those diseases for several reasons. Studies indicated that enhancers regulate chromatin and gene expression to dysregulate transcription in neurodegenerative diseases including Huntington's and Alzheimer's disease (Alcala-Vida et al. 2021). Further, DYRK1A, a kinase implicated in neurodevelopmental disorders (Duchon and Herault 2016), showed enhancer localization and association with CBP/P300 (Li et al. 2018). Another study using RIME identified that an interaction partner of nuclear receptor PPARβ/δ (Legrand et al. 2019) was the neuropathy target esterase PNPLA6, which is implicated in neurodevelopmental, neurodegenerative, and neurological disorders (Hufnagel et al. 2015; Zheng et al. 2018; Sogorb et al. 2016).

3.5 CUT&Tag or Cleavage Under Targets and Tagmentation

3.5.1 CUT&Tag Background

Although ChIP-seq continues to be the gold standard in the field for identifying and mapping the interactions between DNA/chromatin and a wide variety of proteins, a disadvantage of ChIP-seq is the requirement of a large number of starting materials. In the mentioned direction, other methodology has been developed which answers the same questions like ChIP-seq using lesser quantities of starting materials. One such technology is CUT&Tag (Kaya-Okur et al. 2019, 2020), which has few steps similar to TAM-ChIP described in Sect. 3.3.3. Previously, the cross-linking complications and starting material limitations associated with ChIP were addressed by other methods. Some of them are DNA adenine methyltransferase identification (DamID) of 2001 (van Steensel et al. 2001), chromatin endogenous cleavage (ChEC) of 2004 (Schmid et al. 2004), DNA affinity purification sequencing (DAP-seq) of 2017 (Bartlett et al. 2017), and cleavage under targets and release using nuclease (CUT&RUN) of 2017 (Skene and Henikoff 2017).

DamID and ChEC are based on enzymes that get tethered to DNA-binding proteins to exert their actions at the DNA-binding sites. The enzymes involved in DamID and ChEC are DNA adenine methyltransferase, which modifies DNA, and micrococcal nuclease (MNase), which digests DNA, respectively. Direct extraction of DNA from live or permeabilized cells occurs in both methods. DAP-seq identifies transcription factor-binding sites using a protocol that combines affinity-purified transcription factors with NGS of a genomic DNA library (Bartlett et al. 2017). In CUT&RUN (Skene and Henikoff 2017), nuclei from live cells are immobilized on magnetic beads that are coated with lectin and incubated with antibodies against target protein along with protein A-MNase fusion, primed by Ca^{2+}. Next, the protein-DNA complex is isolated, purified, and subject to library preparation. As methods continued to develop based on the advantages and disadvantages of the above processes, CUT&Tag came to be published in 2019.

3.5.2 CUT&Tag Method Outline

Cross-linking by formaldehyde and sonication are not required in CUT&Tag (Fig. 3.3). Here, the process begins with introducing cells to magnetic beads coated with concanavalin A, which is a lectin that has a specific affinity for extracellular glycoproteins containing mannosyl and glucosyl moieties. Hence, cells or nuclei with the above glycans in their extracellular matrices get immobilized on the beads. Next, the cells undergo permeabilization with antibody buffer containing digitonin which solubilizes lipids in cell membranes to facilitate the entry of large molecules like antibodies inside cells.

Subsequently, cells are incubated overnight with primary antibody, washed, and then incubated with the secondary antibody. The next main step is targeted tagmentation, where cells are incubated with protein A-Tn5 transposase. The reaction is stopped after an hour, followed by DNA solubilization using EDTA, SDS, and proteinase K. Then, the DNA undergoes extraction, PCR amplification to affix NGS indices, and sequencing. Important considerations during tagmentation are controlling the concentrations of digitonin to prevent clumping and NaCl to protect accessible chromatin from digestion by Tn5.

3.5.3 Application of CUT&Tag to Study Neurodevelopmental Disorders

CUT&Tag scores over ChIP in some aspects like the requirement of less starting material, exclusion of fixation and sonication, quicker protocol, and need for lesser sequencing. However, ChIP is advantageous over CUT&Tag in identifying and mapping the interactions of those proteins to DNA/chromatin which have low expression or which bind feebly, transiently, or indirectly to DNA/chromatin. The native conditions of CUT&Tag might not capture proteins with the above characteristics, e.g., many transcription factors, for whom cross-linking will be a requirement. Further, CUT&Tag is relatively new and requires time to be optimized for a wide variety of sample types and target proteins. Having a different workflow compared to ChIP-seq, CUT&Tag results might not be easy to directly compare with past ChIP-seq and ENCODE results. It is also not extensively studied as to how

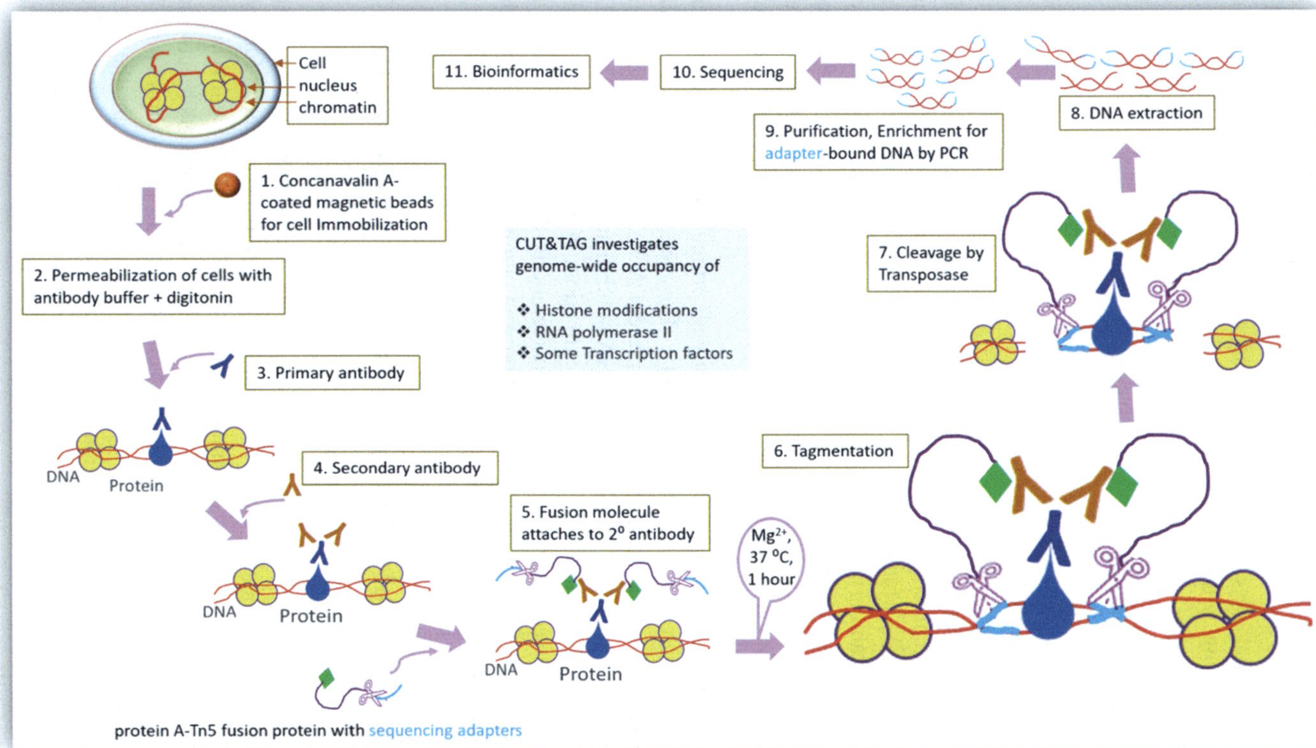

Fig. 3.3 CUT&Tag

CUT&Tag would perform in various regions of silent and inaccessible chromatin, or heterochromatin, because Tn5 only has affinity for open and accessible chromatin. Nonetheless, CUT&Tag has been efficient in mapping the genomic associations of some histone marks, RNA polymerase II, and transcription factors CTCF and NPAT (Kaya-Okur et al. 2019); has been successfully extended to research on crop plants (Tao et al. 2020; Ouyang et al. 2021); and has produced superior quality chromatin accessibility maps upon method modification (Henikoff et al. 2020). Studies on brain and nervous system using CUT&Tag are mentioned below.

3.5.4 Single-Cell CUT&Tag

CUT&Tag could be adapted to single-cell analysis where the workflow from antibody binding to library generation takes place in intact nuclei/cells. An important component of the adaptation involves centrifugation of cells between washes instead of using magnetic beads. Following tagmentation, the sample gets aliquoted as single cells in nanowells for barcoding before sequencing. CUT&Tag successfully enabled single-cell profiling of certain histone marks, and helped distinguish between H1 embryonic stem cells and K562 immortalized myelogenous leukemia cell line (Kaya-Okur et al. 2019).

Despite being new, single-cell CUT&Tag has been performed by other studies to analyze chromatin landscape at the single-cell level (Wu et al. 2021; Bartosovic et al. 2021). In this direction, one study analyzed single cells from complex tissues and differentiated human embryonic stem cells with scalable nanowell and droplet-based single-cell platforms (Wu et al. 2021). The study showed that different blood cell types in humans could be distinguished by single-cell profiling of H3K27 trimethylation, a transcriptionally repressive histone mark caused by the polycomb complex (Wu et al. 2021). The distinctions could produce polycomb landscapes specific to cell types from heterogeneous tissues and provide an orthogonal method to study chromatin landscape for detecting cell states (Wu et al. 2021). The study further applied single-cell CUT&Tag to analyze tumor microenvironment and polycomb-mediated heterogeneity by profiling H3K27me3 in a brain tumor patient (Wu et al. 2021).

Another study on single-cell CUT&Tag used droplet-based platform to prepare single-cell libraries using murine cells to discover epigenomic landscapes in central nervous system (Bartosovic et al. 2021). Using single-cell CUT&Tag profiles, the study elucidated cellular identity and regulatory processes like bivalency of promoters, H3K4me3 occupancy, and connectivity of promoters and enhancers. The study also investigated histone modifications associated with active promoters, e.g., H3M4 trimethylation; enhancers, e.g., H3K27 acetylation; and gene bodies, e.g., H3K36 trimethylation (Bartosovic et al. 2021). Chromatin association of transcription factor OLIG2 and cohesion complex subunit RAD21 was also analyzed at single-cell resolution (Bartosovic et al. 2021). In summary, CUT&Tag is a very new but already popular technology that has also been developed to perform on single-cell platforms. Hence, it is only a matter of time before it is implemented in neurodevelopmental research. Further modifications can help to expand the scope and reduce time and costs leading to more effective application of CUT&Tag (Kaya-Okur et al. 2020), and similar processes involving tagmentation.

3.6 ATAC-seq or Assay for Transposase-Accessible Chromatin with High-Throughput Sequencing

3.6.1 ATAC-seq Background

Tagmentation has been implemented in another important and popular methodology called ATAC-seq which helps map regions of open versus closed chromatin across the genome (Fig. 3.4). ATAC-seq was first reported in 2013, where the technique shed significant insights into personal epigenomes of resting human T cells for health and disease monitoring in clinical timescales (Buenrostro et al. 2013). Buenrostro et al. desribed that the study was successfully performed using 500–50,000 cells to unravel the connections between "*genomic locations of open chromatin, DNA-binding proteins, individual nucleosomes, and higher order compaction at regulatory regions with nucleotide resolution*" (Buenrostro et al. 2013). Genome-wide occupancy maps of 89 transcription factors were derived from the ATAC-seq footprints that helped to systematically reconstruct personalized regulatory networks, which could be potentially implemented in diagnostic applications (Buenrostro et al. 2013).

3.6.2 ATAC-seq Method Outline

Cell content, quality, preparation, and transposition are crucial for successful ATAC-seq. Quantification of cells plays a significant role because it affects the transposition reaction and size distribution profiles of the DNA fragments that are produced. Insufficient or excessive cells result in abnormally higher or lower extent of digestion that diminishes ATAC-seq library quality. Ideally, 25,000–75,000 cells are recommended. In terms of quality, cells need to be in a homogenous, single-cell suspension while being intact and not subjected to harsh treatments.

Following harvest, pure nuclei are obtained by lysing the cells with a nonionic detergent, whose chemistry breaks protein-lipid and lipid-lipid associations only, but does not break protein-protein interactions nor denatures proteins, in general. The chromatin obtained here is treated with Tn5 transposase for the tagmentation reaction where DNA unoccupied by proteins undergoes cleavage and simultaneous tagging by sequencing adapters. In this way, the ATAC-seq library is generated, which is then purified and subjected to PCR amplification with barcoded primers.

Subsequently, the resulting samples might undergo analysis by qPCR or NGS depending on the experimental needs. Like ChIP-seq, bioinformatics analysis of ATAC-seq includes several files, including raw FASTQ files, BAM alignment files, BigWig visualization files, peak calling files, etc.

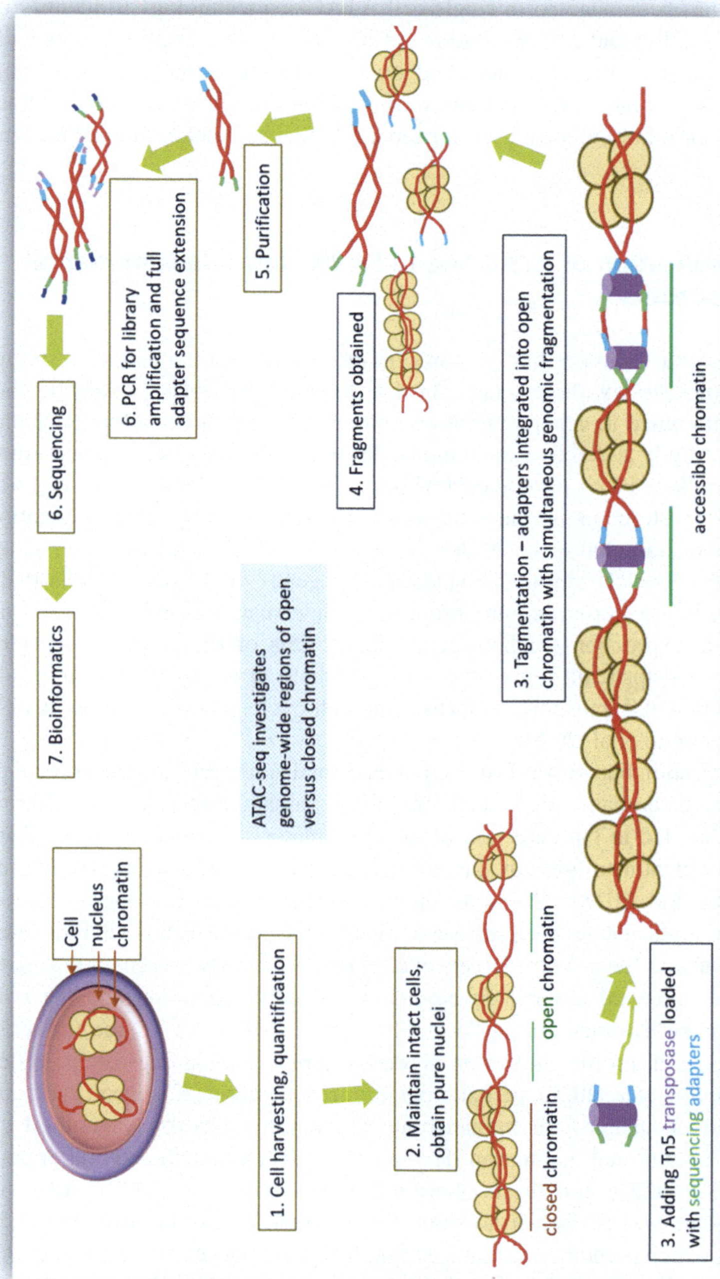

Fig. 3.4 ATAC-seq

3.6.3 Single-Cell ATAC-seq

ATAC-seq has been adapted to single-cell ATAC-seq technology to identify open chromatin in individual cell populations of heterogeneous complex tissues and for lineage tracing of cells. It is achieved by labeling individual cells with unique barcoding for their identification during bioinformatics analysis. Fluidics, microplates, or other tools are used for the individual separation of each nucleus.

3.6.4 Application of ATAC-Seq to Study Neurodevelopmental Disorders

Sensory neuronal subtypes like photoreceptors are lost in retinopathies resulting in permanent blindness without a cure. In this direction, fibroblasts might be chemically reprogrammed to generate photoreceptors, but the process is highly challenging. A recent study added five small molecules to fibroblasts for their transformation into chemically induced photoreceptor-like cells (CiPCs). Pupil reflex and vision were partially restored in murine models of rod degeneration upon transplantation of the CiPCs into subretinal space (Mahato et al. 2020). ATAC-seq was implemented in this study to elucidate chromatin landscape of genes of interest in cells that are intermediate in the reprogramming process, to understand transcription factor binding and gene expression (Mahato et al. 2020). One of the genes showing open chromatin at its upstream regions is Ascl1 (Mahato et al. 2020), which regulates neurodevelopmental transcription factors and cell cycle genes in murine models of brain tumors (Vue et al. 2020).

Data integration between ATAC-seq on germinal zone and subplate/cortical plate of developing human brain, and the Hi-C assay that captures chromatin conformations, led to the detection of distal regulatory elements of genes that are crucial for cortical neurogenesis, including human-gained enhancers (HGEs) (de la Torre-Ubieta et al. 2018). The study shed important insights into the regulation of neural progenitors and gene expression in outer radial glia, which is vital for cerebral cortical expansion (de la Torre-Ubieta et al. 2018). The study presented the relationship between prenatal cortical development and adult cognition and the risk of neuropsychiatric disorders (de la Torre-Ubieta et al. 2018). ATAC-seq was used in several other high-profile studies on neurodevelopment including forebrain development (Trevino et al. 2020), neural progenitor regeneration (Kakebeen et al. 2020), development of an atlas for human brain chromatin accessibility (Fullard et al. 2018), human neuronal maturation (Hickey et al. 2019), human cerebral organoids (Kanton et al. 2020), and rhesus macaque brains (Yin et al. 2020). Analysis of ATAC-seq data was performed to study the neocortical development (Polioudakis et al. 2019), which is another advantage of high-throughput sequencing assays where datasets available from various researchers can be analyzed for data integration and comparison with other experiments without undertaking the wet-lab part of every methodology.

3.7 Conclusions

The continuous development of the state-of-the-art technology to study epigenetics would help unravel the new epigenetic causes, targets, and mechanistic details associated with neurodevelopmental disorders. Several publications have focused on the potential of epigenetics-based therapies against the neurodevelopmental disorders (Urdinguio et al. 2009; Mastrototaro et al. 2017; Millan 2013; Cristancho and Marsh 2020; Ciptasari and van Bokhoven 2020). In addition to the focus on chromatin, epigenetic regulation involving RNA is expected to gain importance in the field of neurodevelopmental disorders (Giuseppina and Alessandro 2018), with further discoveries of RNA modifications (Delatte et al. 2016) and the role of RNA in epigenetic regulations (Long et al. 2020). For instance, the role of RNA is discovered to be essential for the chromatin occupancy and function of polycomb repressive complex 2 (PRC2) in human pluripotent stem cells (Long et al. 2020), and PRC2 is implicated in the development of organs, including the embryonic brain, and congenital disorders (Evans et al. 2020; Deevy and Bracken 2019).

Advancements in epigenetics research and methodology will benefit organoid modeling to study neurodevelopmental disorders (Lewis and Kroll 2018; Lewis et al. 2021) and organ transplantation (Kuscu et al. 2021). With innovations in epigenomics and transcriptomics, focus is also needed on proteomics to obtain a comprehensive understanding of any field of biological research (Timp and Timp 2020). High-throughput assays like large-scale targeted sequencing have detected genes that are involved in neurodevelopmental disorders (Wang et al. 2020). Hence, results from such assays could be integrated with large datasets generated by epigenetic methodology for a deeper understanding of the mechanisms behind neurodevelopmental disorders. It is exciting to see several clinical trials at the crossroads of epigenetics, neurodevelopment, and associated disorders (NCT01 913093, NCT03389178, NCT03918616, NCT04804280). The significance of epigenetics in neurodevelopmental disorders has been established by major discoveries and clinical trials, and hence the immense potential of the field awaits further exploration.

Acknowledgements The author thanks Joseph Fernandez, Theodore DeFrank, Fritz Eibel, Walid Qoronfleh, and colleagues at Active Motif.

Patents Jelinek, M.A., Egan, B.S., Fernandez, J. (2020). *Targeted transposition for use in epigenetic studies* (U.S. Patent No. 10,689,643) U.S. Patent and Trademark Office.

Fernandez J., Jelinek M.A., Egan, B.S. (2018). *Multiplex isolation of protein-associated nucleic acids* (U.S. Patent No. 9,938,524) U.S. Patent and Trademark Office.

Additional Resources Peak calling with MACS2. Introduction to ChIP-Seq using high-performance computing hbctraining.github.io/Intro-to-ChIPseq/lessons/05_peak_calling_macs.html, contributed by Meeta Mistry, Ph.D., and Radhika Khetani, Ph.D. https://zenodo.org/record/5825871#.YhBGyejMJyy

References

Alcala-Vida R et al (2021) Epigenetic mechanisms underlying enhancer modulation of neuronal identity, neuronal activity and neurodegeneration. Neurobiol Dis 147:105155

Allan J et al (1981) Regulation of the higher-order structure of chromatin by histones H1 and H5. J Cell Biol 90(2):279–288

Amir RE et al (2000) Influence of mutation type and X chromosome inactivation on Rett syndrome phenotypes. Ann Neurol 47(5):670–679

Barski A et al (2007) High-resolution profiling of histone methylations in the human genome. Cell 129(4):823–837

Bartlett A et al (2017) Mapping genome-wide transcription-factor binding sites using DAP-seq. Nat Protoc 12(8):1659–1672

Bartosovic M, Kabbe M, Castelo-Branco G (2021) Single-cell CUT&Tag profiles histone modifications and transcription factors in complex tissues. Nat Biotechnol 39(7):825–835

Bell S et al (2019) Mutations in ACTL6B cause neurodevelopmental deficits and epilepsy and lead to loss of dendrites in human neurons. Am J Hum Genet 104(5):815–834

Bomsztyk K et al (2019) PIXUL-ChIP: integrated high-throughput sample preparation and analytical platform for epigenetic studies. Nucleic Acids Res 47(12):e69

Brockers K, Schneider R (2019) Histone H1, the forgotten histone. Epigenomics 11(4):363–366

Buenrostro JD et al (2013) Transposition of native chromatin for fast and sensitive epigenomic profiling of open chromatin, DNA-binding proteins and nucleosome position. Nat Methods 10(12):1213–1218

Buiting K (2010) Prader-Willi syndrome and Angelman syndrome. Am J Med Genet C Semin Med Genet 154C(3):365–376

Cariaga-Martinez AE, Gutierrez KJ, Alelu-Paz R (2018) The vast complexity of the epigenetic landscape during neurodevelopment: an open frame to understanding brain function. Int J Mol Sci 19(5):1333

Chelly J et al (2006) Genetics and pathophysiology of mental retardation. Eur J Hum Genet 14(6): 701–713

Ciptasari U, van Bokhoven H (2020) The phenomenal epigenome in neurodevelopmental disorders. Hum Mol Genet 29(R1):R42–R50

Cotney J et al (2015) The autism-associated chromatin modifier CHD8 regulates other autism risk genes during human neurodevelopment. Nat Commun 6:6404

Cristancho AG, Marsh ED (2020) Epigenetics modifiers: potential hub for understanding and treating neurodevelopmental disorders from hypoxic injury. J Neurodev Disord 12(1):37

de la Torre-Ubieta L et al (2018) The dynamic landscape of open chromatin during human cortical neurogenesis. Cell 172(1–2):289–304.e18

Deevy O, Bracken AP (2019) PRC2 functions in development and congenital disorders. Development 146(19):dev181354

Delatte B et al (2016) RNA biochemistry. Transcriptome-wide distribution and function of RNA hydroxymethylcytosine. Science 351(6270):282–285

Duchon A, Herault Y (2016) DYRK1A, a dosage-sensitive gene involved in neurodevelopmental disorders, is a target for drug development in down syndrome. Front Behav Neurosci 10:104

Egan B et al (2016) An alternative approach to ChIP-Seq normalization enables detection of genome-wide changes in histone H3 lysine 27 trimethylation upon EZH2 inhibition. PLoS One 11(11):e0166438

Esteller M (2007) Rett syndrome: the first forty years: 1966-2006. Epigenetics 2(1):1

Evans MK et al (2020) Ybx1 fine-tunes PRC2 activities to control embryonic brain development. Nat Commun 11(1):4060

Fallah MS et al (2020) Impaired regulation of histone methylation and acetylation underlies specific neurodevelopmental disorders. Front Genet 11:613098

Fullard JF et al (2018) An atlas of chromatin accessibility in the adult human brain. Genome Res 28(8):1243–1252

Gabriele M et al (2018) The chromatin basis of neurodevelopmental disorders: rethinking dysfunction along the molecular and temporal axes. Prog Neuropsychopharmacol Biol Psychiatry 84 (Pt B):306–327

Gilmour DS, Lis JT (1984) Detecting protein-DNA interactions in vivo: distribution of RNA polymerase on specific bacterial genes. Proc Natl Acad Sci U S A 81(14):4275–4279

Gilmour DS, Lis JT (1985) In vivo interactions of RNA polymerase II with genes of Drosophila melanogaster. Mol Cell Biol 5(8):2009–2018

Giuseppina M, Alessandro S (2018) Chapter 9: Emerging role of epigenetics in human neurodevelopmental disorders. Transl Epigenet 6:269–304

Graff J, Mansuy IM (2009) Epigenetic dysregulation in cognitive disorders. Eur J Neurosci 30(1): 1–8

Hebbes TR, Thorne AW, Crane-Robinson C (1988) A direct link between core histone acetylation and transcriptionally active chromatin. EMBO J 7(5):1395–1402

Henikoff S et al (2020) Efficient chromatin accessibility mapping in situ by nucleosome-tethered tagmentation. Elife 9:e63274

Hickey SL, Berto S, Konopka G (2019) Chromatin decondensation by FOXP2 promotes human neuron maturation and expression of neurodevelopmental disease genes. Cell Rep 27(6): 1699–1711.e9

Hoffman EA et al (2015) Formaldehyde crosslinking: a tool for the study of chromatin complexes. J Biol Chem 290(44):26404–26411

Hoffmann A, Spengler D (2019) Chromatin remodeling complex NuRD in neurodevelopment and neurodevelopmental disorders. Front Genet 10:682

Hoffmann A et al (2017) Epigenomics of major depressive disorders and schizophrenia: early life decides. Int J Mol Sci 18(8):1711

Hufnagel RB et al (2015) Neuropathy target esterase impairments cause Oliver-McFarlane and Laurence-Moon syndromes. J Med Genet 52(2):85–94

Inlow JK, Restifo LL (2004) Molecular and comparative genetics of mental retardation. Genetics 166(2):835–881

Johnson DS et al (2007) Genome-wide mapping of in vivo protein-DNA interactions. Science 316(5830):1497–1502

Kakebeen AD et al (2020) Chromatin accessibility dynamics and single cell RNA-Seq reveal new regulators of regeneration in neural progenitors. Elife 9:e52648

Kanton S, Treutlein B, Camp JG (2020) Single-cell genomic analysis of human cerebral organoids. Methods Cell Biol 159:229–256

Karolchik D, Hinrichs AS, Kent WJ (2011) The UCSC genome browser. Curr Protoc Hum Genet Chapter 18: Unit18.6

Kaya-Okur HS et al (2019) CUT&Tag for efficient epigenomic profiling of small samples and single cells. Nat Commun 10(1):1930

Kaya-Okur HS et al (2020) Efficient low-cost chromatin profiling with CUT&Tag. Nat Protoc 15(10):3264–3283

Kelly TK et al (2018) Epigenetic regulation of transcriptional plasticity associated with developmental song learning. Proc Biol Sci 285(1878):20180160

Kim JH et al (2017) Histone lysine methylation and neurodevelopmental disorders. Int J Mol Sci 18(7):1404

Kramer JM, van Bokhoven H (2009) Genetic and epigenetic defects in mental retardation. Int J Biochem Cell Biol 41(1):96–107

Kuscu C, Eason JD, Kuscu C (2021) Technical advancements in epigenomics and applications in transplantation. Curr Opin Organ Transplant 26(1):23–29

Lander ES et al (2001) Initial sequencing and analysis of the human genome. Nature 409(6822): 860–921

Landt SG et al (2012) ChIP-seq guidelines and practices of the ENCODE and modENCODE consortia. Genome Res 22(9):1813–1831

Legrand N et al (2019) PPARbeta/delta recruits NCOR and regulates transcription reinitiation of ANGPTL4. Nucleic Acids Res 47(18):9573–9591

Lewis EM, Kroll KL (2018) Development and disease in a dish: the epigenetics of neurodevelopmental disorders. Epigenomics 10(2):219–231

Lewis EMA et al (2021) Epigenetic regulation during human cortical development: Seq-ing answers from the brain to the organoid. Neurochem Int 147:105039. https://doi.org/10.1016/j.neuint.2021.105039

Li H et al (2009) The sequence alignment/map format and SAMtools. Bioinformatics 25(16): 2078–2079

Li J, Zhao G, Gao X (2013) Development of neurodevelopmental disorders: a regulatory mechanism involving bromodomain-containing proteins. J Neurodev Disord 5(1):4

Li S et al (2018) DYRK1A interacts with histone acetyl transferase p300 and CBP and localizes to enhancers. Nucleic Acids Res 46(21):11202–11213

Long Y et al (2020) RNA is essential for PRC2 chromatin occupancy and function in human pluripotent stem cells. Nat Genet 52(9):931–938

Lorentsen KJ et al (2018) Bcl11b is essential for licensing Th2 differentiation during helminth infection and allergic asthma. Nat Commun 9(1):1679

Luger K et al (1997) Crystal structure of the nucleosome core particle at 2.8 A resolution. Nature 389(6648):251–260

Mahato B et al (2020) Pharmacologic fibroblast reprogramming into photoreceptors restores vision. Nature 581(7806):83–88

Mastrototaro G, Zaghi M, Sessa A (2017) Epigenetic mistakes in neurodevelopmental disorders. J Mol Neurosci 61(4):590–602

Mikkelsen TS et al (2007) Genome-wide maps of chromatin state in pluripotent and lineage-committed cells. Nature 448(7153):553–560

Millan MJ (2013) An epigenetic framework for neurodevelopmental disorders: from pathogenesis to potential therapy. Neuropharmacology 68:2–82

Mistry M et al (2022) hbctraining/Intro-to-ChIPseq-flipped: ChIP-seq data analysis lessons from HCBC (first release). Zenodo. https://doi.org/10.5281/zenodo.5825871

Miyake K et al (2012) Epigenetics in autism and other neurodevelopmental diseases. Adv Exp Med Biol 724:91–98

Mohammed H et al (2013) Endogenous purification reveals GREB1 as a key estrogen receptor regulatory factor. Cell Rep 3(2):342–349

Mohammed H et al (2015) Progesterone receptor modulates ERalpha action in breast cancer. Nature 523(7560):313–317

Mohammed H et al (2016) Rapid immunoprecipitation mass spectrometry of endogenous proteins (RIME) for analysis of chromatin complexes. Nat Protoc 11(2):316–326

Morris-Rosendahl DJ, Crocq MA (2020) Neurodevelopmental disorders-the history and future of a diagnostic concept. Dialogues Clin Neurosci 22(1):65–72

Mossink B et al (2021) The emerging role of chromatin remodelers in neurodevelopmental disorders: a developmental perspective. Cell Mol Life Sci 78(6):2517–2563

Nakato R, Sakata T (2021) Methods for ChIP-seq analysis: a practical workflow and advanced applications. Methods 187:44–53

O'Neill LP, Turner BM (2003) Immunoprecipitation of native chromatin: NChIP. Methods 31(1): 76–82

Orlando V (2000) Mapping chromosomal proteins in vivo by formaldehyde-crosslinked-chromatin immunoprecipitation. Trends Biochem Sci 25(3):99–104

Ouyang W et al (2021) Rapid and low-input profiling of histone marks in plants using nucleus CUT&Tag. Front Plant Sci 12:634679

Papachristou EK et al (2018) A quantitative mass spectrometry-based approach to monitor the dynamics of endogenous chromatin-associated protein complexes. Nat Commun 9(1):2311

Park PJ (2009) ChIP-seq: advantages and challenges of a maturing technology. Nat Rev Genet 10(10):669–680

Polioudakis D et al (2019) A single-cell transcriptomic atlas of human neocortical development during mid-gestation. Neuron 103(5):785–801 e8

Quievryn G, Zhitkovich A (2000) Loss of DNA-protein crosslinks from formaldehyde-exposed cells occurs through spontaneous hydrolysis and an active repair process linked to proteosome function. Carcinogenesis 21(8):1573–1580

Raisner R et al (2018) Enhancer activity requires CBP/P300 bromodomain-dependent histone H3K27 acetylation. Cell Rep 24(7):1722–1729

Robertson G et al (2007) Genome-wide profiles of STAT1 DNA association using chromatin immunoprecipitation and massively parallel sequencing. Nat Methods 4(8):651–657

Rockwell AL, Hongay CF (2019) The m(6)A dynamics of profilin in neurogenesis. Front Genet 10: 987

Schmid M, Durussel T, Laemmli UK (2004) ChIC and ChEC; genomic mapping of chromatin proteins. Mol Cell 16(1):147–157

Schrauwen I et al (2018) De novo variants in GREB1L are associated with non-syndromic inner ear malformations and deafness. Hum Genet 137(6–7):459–470

Sims D et al (2014) Sequencing depth and coverage: key considerations in genomic analyses. Nat Rev Genet 15(2):121–132

Skene PJ, Henikoff S (2017) An efficient targeted nuclease strategy for high-resolution mapping of DNA binding sites. Elife 6:e21856

Sogorb MA et al (2016) Roles of NTE protein and encoding gene in development and neurodevelopmental toxicity. Chem Biol Interact 259(Pt B):352–357

Sokpor G et al (2017) Chromatin remodeling BAF (SWI/SNF) complexes in neural development and disorders. Front Mol Neurosci 10:243

Solomon MJ, Varshavsky A (1985) Formaldehyde-mediated DNA-protein crosslinking: a probe for in vivo chromatin structures. Proc Natl Acad Sci U S A 82(19):6470–6474

Solomon MJ, Larsen PL, Varshavsky A (1988) Mapping protein-DNA interactions in vivo with formaldehyde: evidence that histone H4 is retained on a highly transcribed gene. Cell 53(6): 937–947

Sun Z et al (2019) EGR1 recruits TET1 to shape the brain methylome during development and upon neuronal activity. Nat Commun 10(1):3892

Tao X et al (2020) Efficient chromatin profiling of H3K4me3 modification in cotton using CUT&Tag. Plant Methods 16:120

Timp W, Timp G (2020) Beyond mass spectrometry, the next step in proteomics. Sci Adv 6(2): eaax8978

Timpano S, Picketts DJ (2020) Neurodevelopmental disorders caused by defective chromatin remodeling: phenotypic complexity is highlighted by a review of ATRX function. Front Genet 11:885

Trevino AE et al (2020) Chromatin accessibility dynamics in a model of human forebrain development. Science 367(6476):eaay1645

Trivier E et al (1996) Mutations in the kinase Rsk-2 associated with Coffin-Lowry syndrome. Nature 384(6609):567–570

Urdinguio RG, Sanchez-Mut JV, Esteller M (2009) Epigenetic mechanisms in neurological diseases: genes, syndromes, and therapies. Lancet Neurol 8(11):1056–1072

van Steensel B, Delrow J, Henikoff S (2001) Chromatin profiling using targeted DNA adenine methyltransferase. Nat Genet 27(3):304–308

Varshavsky A (2008) Discovery of cellular regulation by protein degradation. J Biol Chem 283(50): 34469–34489

Vue TY et al (2020) ASCL1 regulates neurodevelopmental transcription factors and cell cycle genes in brain tumors of glioma mouse models. Glia 68(12):2613–2630

Wang T et al (2020) Large-scale targeted sequencing identifies risk genes for neurodevelopmental disorders. Nat Commun 11(1):4932

Wenderski W et al (2020) Loss of the neural-specific BAF subunit ACTL6B relieves repression of early response genes and causes recessive autism. Proc Natl Acad Sci U S A 117(18): 10055–10066

Wu SJ et al (2021) Single-cell CUT&Tag analysis of chromatin modifications in differentiation and tumor progression. Nat Biotechnol 39(7):819–824

Yin S et al (2020) Transcriptomic and open chromatin atlas of high-resolution anatomical regions in the rhesus macaque brain. Nat Commun 11(1):474

Zaghi M, Broccoli V, Sessa A (2019) H3K36 methylation in neural development and associated diseases. Front Genet 10:1291

Zahir FR, Brown CJ (2011) Epigenetic impacts on neurodevelopment: pathophysiological mechanisms and genetic modes of action. Pediatr Res 69(5 Pt 2):92R–100R

Zamora-Moratalla A, Martínez de Lagrán M, Dierssen M (2021) Neurodevelopmental disorders: 2021 update. Free Neuropathol 2(6)

Zheng R et al (2018) A novel PNPLA6 compound heterozygous mutation identified in a Chinese patient with Boucher-Neuhauser syndrome. Mol Med Rep 18(1):261–267

Methionine Is a Major Methyl Donor Whose Dietary Intake Likely Plays a Causative Role for Neurodevelopmental Disorders via Epigenomic Profile Alterations

4

Ghada Mubarak and Farah R. Zahir

Abstract

Methionine is an essential amino acid that plays a significant role in one-carbon metabolism. One-carbon metabolism pathways include the folate cycle, transsulfuration pathway, and methionine metabolism cycle as major components. They are deeply interconnected and yield a single-carbon molecule—the methyl group CH3—via the universal methyl donor, *S*-adenosylmethionine (SAM). CH3 is an integral component of epigenomic regulation via its role in the methylation of nuclear material. Thus, methionine exposure levels are expected to impact epigenomic regulation. Epigenomic imbalances are a known significant cause for neurodevelopmental disorders (ND). Extrapolating, methionine imbalance should lead to ND causation via epigenomic deregulation. In this chapter we collect evidence for such a mechanism of disease, and show that the body does seem to be methionine sensitive, and that methionine metabolism imbalances indeed cause ND. We also discuss the modulation of methionine dietary intake as a potential therapeutic.

While there exists a significant amount of literature for other major methyl donors such as folate, betaine, and choline, it is only recently that evidence for methionine impact on health is becoming known. However, as the more direct precursor to SAM, the body appears more sensitive to methionine exposure. This

G. Mubarak
Sidra Medicine, Doha, Qatar
e-mail: gmubarak@sidra.org

F. R. Zahir (✉)
Department of Medical Genetics, BC Children's and Women's Hospital, University of British Columbia, Vancouver, BC, Canada
e-mail: farahz@bcchr.ca

M. W. Qoronfleh et al. (eds.), *Proteins Associated with Neurodevelopmental Disorders*, Nutritional Neurosciences, https://doi.org/10.1007/978-981-15-9781-7_4

chapter is expected to fill a vital void in our appreciation for methionine in terms of epigenomic regulation and ND, in its capacity as a major methyl donor.

Keywords

Epigenome · Neurodevelopmental disorders · Dietary intake · One-carbon metabolism methionine · Methyl donor · *S*-adenosylmethionine · SAM

4.1 Introduction

Methionine is an essential amino acid that supplies vital building blocks for a number of biological processes via the methionine metabolism cycle (MMC) and its branches. The MMC is one of the three key components in one-carbon metabolism (the others being the folate cycle and the transsulfuration pathway) that produce the methyl moiety as a key output. The methyl group—CH3—is necessary for the methylation of several compounds such as DNA, protein, amino acids, creatine, and phospholipids (Clare et al. 2019).

Importantly, methylation of nuclear material is a fundamental epigenomic regulatory process. Methylation can occur on DNA, and methylated DNA can be further modified (e.g., hydroxymethylated), as well as on the tails of histone proteins. Further, it is recognized that epigenomic regulation is a significant cause for neurodevelopmental disorders (ND) (Zahir and Brown 2011). Therefore, methionine, as a methyl donor, may be involved in ND causation via perturbation of its breakdown and subsequent impact on the epigenome (Fig. 4.1). It is in this context that we discuss methionine here.

Disruption of methionine metabolism may be categorized due to two main causes: intrinsic and extrinsic (Fig. 4.1). Intrinsic causes include mainly disruption of the MMC, usually due to genetic defects in genes that encode key pathway

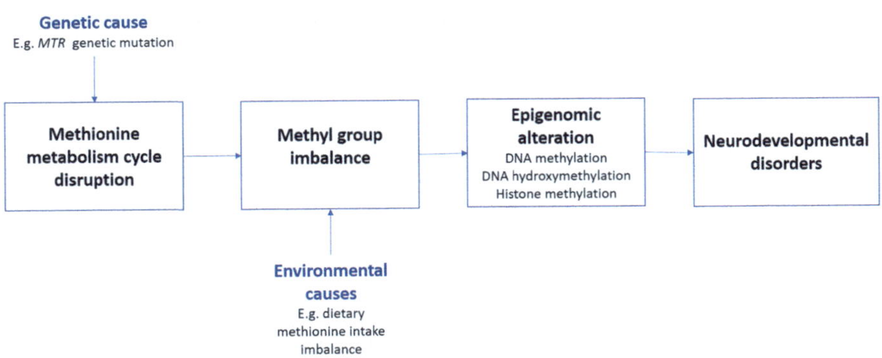

Fig. 4.1 Schematic illustration of intrinsic (e.g., genetic factors that lead to aberrant enzymatic action) or extrinsic (e.g., dietary imbalance of methionine intake) factors that disrupt the methionine metabolism cycle leading to epigenomic alteration-induced neurodevelopmental disease (ND)

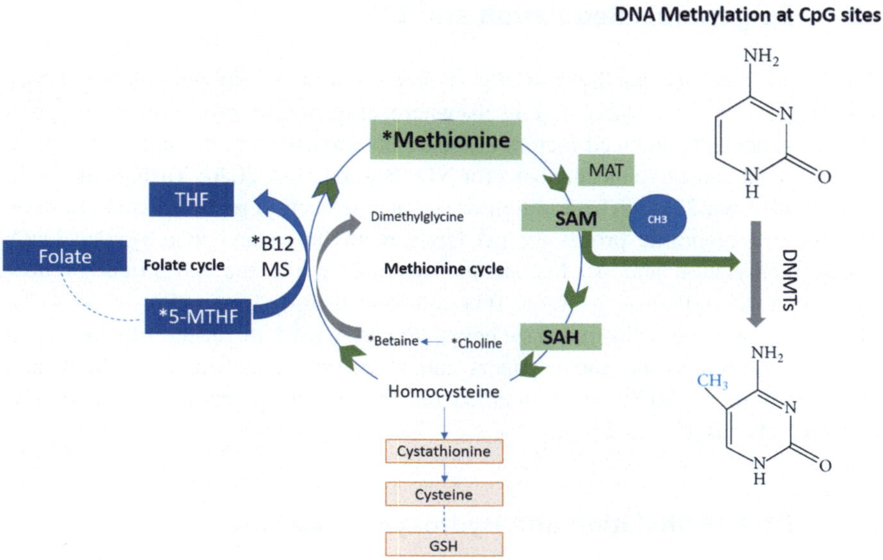

Fig. 4.2 Methionine metabolism cycle (MMC) and its link to epigenomic alteration (DNA methylation as an example). The color represents the three main pathways that comprise the one-carbon metabolism cycle: green—MMC, blue—folate cycle, and orange—transsulfuration pathway. Abbreviations: *SAM* S-adenosyl-methionine, *SAH* S-adenosyl-homocysteine, *MAT* methionine adenosine transferase, *CH3* methyl group, *MT* methyl transferases, *DNMTs* DNA methyltransferases, *5-MTHF* 5-methyl tetrahydrofolate, *B12* vitamin B12, *MS* methionine synthase, *THF* tetrahydrofolate, *GSH* the antioxidant glutathione, (*) indicates methyl group donors obtained from diet

enzymes. Extrinsic causes are mainly exposure to methionine as a dietary substance. We discuss both categories in depth in this chapter.

It is important to note that while methionine is a major donor of the CH3 moiety, there are others. They include folate, betaine, and choline, which are all connected via the one-carbon metabolism pathway. Figure 4.2 gives a schematic highlighting relevant details of the one-carbon metabolism pathway, viz., the connection between folate, betaine, choline, and the MMC. Remarkably, these compounds are all obtained from diet, and studies have shown that deficiencies in any of these substances have been associated with disturbance to human health (Shea and Rogers 2014). An especially well-known example is that deficiency in folic acid during pregnancy is proved to be associated with neural tube defects, and it is known that supplementation with folic acid before conceiving and during the first trimester of pregnancy is preventative (Li et al. 2019). Although the mechanism of action remains unclear, what is apparent is that dietary intake of methyl donors can and does impact neural development.

Therefore, as these substances are all active in one-carbon metabolism, expectedly there is an extensive interplay between their metabolic pathways (Fig. 4.2). While we will occasionally necessarily allude to them in our discussion, we will focus this chapter on methionine alone as a methyl donor.

4.2 Epigenomic Regulation and CH3

The chemical moiety "methyl"—CH3—is integral to major epigenomic regulatory processes. As we have elaborated on elsewhere, epigenomics processes, both due to inherent genetically induced factors and external environmentally induced factors, are significant causative mechanisms for ND (Bastaki et al. 2020; Yasin et al. 2020; Zahir and Brown 2011). The addition or removal of methyl group/s occurs in three categorical epigenomic processes: (a) DNA methylation and thereby also DNA hydroxymethylation and (b) histone methylation. While the full extent of how these processes influence genomic function (and thereby cause ND) is yet to be fully understood, we will summarize below how they work in general (the interested reader is referred to the above papers and to Fahrner and Bjornsson 2014, and Haghshenas et al. 2020, for a detailed discussion on epigenomic perturbation-induced ND causation and profiling).

4.2.1 DNA Methylation and Hydroxymethylation

The addition or removal of a methyl group to the 5-carbon position of the DNA ribose sugar-phosphodiester backbone is termed "DNA methylation." This usually occurs on cytosine, forming methyl-cytosine (mC), when the cytosine is in a CpG (cytosine linked to guanosine via a phosphodiester bond) configuration (mCpG). However, it was recently found that mC may also occur when cytosine is in non-CpG configuration (mCpH where H is A, T, or G), as a frequently occurring epigenomic process in neuronal tissue, that may play a significant role in postnatal brain development and function (Clemens and Gabel 2020; de Mendoza et al. 2021; Guo et al. 2014; Lister et al. 2013; Price et al. 2019). The presence/absence of the methyl moiety at specific locations on the DNA strand acts as a chemical marker that determines whether it is in a turned "on" (i.e., genes are able to be transcribed) or "off" (i.e., genes are not able to be transcribed) state (Cerrato et al. 2020).

DNA hydroxymethylation occurs due to the oxidation of mC yielding 5-hydroxymethyl cytosine (hmC). Though the DNA hydroxymethylation mark was only recently discovered (Kriaucionis and Heintz 2009), it is already understood to have a prominent role in brain function, due to being found specifically as a leading post-developmental mark in neuronal tissue [as reviewed by Kinde et al. 2015] (Kinde et al. 2015). Interestingly hmC has been found to predominantly occur in brain tissue at CpG sites, i.e., as hmCpG. It is less clear whether hmCpH is found in vivo, since, though studies to date have not unearthed significant evidence, this may be due to lack of appropriate experimental technique. What is clear however is that hydroxymethylation of cytosine, i.e., hmC, appears to be a significant epigenomic regulatory process in brain development and function (Efimova et al. 2020; Kochmanski and Bernstein 2020; Rustad et al. 2019; Torres et al. 2019; Yang et al. 2020). While a thorough discussion of the impact of hmCpG and hmCpH is beyond the scope of this chapter, we note this epigenomic mark as being necessarily linked to DNA methylation, and hence CH3 availability.

4.2.2 Histone Methylation

Posttranslational histone modification is important for the regulation of chromatin structure. Histone methylation is one of several possible histone tail modifications. Histone methylation occurs on select residues of the histone amino acid tails, and plays a key role in silencing or activating transcription based upon the pattern and location of methylation.

Importantly histone methylation has been shown to have a pathophysiological impact on cognition; postmortem prefrontal cortex neurons of autistic subjects showed altered H3K4me3 (trimethylated histone3, lysine4) peaks with significant enrichments in genes and loci implicated in neurodevelopmental diseases (Shulha et al. 2012). Changes in hippocampal H3K4 (histone3, lysine4) methylation were observed in rodent fear conditioning and showed association with memory formation (Gupta et al. 2010). Association between histone methylation and DNA methylation in specific genes in adult mice brain indicated a possible role in long-term memory formation (Gupta et al. 2010).

Consequential to this chapter's thesis, a recent study showed that dietary methionine level exposure drives histone methylation levels for key genes, impacting their function (Mentch et al. 2015). Another study on human cancer cells and a mouse model also demonstrated that methionine restriction is able to alter histone methylation patterns and thereby impact gene expression. These remarkable findings indicate the close connection between dietary intake of methionine and health, via epigenomic regulatory mechanisms—an emerging area of research we will explore here.

4.3 The MMC and CH3

Methionine is a sulfur-containing amino acid vital for protein synthesis, as well as DNA and histone methylation (Mladenović et al. 2019). It is metabolized via the MMC as follows: methionine is converted via methionine adenosyl transferase (MAT) to *S*-adenosyl methionine (SAM). SAM is a universal methyl donor in all mammalian cells (Mato et al. 2013), and hence is a foundational metabolite for several biological processes. The removal of CH3 from SAM (demethylation of SAM) results in *S*-adenosylhomocysteine (SAH), which is then converted to homocysteine. However, the SAH-to-homocysteine reaction is reversible. Consistently high levels of homocysteine can convert back to SAH (Fig. 4.2). This is of note as SAH can act as an inhibitor to DNA methyl transferases, which may lead to DNA hypomethylation (James et al. 2002). Homocysteine is then either (a) remethylated back to methionine by methylcobalamin (vitamin B12) via the folate cycle and through methionine synthase or (b) remethylated back to methionine by betaine via a folate-independent reaction or (c) further converted to cystathionine (Ctt) and cysteine that are used later in the generation of glutathione (GSH) (Fig. 4.2) (Zhang 2018). Importantly, through methionine salvage pathways, SAM can be converted to *S*-methyl-5′-thioadenosine (MTA) and back to methionine (Tang et al. 2017). As may be surmised from the above-simplified explanation of the MMC, methionine plays an integral role in the generation of CH3 in the body as well as in the sequestration of excess CH3. It is distinguished from the other major methyl donors by its more immediate connection to the universal methyl donor SAM.

4.4 Disruption of the Methionine Cycle Due to Genetic Defects Leads to ND

Disruption of the methionine cycle may occur due to genetic defects that impact the activity of key enzymes in the MMC, leading to reduced CH3 availability. As we have hypothesized, this is expected to result in altered epigenomic methylation profiles that could be causative for ND. Indeed, this is the case and has been reported. Table 4.1 lists pertinent examples for where there is evidence of altered epigenomic landscape due to MMC disruption. Below we also detail reported genetic defects affecting the MMC that cause ND.

4.4.1 *MTR* and *MTRR*

Variations in the *MTRR* gene have been shown to disrupt the MMC directly. *MTRR* encodes methionine synthase reductase, a necessary enzyme that functions to regenerate another enzyme termed methionine synthase. Methionine synthase is encoded by the *MTR* gene. Genetic variants in both *MTRR* and *MTR* have been associated with the risk for autism spectrum disorder (ASD) and attention deficit hyperactivity disorder (ADHD) by several studies (Dutta et al. 2011; James et al. 2006; Saha et al. 2017, 2018) as reviewed by Lintas et al. (Lintas 2019).

4.4.2 *MTHFR*

Mutations that affect the activity of the methylene tetrahydrofolate reductase enzyme (MTHFR) have been shown as causative for ND in children (Lintas 2019; Mitchell et al. 2014; Schaevitz and Berger-Sweeney 2012; Shaik Mohammad et al. 2016). Furthermore, a well-studied sequence variant of *MTHFR* which yields active enzyme, albeit with altered activity, is the C677T variant. It has been shown to cause reduced DNA methylation in homozygous individuals compared to homozygous wild type (Weiner et al. 2014), clearly linking disruption of the MMC to aberrant DNA methylation.

4.4.3 *SIRT1*

SIRT1 is one of the seven mammalian sirtuin protein-encoding genes (*SIRT 1* to *7*). SIRT1 is thought to be localized to the nucleus or the cytoplasm in a function-

Table 4.1 Reported genetic defects in MMC enzymes/factors that have been shown to result in epigenomic methylation defects. N.B., this is not an exhaustive list

Gene	Encoded enzyme/protein	Comment
MTR	Methionine synthase	The variant rs1805087 (MTR A2756G) showed increased DNMT1 promoter methylation (Weiner et al. 2014), and was associated with increased global methylation in homozygous form (Coppedè et al. 2019)
MTRR rs1801394	Methionine synthase reductase	Reduced enzyme activity may alter methionine/homocysteine ratio in silico (Olteanu et al. 2002; Olteanu and Banerjee 2001; Saha et al. 2018). Significant association with idiopathic intellectual disability; however the mechanism remains unclear (Dutta et al. 2011)
MTHFR rs1801133	Methylene tetrahydrofolate reductase enzyme	Reduced DNA methylation (Weiner et al. 2014)
MAT1A	Isoforms of methionine adenosyltransferase	Possible effect on histone methylation (Chien et al. 2015)
SIRT1	Sirtuin proteins	Mouse studies showed profound impact of sirt1 on histone methylation (Kishi et al. 2011, 2010; Libert et al. 2011; Tang et al. 2017)
GNMT Rs121907888r Rs864321678	GNMT enzyme that is important for transmethylation reactions	Global DNA hypermethylation in the liver of knock-off mice (Luka et al. 2006)
SIN3 Drosophila study	SIN3 regulates the expression of methionine metabolic genes and histone modification at promoter regions	Knockdown of *sin3A* results in an increase in global histone H3K4me methylation (Liu and Pile 2017)

dependent way, and is classed as an NAD-dependent deacetylase (Tang et al. 2017). Along with known roles for obesity and aging (Tang et al. 2017), the role of SIRT1 in neural development and brain function is gaining prominence, as it has been shown to be an effector for a number of disorders (Herskovits and Guarente 2014). Genetic variants in *SIRT1* were associated with schizophrenia (Kishi et al. 2011) and major depressive disorder (Kishi et al. 2010). In the Japanese population it was associated with anxiety and mood disorder (Libert et al. 2011). However, how SIRT1 may impact the MMC has only recently been unearthed; in an elegant mouse study, Tang et al. show that Sirt1 has a profound impact on histone methylation (Tang et al. 2017). *Sirtl* controls the enzyme methionine adenosyltransferase 2a (MAT2a) that converts methionine to SAM in mouse embryonic stem cells (mESCs) (Tang et al. 2017). A knockout of *Sirtl* in mice reduced the conversion of methionine to SAM and a subsequently reduced histone methylation (Tang et al. 2017). These early findings in mice suggest that as-yet unknown connections of sirtuins to the methionine cycle may be likely in humans, adding to our growing understanding of how genetic defects of key metabolic enzymes can lead to methionine depletion defects.

4.4.4 Hypermethioninemia Reported to Cause Neurological Disease

Hypermethioninemia due to genetic defects is also known, with several reports as collated by Harvey Mudd (2011). A pertinent example is that of *MAT1A* which encodes MAT. Individuals presented a wide range of clinical manifestations depending on the specific defect in the gene—with individuals carrying a more elevated blood methionine level showing a more severe neurological presentation (Chamberlin et al. 2000). We note that this occurrence is not on a normal genomic background and is likely linked to aberrant MMC functioning. Indeed, in this case failure to convert methionine to SAM and subsequent low levels of SAM in the brain is thought to be the main reason for the neurological manifestation (Furujo et al. 2012).

4.4.5 Mitochondrial DNA Depletion Disrupts the Methionine Cycle

In an interesting study, Lozoya et al. detected unexpected links between progressive mtDNA depletion, and MMC, and impact on global genomic DNA epigenomic profiles. The observed changes are complex and include alteration of several aspects of the cycle, including increased methionine degradation, and changes in recycling of homocysteine among others (Lozoya et al. 2018). While the finding being preliminary and thus precluding further comment, they suggest that further exploration is needed in how methionine may be connected to cellular energy metabolism and thus to ND.

4.5 Several NDs Show Aberrant and Even Diagnostic Methylation Epigenome Profiles

The premise of this chapter is that imbalanced methionine exposure may alter the epigenomic landscape due to the disruption of proper CH3 availability. The existence of ND where no genetic defect is reported, but altered epigenomic profiles are found, lends weight to the notion that environmental exposure to a major methyl donor such as methionine may contribute to ND as well. In this section we explore the emerging understanding of extant causative/diagnostic epigenomic profiles for ND and link this to methionine.

In the past decade, landmark studies have shown that methylation capacity may be reduced in ND, though the majority of current data is focused on ASD (James et al. 2002; Melnyk et al. 2012). DNA hypomethylation was expressed by the reduction of SAM-to-SAH ratio in autistic children (Melnyk et al. 2012). It is worth mentioning that in this cohort, the intracellular antioxidant glutathione (GSH) as well as detoxification mechanism was found to be also reduced; however decreased methylation capacity was not proven to be a consequence of exposure to oxidative stress (Melnyk et al. 2012). In a meta-analysis, blood levels of core biomarkers of methylation capacity (methionine, SAM, and SAM/SAH) were shown to be significantly decreased in autistic children compared to a matched control, while SAH levels were significantly increased in multiple autism studies regardless of sample source (Guo et al. 2020).

The above data indicating possible global nuclear material methylation defects as causative for ND were confirmed by more direct investigations. Siu et al. are among the first to show that DNA methylation profiles which are diagnostic are extant—they profiled peripheral blood DNA methylation and were able to distinguish between genetically defined ASD cases, heterogenous ASD cases of undefined genetic cause, and healthy controls (Siu et al. 2019), further corroborating evidence for epigenomic methylation profiles that could possibly be causative for ND.

Related to the above is the more abundant literature supporting characteristic epigenomic profiles for ND, albeit due to a known genetic cause. Aref-Eshghi et al. investigated DNA methylation epi-signatures of 42 NDs, and concluded that individuals with certain congenital syndromes have specific DNA methylation patterns in their peripheral blood (Aref-Eshghi et al. 2020). For example, disruption of genes directly controlling DNA methylation such as *DMNT1*, *DNMT3A*, and *DNMT3B* was associated with distinct DNA methylation patterns in the case of TBRS (Tatton-Brown-Rahman) and ADCADN (cerebellar ataxia, deafness, and narcolepsy, AD) syndromes (Aref-Eshghi et al. 2020). Interestingly, the authors were able to predict the diagnosis of 9 out of 965 individuals with idiopathic ND using their epi-signature classification method, and confirm two diagnoses, Rahman syndrome (RMNS) and Kleefstra syndrome 1 (Aref-Eshghi et al. 2020), indicating that the epi-signature is sufficient as a diagnostic marker.

Additionally, the HVDAS syndrome (Helsmoortel-van der Aa syndrome) that includes both ASD and intellectual disability (ID) (Helsmoortel et al. 2014) was demonstrated to have two distinct epi-signature profiles caused by a single-gene mutation (Bend et al. 2019), demonstrating that DNA methylation epi-signatures may be useful to distinguish clinically overlapping syndromes.

The above studies, though preliminary, are shedding light on the fact that distinctive epigenomic profiles diagnostic for ND do exist. While the majority of studies currently published focus on profiling the aberrant epigenome for ND of known genetic cause, the fact that children with idiopathic ASD demonstrate changes in key methyl donors in their blood lends support to the notion that aberrant epigenomic methylation profiles sans genomic defect may also cause ND. As this area of research grows, we expect more definitive and comprehensive findings that better define causative methylation profiles for ND. This is critical as we attempt to understand the connection between methionine—a methyl donor—and its impact on nuclear material methylation and ND causation.

4.6 Nutritional Deficiency of Major Methyl Donors Impacts ND Causation

4.6.1 During Fetal, Pre-, Peri-, and Postnatal Development

There is a growing understanding that maternal nutrient intake during gestation is able to bring about epigenetic changes in the fetus (sometimes lasting) with proven potential to impact phenotype. One of the most prominent studies launched in the past two decades to investigate the role of environmentally induced epigenomics with respect to ND is the CHARGE (Childhood Autism Risks from Genetics and Environment) study (Irva et al. 2006). Specifically, the CHARGE study investigates ASD risk, examining a variety of environmental exposures at pre- and postnatal stages as well as early childhood. This endeavor has resulted in a large number of publications, some of which discuss the impacts of maternal gestational nutritional intake on fetal development (Goodrich et al. 2018; Schmidt et al. 2014, 2021). While the CHARGE study exemplifies this promising shift in research focus, there are other studies that also probe the same area, with more direct relevance to the focus of this chapter.

The MANOE (Maternal Nutrition and Offspring's Epigenome) study is notable in this regard, as it specifically looks at how gestational intake of methyl donors may impact the child's epigenomic profile (Pauwels et al. 2016). This is a prospective, observational cohort study initiated in April 2012 that as far as we are aware is still ongoing. Based in Belgium, it relies on self-reporting by women enrolled using a validated food frequency questionnaire (Pauwels et al. 2014). Initial results analyzing 114 mother-child pairs showed that mothers had on average a higher intake over daily requirement of methionine, during pregnancy. Further, the maternal first-trimester methionine intake was associated with hypomethylation of the *DNMT1* gene in the infant, upon testing infant buccal epithelial cells at 6 months of age (Pauwels et al. 2017b). This is remarkable as *DNMT1* encodes a major epigenomic regulatory enzyme, which has bearing on global epigenomic programming. Interestingly, a subsequent publication showed a significant negative correlation between maternal methionine intake and whole DNA methylation and hydroxymethylation levels in women's blood; a high intake of methionine prepregnancy and in the first trimester resulted in reduced hydroxymethylation levels in weeks 11–13 and 18–22, respectively. While this pattern was observed for both choline and betaine intake, conversely, folic acid intake showed a positive relation to global DNA methylation and hydroxymethylation levels (Pauwels et al. 2016). A third study investigating the same MANOE cohort found that breastfeeding impacts DNA methylation patterns for genes related to fat metabolism (*RXRA*) and appetite control (*LEP*), in the child (Pauwels et al. 2019). While the breastfeeding mother's nutritional intake is not specifically discussed in this chapter, it is presumed that they would have a high methyl donor intake as they are the same participants discussed in other MANOE publications. Finally, another study from the same group shows difference in cord blood methylation levels associated with maternal methyl donor intake that included methionine (Pauwels et al. 2017a).

These expanding areas of research have been systematically reviewed by others. We point the reader to two important recent reviews: the first by Cheng et al. (2019) who conducted a systematic literature review for all reports of perinatal environmental exposures that increase the risk for ASD that were published between 2005 and 2018. They find a variety of environmental exposures that are important, including several foods and nutritional supplements. Among them are polyunsaturated fatty acids which are metabolically linked to the MMC (James et al. 2018). The second review we highlight is a very useful narrative review of studies in the literature carefully selected to include data in three phases: (a) studies that show preconception and/or pregnancy nutritional outcome on the child's DNA methylation profile, (b) associations between DNA methylation status of genes found in the above first-phase data and cognitive outcomes, and (c) studies that explicitly link maternal nutritional exposure to a DNA methylation profile tied to the child's phenotype. It is striking that the literature search performed by three independent investigators in PubMed and Google scholar, during the first quarter of 2017, yielded 34 papers that fulfill the first phase, 31 that fulfill the second, and fully 8 papers that fulfill the most exacting phase 3 (James et al. 2018). Notably, they find that several of the papers fulfilling their exacting inclusion and exclusion criteria demonstrate offspring effects due to maternal gestational intake of one-carbon metabolism components which include the methyl donors—folate, betaine, choline, and methionine.

4.6.2 During Childhood

While information particular to methionine intake in children is sparse, more information is available for other methyl donors. Choline and betaine were shown to be elevated in ASD cases (Orozco et al. 2019). They are known to act as methyl group donors during folate deficiency to remethylate homocysteine back to methionine (Zeisel and Blusztajn 1994). A study of urinary folate-methionine cycle metabolites in autistic children showed a significantly lower *N-N*-dimethylglycine metabolite profile, despite normal levels of choline and betaine (Belardo et al. 2019). The authors attributed their results to a possible impact due to vitamin B12 deficiency-induced impairment of the one-carbon metabolism pathway (Belardo et al. 2019). This highlights the importance of vitamin B12 as a cofactor for methionine recycling in the MMC. Furthering the above idea, a study on human brain tissue showed that vitamin B12, especially its methylated form MeCbl (methylcobalamin), is decreased with age in the frontal cortex of control human subjects, whereas serum Vit B12 levels did not show similar decrease with age, suggesting that brain vitamin B12 is distinctly regulated from the rest of the body (Zhang et al. 2016), and hence further implying correct MMC activity for brain function.

We have also discussed, in a previous section in this chapter, the altered profile of metabolites in the methionine cycle as biomarkers in children with ASD (vide supra), supporting the notion that proper methionine intake is correlated to ND in children.

4.7 Therapeutic and Prophylactic Use of Methionine for ND

4.7.1 Methionine Supplementation

Several studies on animals showed that availability of methyl donor micronutrients may influence brain function and behavior in the offspring. For example, supplementing C58 mice with methyl donors altered neurodevelopmental trajectory and increased global DNA methylation in the cortex and cerebellum of the adult offspring (Muehlmann et al. 2020; Pizzorusso and Tognini 2020). Interestingly, in rat, a paternal diet high in methyl donor nutrient rate was shown to influence behavioral and cognitive function in the offspring (McCoy et al. 2018). It was demonstrated that supplementing mouse mothers on a high-fat diet with folate during gestation rescued cognitive and behavioral outcomes affected by obesity, and was also able to restore prefrontal cortex DNA methylation levels, for male offspring (McKee et al. 2018; Pizzorusso and Tognini 2020). Furthermore, methionine administration significantly increased hypermethylation of *Bdnf* (which encodes a key neuronal tissue growth and maintenance factor), and enhanced epilepsy-negative outcomes, in an epileptic rodent model (Parrish et al. 2015). Despite the promising results of these animal model studies, more work is required to understand the potential impacts for methionine supplementation in humans, especially in the context of ND.

Conversely, and concerningly, there is some evidence for negative health impacts due to increased methionine exposure. An alarming publication in 2003 reported brain edema and MRI abnormalities in ten babies fed on an infant formulae with excessively high methionine levels between 1999 and 2001 (Harvey Mudd et al. 2003). The babies were tested for common genetic causes of MMC aberration and found to be normal indicating that the insult was likely the result of diet. Therefore, in this case, hypermethioninemia arose on a normal genomic background.

4.7.2 Methionine Restriction

The key question arises as to whether dietary methionine intake level adjustments may be used therapeutically. It appears that while some presentation of hypermethioninemia may be resolved after normalization of plasma methionine, determining the correct dosage is difficult. For instance, despite methionine restriction being reported as helpful in the management of some cases, concerns arose that this would lead to depletion of the universal methyl donor SAM (Mudd 2011) that may then lead to several other as-yet unknown adverse effects. For more details, the interested reader is referred to an excellent review that discusses all cases of hypermethioninemia due to both genetic and nongenetic causes (Mudd 2011).

Briefly, the current indication is that methionine is dosage sensitive and optimal health requires a carefully controlled methionine availability. Finally, we note that more extensive literature is available that comprehensively addresses the effects of other methyl donors on ND causation and prevention. The interested reader is referred to these reviews (Gao et al. 2016; James et al. 2018; Li et al. 2019; Lintas 2019).

4.8 Conclusion

In this chapter we have explored the link between methionine as a key methyl donor; epigenomic perturbation, viz., DNA methylation, hydroxymethylation, and histone methylation; and ND. While more information is extant on other methyl donors, notably for folate, it is only recently the unique role of methionine as a methyl donor has begun to be explored.

The other known major methyl donors—folate, betaine, and choline—feed into the MMC via complex pathways. Methionine, however, is a more direct precursor of the universal methyl donor *S*-adenosylmethionine (SAM), and hence it may be hypothesized that the imbalance of methionine would have a more pronounced impact on methyl group availability and thus upon epigenomic regulation. Indeed, this does seem to be the case, as, in contrast to other methyl donors, the body seems to be adversely affected due to both methionine restriction and methionine abundance (Chien et al. 2015; Furujo et al. 2012; Mladenović et al. 2019; Mudd 2011; Pizzorusso and Tognini 2020). This heightened biological sensitivity is expected to be a direct precursor in such a foundational pathway, as methionine is to the MMC and indeed thus also to one-carbon metabolism as a whole.

In keeping with the above, recent evidence is indeed unearthing the importance of methyl groups and methyl group precursor concentration for cognitive development and function, especially in complex conditions such as ASD (Melnyk et al. 2012; Guo et al. 2020). Excitingly, groundbreaking work is beginning to prove that DNA methylation epi-signatures (irrespective of genetic background) can be diagnostic for ND (Siu et al. 2019; Aref-Eshghi et al. 2020). Concomitantly, environmentally induced epigenomic profile alterations in children have been demonstrated through maternal nutrition pre-, peri-, and post-pregnancy (Pauwels et al. 2016, 2017a, b). Taken together, these findings raise the possibility that a key methyl donor such as methionine may have integral roles in neurodevelopment and function. This concept is further enhanced by the fact that preliminary animal model work exploring both maternal (Muehlmann et al. 2020; Pizzorusso and Tognini 2020) and paternal (McCoy et al. 2018) methyl donor intake produced an altered neurodevelopment trajectory, and increased DNA methylation in offspring, indicating intergenerational effects for this fundamental substance.

Critically however, the few studies currently available document both a positive and a negative impact on brain development and function due to methionine exposure, depending on the level. Hence, we draw attention to the notion that methionine is likely to be dose dependent, and more research in this area is sorely needed. To end, this essential amino acid, though having arrived late in the current discussions surrounding methyl donor exposure, epigenomics, and disease, nevertheless commands a great deal of attention as a possible vital player for both ND causation and therapeutics.

References

Aref-Eshghi E, Kerkhof J, Pedro VP, Barat-Houari M, Ruiz-Pallares N, Andrau JC et al (2020) Evaluation of DNA methylation episignatures for diagnosis and phenotype correlations in 42 Mendelian neurodevelopmental disorders. Am J Hum Genet 106(3):356–370. https://doi.org/10.1016/j.ajhg.2020.01.019

Bastaki KN, Alwan S, Zahir FR (2020) In: Essa MM, Qoronfleh MW (eds) Maternal prenatal exposures in pregnancy and autism spectrum disorder: an insight into the epigenetics of drugs and diet as key environmental influences BT—personalized food intervention and therapy for autism spectrum disorder management. Springer International Publishing, Cham, pp 143–162. https://doi.org/10.1007/978-3-030-30402-7_5

Belardo A, Gevi F, Zolla L (2019) The concomitant lower concentrations of vitamins B6, B9 and B12 may cause methylation deficiency in autistic children. J Nutr Biochem 70:38–46. https://doi.org/10.1016/j.jnutbio.2019.04.004

Bend EG, Aref-Eshghi E, Everman DB, Rogers RC, Cathey SS, Prijoles EJ et al (2019) Gene domain-specific DNA methylation episignatures highlight distinct molecular entities of ADNP syndrome. Clin Epigenetics 11(1):64. https://doi.org/10.1186/s13148-019-0658-5

Cerrato F, Sparago A, Ariani F, Brugnoletti F, Calzari L, Coppedè F et al (2020) DNA methylation in the diagnosis of monogenic diseases. Genes 11(4):355. https://doi.org/10.3390/genes11040355

Chamberlin ME, Ubagai T, Mudd SH, Thomas J, Pao VY, Nguyen TK et al (2000) Methionine adenosyltransferase I/III deficiency: novel mutations and clinical variations. Am J Hum Genet 66(2):347–355. https://doi.org/10.1086/302752

Cheng J, Eskenazi B, Widjaja F, Cordero JF, Hendren RL (2019) Improving autism perinatal risk factors: a systematic review. Med Hypotheses 127:26–33. https://doi.org/10.1016/j.mehy.2019.03.012

Chien Y-H, Abdenur JE, Baronio F, Bannick AA, Corrales F, Couce M et al (2015) Mudd's disease (MAT I/III deficiency): a survey of data for MAT1A homozygotes and compound heterozygotes. Orphanet J Rare Dis 10(1):99. https://doi.org/10.1186/s13023-015-0321-y

Clare CE, Brassington AH, Kwong WY, Sinclair KD (2019) One-carbon metabolism: linking nutritional biochemistry to epigenetic programming of long-term development. Annu Rev Anim Biosci 7(1):263–287. https://doi.org/10.1146/annurev-animal-020518-115206

Clemens AW, Gabel HW (2020) Emerging insights into the distinctive neuronal methylome. Trends Genet 36(11):816–832. https://doi.org/10.1016/j.tig.2020.07.009

Coppedè F, Stoccoro A, Tannorella P, Migliore L (2019) Plasma homocysteine and polymorphisms of genes involved in folate metabolism correlate with DNMT1 gene methylation levels. Metabolites 9(12):298. https://doi.org/10.3390/metabo9120298

de Mendoza A, Poppe D, Buckberry S, Pflueger J, Albertin CB, Daish T et al (2021) The emergence of the brain non-CpG methylation system in vertebrates. Nat Ecol Evol 5(3):369–378. https://doi.org/10.1038/s41559-020-01371-2

Dutta S, Shaw J, Chatterjee A, Sarkar K, Usha R, Chatterjee A et al (2011) Importance of gene variants and co-factors of folate metabolic pathway in the etiology of idiopathic intellectual disability. Nutr Neurosci 14(5):202–209. https://doi.org/10.1179/1476830511Y.0000000016

Efimova OA, Koltsova AS, Krapivin MI, Tikhonov AV, Pendina AA (2020) Environmental epigenetics and genome flexibility: focus on 5-hydroxymethylcytosine. Int J Mol Sci 21(9):3223. https://doi.org/10.3390/ijms21093223

Fahrner JA, Bjornsson HT (2014) Mendelian disorders of the epigenetic machinery: tipping the balance of chromatin states. Annu Rev Genomics Hum Genet 15:269–293. https://doi.org/10.1146/annurev-genom-090613-094245

Furujo M, Kinoshita M, Nagao M, Kubo T (2012) Methionine adenosyltransferase I/III deficiency: neurological manifestations and relevance of S-adenosylmethionine. Mol Genet Metab 107(3):253–256. https://doi.org/10.1016/j.ymgme.2012.08.002

Gao Y, Sheng C, Xie RH, Sun W, Asztalos E, Moddemann D et al (2016) New perspective on impact of folic acid supplementation during pregnancy on neurodevelopment/autism in the offspring children—a systematic review. PLoS One 11(11):1–16. https://doi.org/10.1371/journal.pone.0165626

Goodrich AJ, Volk HE, Tancredi DJ, McConnell R, Lurmann FW, Hansen RL, Schmidt RJ (2018) Joint effects of prenatal air pollutant exposure and maternal folic acid supplementation on risk of autism spectrum disorder. Autism Res 11(1):69–80. https://doi.org/10.1002/aur.1885

Guo JU, Su Y, Shin JH, Shin J, Li H, Xie B et al (2014) Distribution, recognition and regulation of non-CpG methylation in the adult mammalian brain. Nat Neurosci 17(2):215–222. https://doi.org/10.1038/nn.3607

Guo BQ, Ding SB, Li HB (2020) Blood biomarker levels of methylation capacity in autism spectrum disorder: a systematic review and meta-analysis. Acta Psychiatr Scand 141(6):492–509. https://doi.org/10.1111/acps.13170

Gupta S, Kim SY, Artis S, Molfese DL, Schumacher A, Sweatt JD et al (2010) Histone methylation regulates memory formation. J Neurosci 30(10):3589–3599. https://doi.org/10.1523/JNEUROSCI.3732-09.2010

Haghshenas S, Bhai P, Aref-Eshghi E, Sadikovic B (2020) Diagnostic utility of genome-wide DNA methylation analysis in Mendelian neurodevelopmental disorders. Int J Mol Sci 21(23):9303. https://doi.org/10.3390/ijms21239303

Harvey Mudd S, Braverman N, Pomper M, Tezcan K, Kronick J, Jayakar P et al (2003) Infantile hypermethioninemia and hyperhomocysteinemia due to high methionine intake: a diagnostic trap. Mol Genet Metab 79(1):6–16. https://doi.org/10.1016/S1096-7192(03)00066-0

Helsmoortel C, Vulto-van Silfhout AT, Coe BP, Vandeweyer G, Rooms L, van den Ende J et al (2014) A SWI/SNF-related autism syndrome caused by de novo mutations in ADNP. Nat Genet 46(4):380–384. https://doi.org/10.1038/ng.2899

Herskovits AZ, Guarente L (2014) SIRT1 in neurodevelopment and brain senescence. Neuron 81(3):471–483. https://doi.org/10.1016/j.neuron.2014.01.028

Irva H-P, Croen LA, Hansen R, Jones CR, van de Water J, Pessah IN (2006) The CHARGE study: an epidemiologic investigation of genetic and environmental factors contributing to autism. Environ Health Perspect 114(7):1119–1125. https://doi.org/10.1289/ehp.8483

James SJ, Melnyk S, Pogribna M, Pogribny IP, Caudill MA (2002) Elevation in S-adenosylhomocysteine and DNA hypomethylation: potential epigenetic mechanism for homocysteine-related pathology. J Nutr 132(8):2361S–2366S. https://doi.org/10.1093/jn/132.8.2361S

James SJ, Melnyk S, Jernigan S, Cleves MA, Halsted CH, Wong DH et al (2006) Metabolic endophenotype and related genotypes are associated with oxidative stress in children with autism. Am J Med Genet B Neuropsychiatr Genet 141B(8):947–956. https://doi.org/10.1002/ajmg.b.30366

James P, Sajjadi S, Tomar AS, Saffari A, Fall CHD, Prentice AM et al (2018) Candidate genes linking maternal nutrient exposure to offspring health via DNA methylation: a review of existing evidence in humans with specific focus on one-carbon metabolism. Int J Epidemiol 47(6):1910–1937. https://doi.org/10.1093/ije/dyy153

Kinde B, Gabel HW, Gilbert CS, Griffith EC, Greenberg ME (2015) Reading the unique DNA methylation landscape of the brain: non-CpG methylation, hydroxymethylation, and MeCP2. Proc Natl Acad Sci U S A 112(22):6800–6806. https://doi.org/10.1073/pnas.1411269112

Kishi T, Yoshimura R, Kitajima T, Okochi T, Okumura T, Tsunoka T et al (2010) SIRT1 gene is associated with major depressive disorder in the Japanese population. J Affect Disord 126(1):167–173. https://doi.org/10.1016/j.jad.2010.04.003

Kishi T, Fukuo Y, Kitajima T, Okochi T, Yamanouchi Y, Kinoshita Y et al (2011) SIRT1 gene, schizophrenia and bipolar disorder in the Japanese population: an association study. Genes Brain Behav 10(3):257–263. https://doi.org/10.1111/j.1601-183X.2010.00661.x

Kochmanski J, Bernstein AI (2020) The impact of environmental factors on 5-hydroxymethylcytosine in the brain. Curr Environ Health Rep 7(2):109–120. https://doi.org/10.1007/s40572-020-00268-3

Kriaucionis S, Heintz N (2009) The nuclear DNA base 5-hydroxymethylcytosine is present in Purkinje neurons and the brain. Science (New York, NY) 324(5929):929–930. https://doi.org/10.1126/science.1169786

Li M, Francis E, Hinkle SN, Ajjarapu AS, Zhang C (2019) Preconception and prenatal nutrition and neurodevelopmental disorders: a systematic review and meta-analysis. Nutrients 11(7):1628. https://doi.org/10.3390/nu11071628

Libert S, Pointer K, Bell EL, Das A, Cohen DE, Asara JM et al (2011) SIRT1 activates MAO-A in the brain to mediate anxiety and exploratory drive. Cell 147(7):1459–1472. https://doi.org/10.1016/j.cell.2011.10.054

Lintas C (2019) Linking genetics to epigenetics: the role of folate and folate-related pathways in neurodevelopmental disorders. Clin Genet 95(2):241–252. https://doi.org/10.1111/cge.13421

Lister R, Mukamel EA, Nery JR, Urich M, Puddifoot CA, Johnson ND et al (2013) Global epigenomic reconfiguration during mammalian brain development. Science (New York, NY) 341(6146):1237905. https://doi.org/10.1126/science.1237905

Liu M, Pile LA (2017) The transcriptional corepressor SIN3 directly regulates genes involved in methionine catabolism and affects histone methylation, linking epigenetics and metabolism. J Biol Chem 292(5):1970–1976. https://doi.org/10.1074/jbc.M116.749754

Lozoya OA, Martinez-Reyes I, Wang T, Grenet D, Bushel P, Li J et al (2018) Mitochondrial nicotinamide adenine dinucleotide reduced (NADH) oxidation links the tricarboxylic acid (TCA) cycle with methionine metabolism and nuclear DNA methylation. PLoS Biol 16(4): 1–23. https://doi.org/10.1371/journal.pbio.2005707

Luka Z, Capdevila A, Mato JM, Wagner C (2006) A glycine N-methyltransferase knockout mouse model for humans with deficiency of this enzyme. Transgenic Res 15(3):393–397. https://doi.org/10.1007/s11248-006-0008-1

Mato JM, Martínez-Chantar ML, Lu SC (2013) S-adenosylmethionine metabolism and liver disease. Ann Hepatol 12(2):183–189. https://doi.org/10.1016/S1665-2681(19)31355-9

McCoy CR, Jackson NL, Brewer RL, Moughnyeh MM, Smith DL Jr, Clinton SM (2018) A paternal methyl donor depleted diet leads to increased anxiety- and depression-like behavior in adult rat offspring. Biosci Rep 38(4):BSR20180730. https://doi.org/10.1042/BSR20180730

McKee SE, Zhang S, Chen L, Rabinowitz JD, Reyes TM (2018) Perinatal high fat diet and early life methyl donor supplementation alter one carbon metabolism and DNA methylation in the brain. J Neurochem 145(5):362–373. https://doi.org/10.1111/jnc.14319

Melnyk S, Fuchs GJ, Schulz E, Lopez M, Kahler SG, Fussell JJ et al (2012) Metabolic imbalance associated with methylation dysregulation and oxidative damage in children with autism. J Autism Dev Disord 42(3):367–377. https://doi.org/10.1007/s10803-011-1260-7

Mentch SJ, Mehrmohamadi M, Huang L, Liu X, Gupta D, Mattocks D et al (2015) Histone methylation dynamics and gene regulation occur through the sensing of one-carbon metabolism. Cell Metab 22(5):861–873. https://doi.org/10.1016/j.cmet.2015.08.024

Mitchell ES, Conus N, Kaput J (2014) B vitamin polymorphisms and behavior: evidence of associations with neurodevelopment, depression, schizophrenia, bipolar disorder and cognitive decline. Neurosci Biobehav Rev 47:307–320. https://doi.org/10.1016/j.neubiorev.2014.08.006

Mladenović D, Radosavljević T, Hrnčić D, Rasic-Markovic A, Stanojlović O (2019) The effects of dietary methionine restriction on the function and metabolic reprogramming in the liver and brain—implications for longevity. Rev Neurosci 30(6):581–593. https://doi.org/10.1515/revneuro-2018-0073

Mudd SH (2011) Hypermethioninemias of genetic and non-genetic origin: a review. Am J Med Genet C: Semin Med Genet 157(1):3–32. https://doi.org/10.1002/ajmg.c.30293

Muehlmann AM, Bliznyuk N, Duerr I, Yang TP, Lewis MH (2020) Early exposure to a methyl donor supplemented diet and the development of repetitive motor behavior in a mouse model. Dev Psychobiol 62(1):77–87. https://doi.org/10.1002/dev.21914

Olteanu H, Banerjee R (2001) Human methionine synthase reductase, a soluble P-450 reductase-like dual flavoprotein, is sufficient for NADPH-dependent methionine synthase activation*. J Biol Chem 276(38):35558–35563. https://doi.org/10.1074/jbc.M103707200

Olteanu H, Munson T, Banerjee R (2002) Differences in the efficiency of reductive activation of methionine synthase and exogenous electron acceptors between the common polymorphic variants of human methionine synthase reductase. Biochemistry 41(45):13378–13385. https://doi.org/10.1021/bi020536s

Orozco JS, Hertz-Picciotto I, Abbeduto L, Slupsky CM (2019) Metabolomics analysis of children with autism, idiopathic-developmental delays, and Down syndrome. Transl Psychiatry 9(1): 243. https://doi.org/10.1038/s41398-019-0578-3

Parrish RR, Buckingham SC, Mascia KL, Johnson JJ, Matyjasik MM, Lockhart RM, Lubin FD (2015) Methionine increases BDNF DNA methylation and improves memory in epilepsy. Ann Clin Transl Neurol 2(4):401–416. https://doi.org/10.1002/acn3.183

Pauwels S, Doperé I, Huybrechts I, Godderis L, Koppen G, Vansant G (2014) Validation of a food-frequency questionnaire assessment of methyl-group donors using estimated diet records and plasma biomarkers: the method of triads. Int J Food Sci Nutr 65(6):768–773. https://doi.org/10.3109/09637486.2014.917149

Pauwels S, Duca RC, Devlieger R, Freson K, Straetmans D, Van Herck E et al (2016) Maternal methyl-group donor intake and global DNA (hydroxy)methylation before and during pregnancy. Nutrients 8(8):474. https://doi.org/10.3390/nu8080474

Pauwels S, Ghosh M, Duca RC, Bekaert B, Freson K, Huybrechts I et al (2017a) Dietary and supplemental maternal methyl-group donor intake and cord blood DNA methylation. Epigenetics 12(1):1–10. https://doi.org/10.1080/15592294.2016.1257450

Pauwels S, Ghosh M, Duca RC, Bekaert B, Freson K, Huybrechts I et al (2017b) Maternal intake of methyl-group donors affects DNA methylation of metabolic genes in infants. Clin Epigenetics 9(1):16. https://doi.org/10.1186/s13148-017-0321-y

Pauwels S, Symons L, Vanautgaerden E-L, Ghosh M, Duca RC, Bekaert B et al (2019) The influence of the duration of breastfeeding on the infant's metabolic epigenome. Nutrients 11(6):1408. https://doi.org/10.3390/nu11061408

Pizzorusso T, Tognini P (2020) Interplay between metabolism, nutrition and epigenetics in shaping brain DNA methylation, neural function and behavior. Genes 11(7):1–18. https://doi.org/10.3390/genes11070742

Price AJ, Collado-Torres L, Ivanov NA, Xia W, Burke EE, Shin JH et al (2019) Divergent neuronal DNA methylation patterns across human cortical development reveal critical periods and a unique role of CpH methylation. Genome Biol 20(1):196. https://doi.org/10.1186/s13059-019-1805-1

Rustad SR, Papale LA, Alisch RS (2019) In: Binder EB, Klengel T (eds) DNA methylation and hydroxymethylation and behavior BT—behavioral neurogenomics. Springer International Publishing, Cham, pp 51–82. https://doi.org/10.1007/7854_2019_104

Saha T, Chatterjee M, Sinha S, Rajamma U, Mukhopadhyay K (2017) Components of the folate metabolic pathway and ADHD core traits: an exploration in eastern Indian probands. J Hum Genet 62(7):687–695. https://doi.org/10.1038/jhg.2017.23

Saha T, Chatterjee M, Verma D, Ray A, Sinha S, Rajamma U, Mukhopadhyay K (2018) Genetic variants of the folate metabolic system and mild hyperhomocysteinemia may affect ADHD associated behavioral problems. Prog Neuropsychopharmacol Biol Psychiatry 84:1–10. https://doi.org/10.1016/j.pnpbp.2018.01.016

Schaevitz LR, Berger-Sweeney JE (2012) Gene–environment interactions and epigenetic pathways in autism: the importance of one-carbon metabolism. ILAR J 53(3–4):322–340. https://doi.org/10.1093/ilar.53.3-4.322

Schmidt RJ, Tancredi DJ, Krakowiak P, Hansen RL, Ozonoff S (2014) Maternal intake of supplemental iron and risk of autism spectrum disorder. Am J Epidemiol 180(9):890–900. https://doi.org/10.1093/aje/kwu208

Schmidt RJ, Kogan V, Shelton JF, Delwiche L, Hansen RL, Ozonoff S (2021) Combined prenatal pesticide exposure and folic acid intake in relation to autism spectrum disorder. Environ Health Perspect 125(9):97007. https://doi.org/10.1289/EHP604

Shaik Mohammad N, Sai Shruti P, Bharathi V, Krishna Prasad C, Hussain T, Alrokayan SA et al (2016) Clinical utility of folate pathway genetic polymorphisms in the diagnosis of autism spectrum disorders. Psychiatr Genet 26(6):281–286

Shea TB, Rogers E (2014) Lifetime requirement of the methionine cycle for neuronal development and maintenance. Curr Opin Psychiatry 27(2):138–142. https://doi.org/10.1097/YCO.0000000000000046

Shulha HP, Cheung I, Whittle C, Wang J, Virgil D, Lin CL et al (2012) Epigenetic signatures of autism: trimethylated H3K4 landscapes in prefrontal neurons. Arch Gen Psychiatry 69(3):314–324. https://doi.org/10.1001/archgenpsychiatry.2011.151

Siu MT, Butcher DT, Turinsky AL, Cytrynbaum C, Stavropoulos DJ, Walker S et al (2019) Functional DNA methylation signatures for autism spectrum disorder genomic risk loci: 16p11.2 deletions and CHD8 variants. Clin Epigenetics 11(1):103. https://doi.org/10.1186/s13148-019-0684-3

Tang S, Fang Y, Huang G, Xu X, Padilla-Banks E, Fan W et al (2017) Methionine metabolism is essential for SIRT 1-regulated mouse embryonic stem cell maintenance and embryonic development. EMBO J 36(21):3175–3193. https://doi.org/10.15252/embj.201796708

Torres RF, Kouro R, Kerr B (2019) Writers and readers of DNA methylation/hydroxymethylation in physiological aging and its impact on cognitive function. Neural Plast 2019:5982625. https://doi.org/10.1155/2019/5982625

Weiner AS, Boyarskikh UA, Voronina EN, Mishukova OV, Filipenko ML (2014) Methylenetetrahydrofolate reductase C677T and methionine synthase A2756G polymorphisms influence on leukocyte genomic DNA methylation level. Gene 533(1):168–172. https://doi.org/10.1016/j.gene.2013.09.098

Yang J, Bashkenova N, Zang R, Huang X, Wang J (2020) The roles of TET family proteins in development and stem cells. Development (Cambridge, England) 147(2):dev183129. https://doi.org/10.1242/dev.183129

Yasin H, Stowe R, Wong CK, Jithesh PV, Zahir FR (2020) First whole transcriptome RNAseq on CHD8 haploinsufficient patient and meta-analyses across cellular models uncovers likely key pathophysiological target genes. Cureus 12(11):e11571–e11571. https://doi.org/10.7759/cureus.11571

Zahir FR, Brown CJ (2011) Epigenetic impacts on neurodevelopment: pathophysiological mechanisms and genetic modes of action. Pediatr Res 69(8):92–100. https://doi.org/10.1203/PDR.0b013e318213565e

Zeisel SH, Blusztajn JK (1994) Choline and human nutrition. Annu Rev Nutr 14(1):269–296. https://doi.org/10.1146/annurev.nu.14.070194.001413

Zhang N (2018) Role of methionine on epigenetic modification of DNA methylation and gene expression in animals. Anim Nutr 4(1):11–16. https://doi.org/10.1016/j.aninu.2017.08.009

Zhang Y, Hodgson NW, Trivedi MS, Abdolmaleky HM, Fournier M, Cuenod M et al (2016) Decreased brain levels of vitamin B12 in aging, autism and schizophrenia. PLoS One 11(1):1–19. https://doi.org/10.1371/journal.pone.0146797

Attention-Deficit Hyperactivity Disorder: Genetic, Pharmacogenetic, and Metabolomic Insights

5

Salma N. Younes, Rana Al-Jurf, Sara Hammuda, Gheyath K. Nasrallah, Amal Al-Jurf, Ayah Ziyada, Palli Valapila Abdulrouf, M. Walid Qoronfleh, Muthanna Samara, and Nader Al-Dewik

Abstract

ADHD is a neurodevelopmental disorder that affects children, adolescents, and adults at a high rate around the globe, resulting in significant impairment. Inattention, impulsivity, restlessness, and hyperactivity are all hallmarks of ADHD. Symptoms may persist into adulthood in 55–66% of all cases. The causes of ADHD remain unclear, but it is believed to be a complex disease with a variety of contributing variables, including heredity, neurodevelopmental problems, severe brain traumas, neuroinflammation, consanguineous marriages, prematurity, and exposure to environmental toxins. Numerous genetic polymorphisms

S. N. Younes (✉)
Department of Research, Women's Wellness and Research Center (WWRC), Hamad Medical Corporation (HMC), Doha, Qatar
e-mail: V-SYounes@hamad.qa

R. Al-Jurf · G. K. Nasrallah
Department of Biomedical Science, College of Health Sciences, Member of Q.U. Health, Qatar University, Doha, Qatar
e-mail: 199660690@student.qu.edu.qa; gheyath.nasrallah@qu.edu.qa

S. Hammuda · M. Samara
Department of Psychology, Kingston University London, Kingston upon Thames, London, UK
e-mail: M.Samara@kingston.ac.uk

A. Al-Jurf
Department of Special Education and Gifted Students, Ministry of Education, Doha, Qatar
e-mail: A.aljurf@edu.gov.qa

A. Ziyada
Genomics and Precision Medicine (GPM), College of Health and Life Science (CHLS), Hamad Bin Khalifa University (HBKU), Doha, Qatar
e-mail: ayzi23678@hbku.edu.qa

M. W. Qoronfleh et al. (eds.), *Proteins Associated with Neurodevelopmental Disorders*, Nutritional Neurosciences, https://doi.org/10.1007/978-981-15-9781-7_5

linked with ADHD have been discovered in the twenty-first century. These findings have already given a starting point for the study of ADHD biology and innovative treatment options. Pharmacotherapy using methylphenidate (MPH) seems to be the first-line treatment option for adults with ADHD. Moreover, research has been done on genes that influence the response to MPH among ADHD-affected individuals. Furthermore, a few peripheral biomarkers have been discovered in ADHD adults. In this chapter, the authors summarize current evidence on genetic, pharmacogenetic, and biochemical (metabolomics) investigations in ADHD. Also, the authors address the neurobiology of ADHD, with a focus on functional or structural alterations in the brain of ADHD-affected individuals and their connections with complicated chromosomal variants using imaging genetics methods. In addition, the biological mechanisms involved in ADHD have been summarized. Finally, the scope for additional research for a better understanding of the pathophysiology of ADHD in the context of disrupted signaling pathways is reviewed, which could eventually lead to the discovery of possible therapeutic targets and novel treatment strategies.

Keywords

Attention-deficit hyperactivity disorder · ADHD · Pathways · Developmental mechanisms · Genes · Pharmacogenetics · Metabolomics

P. V. Abdulrouf
Department of Research, Women's Wellness and Research Center (WWRC), Hamad Medical Corporation (HMC), Doha, Qatar

Department of Pharmacy, Women's Wellness and Research Center (WWRC), Hamad Medical Corporation, Doha, Qatar
e-mail: PABDULROUF@hamad.qa

M. W. Qoronfleh
Q3CG Research Institute (QRI), Research and Policy Division, Ypsilanti, MI, USA

21HealthStreet, Consulting Services, 1 Christian Fields, London, UK

Applied Biomedicine, Inc., Doha, Qatar

N. Al-Dewik (✉)
Department of Research, Women's Wellness and Research Center (WWRC), Hamad Medical Corporation (HMC), Doha, Qatar

Interim Translational Research Institute (iTRI), Hamad Medical Corporation (HMC), Doha, Qatar

Faculty of Health and Social Care Sciences, Kingston University, St. George's University of London, London, UK

Clinical and Metabolic Genetics, Department of Pediatrics, Hamad General Hospital (HGH), Hamad Medical Corporation (HMC), Doha, Qatar

Genomics and Precision Medicine (GPM), College of Health and Life Science (CHLS), Hamad Bin Khalifa University (HBKU), Doha, Qatar
e-mail: Naldewik@hamad.qa; Nader.Al-Dewik@kingston.ac.uk

Abbreviations

5-HT1A	Serotonin 1A receptor
5-HT1B	Serotonin 1B receptor
5-HT1D	Serotonin 1D receptor
5-HT1E	Serotonin 1E receptor
5-HT1F	Serotonin 1F receptor
5-HT2A	Serotonin 2A receptor
5-HT2B	Serotonin 2B receptor
5-HT2C	Serotonin 2C receptor
5-HT3A	Serotonin 3A receptor
5-HT3B	Serotonin 3B receptor
5-HT4	Serotonin 4 receptor
5-HT5A	Serotonin 5A receptor
5-HT6	Serotonin 6 receptor
5-HT7	Serotonin 7 receptor
ADRA2A	Adrenergic alpha 2A receptor
ADRA2C	Adrenergic alpha 2C receptor
ANK3	Ankyrin
ANKK1	Ankyrin repeat and kinase domain containing 1
bAREase	Arylesterase
BCHE	Acetylcholine-metabolizing butyrylcholinesterase
BDNF	Brain-derived neurotrophic factor
CACNA1A	Calcium channel, voltage-dependent P/Q type, alpha 1A subunit
CACNA1C	a1C subunit of an L-type voltage-dependent calcium channel
CDH13	Cadherin 13
cHcy	Homocysteine
CLOCK	Circadian locomotor output cycles protein kaput
CLPX1-4	Caseinolytic mitochondrial matrix peptidase chaperone subunit
CMTM8	CKLF-like MARVEL transmembrane domain containing 8
CNTFR/CNTF	Ciliary neurotrophic factor/receptor
CTNNA2	aN-catenin protein
DBH	Dopamine beta hydroxylase
DDC	Dopamine decarboxylase
DFNB31	Deafness, autosomal recessive 31
DGKH	Diacylglycerol kinase eta
DIRAS2	Brain-expressed GTP-binding RAS-like 2
DISC1	Disrupted in schizophrenia 1
dMDA	Malondialdehyde
DRD1	Dopamine D1 receptor
DRD2	Dopamine D2 receptor
DRD3	Dopamine D3 receptor
DRD5	Dopamine D5 receptor
EGFR	Epidermal growth factor receptor

eNO	Nitric oxide metabolite
fOSI	Oxidative stress index
FOXP2	Forkhead box P2
gPON	Paraoxonase
GRM7	Glutamate receptor, metabotropic 7
hSOD	Superoxide dismutase
KCNIP4	Kv channel-interacting protein 4
LPHN3	Latrophilin 3
lTAS	Total antioxidant status
MAOA	Monoamine oxidase A
MAOB	Monoamine oxidase B
mTOS	Total oxidant status
MYO5B	Myosin 5B
NCAM1	Neural cell adhesion molecule 1
NGF	Nerve growth factor
NGFR	Nerve growth factor receptor
nOL	Oleic acid; pPAOL
NOS1/3	Nitric oxide synthase 1/3
NPAS3	Neuronal PAS domain protein 3
NR3C2	Mineralocorticoid receptor
NSF	*N*-ethylmaleimide-sensitive factor
NTF3	Neurotrophin 3
NTF4/5	Neurotrophin 4/5
NTRK1	Neurotrophic tyrosine kinase, receptor, type 1
NTRK2	Neurotrophic tyrosine kinase, receptor, type 2
NTRK3	Neurotrophic tyrosine kinase, receptor, type 3
OPRM1	u-Opioid receptor
PPP2R2C	Phosphatase 2, regulatory subunit B, gamma
PR KG1	Protein kinase G
qBDNF	Brain-derived neurotrophic factor
RAB3A	Member RAS oncogene family
rHVA	Homovanillic acid; s5-HIAA
SLC39A3	Solute carrier family 39 (zinc transporter), member 3
SLC6A2	Norepinephrine transporter
SLC6A3	Dopamine transporter
SLC6A4	Serotonin transporter
SNAP-25	Synaptosomal-associated protein-25
SNPH	Syntaphilin
SPOCK3	Ca(2+)-binding extracellular heparan/chondroitin sulfate proteoglycan
STX1A	Syntaxin 1A
STXBP1	Syntaxin-binding protein 1
SYNIII	Synapsin III
SYP	Synaptophysin
SYT1	Synaptotagmin 1

SYT2	Synaptotagmin 2
tCSF	Cerebrospinal fluid
TH	Tyrosine hydroxylase
TPH1	Tryptophan hydroxylase 1
TPH2	Tryptophan hydroxylase 2
TSPAN8	Tetraspanin-8
TTC12	Tetratricopeptide repeat domain 12
UTR	Untranslated region
VAMP1	Synaptobrevin-1
VAMP-2	Synaptobrevin-2
VNTR	Variable tandem repeat polymorphism
ZNF804A	Zinc finger protein 804A

5.1 Introduction

Attention-deficit/hyperactivity disorder (ADHD) is a common clinically heterogeneous neurodevelopmental disorder characterized by excessive inattention, impulsivity, and hyperactivity (Faraone et al. 2015), affecting 5–7% of children worldwide (Polanczyk and Rohde 2007; Simon et al. 2009; Nigg 2013; Wilens and Spencer 2010).

Although it was long thought to be a childhood disease, 65% of afflicted children continue to exhibit symptoms throughout adulthood (Faraone et al. 2006). ADHD prevalence in childhood was estimated to be 5.3% (Polanczyk and Rohde 2007; Polanczyk et al. 2014; Willcutt 2012), and 2.5–4.9% in adulthood (Ramos-Quiroga et al. 2014a; Simon et al. 2009). Although many clinical characteristics of ADHD in children and adults are comparable, adults exhibit less hyperactivity and impulsivity and more inattentive symptoms (Volkow and Swanson 2013; Buitelaar et al. 2012; Haavik et al. 2010). Furthermore, certain structural brain discoveries distinguish ADHD in children from ADHD in adults (Shaw et al. 2013). The above findings point towards distinct etiological routes for children and adults.

ADHD might occur alone or in conjunction with other neurological, psychiatric, and neurodevelopmental disorders (Scott et al. 2016; Yurtbasi et al. 2018). ADHD has a broad and complex negative effect on society since it impacts not only all areas of a child's life but also those of siblings and parents, creating significant disruptions to normal family functioning (Quintero et al. 2018; Martinez-Raga et al. 2017). In addition, there are financial constraints associated with treatment expenses and decreased job opportunities (Ogundele 2018). Depending on the degree of the ADHD condition, it might impair a child's academic performance. If untreated or misdiagnosed, it might continue into adulthood, impacting both personal and professional life (Ogundele 2018).

Many factors, including genetic susceptibility, neurodevelopmental problems, aberrant neuronal maturation, brain damage, environmental exposures, and

Fig. 5.1 ADHD and risk factors. ADHD symptoms have been linked to various factors, including gene variants; environmental, psychological, and individual factors such as age; abnormalities in various neurological pathways, including dopaminergic and serotonergic; and comorbidities with multiple disorders

consanguinity, might contribute to ADHD symptoms. Prenatal, postnatal, and perinatal exposure to environmental risk factors has been linked to ADHD symptoms. Factors during pregnancy and the postpartum period, history of preterm delivery, low birth weight, a mother's smoking history, trauma, and obesity are all known risk factors for ADHD. Trauma, parenting style, chemical coloring and flavoring food components, and pesticide and pollutant exposure are all postnatal risk factors that might exacerbate ADHD symptoms (Posner et al. 2020) (Fig. 5.1).

It is believed that ADHD symptoms have been caused by aberrant anatomical and functional problems in the brain (Cortese and Castellanos 2012). Brain alterations, including morphological, microstructural, functional, biochemical, and molecular alterations, have been investigated in multiple neuroimaging studies conducted among individuals affected with ADHD (Klein et al. 2017; Weyandt et al. 2013; Samea et al. 2019). Moreover, many neural networks implicating attention, executive, and reward functions have been identified, along with numerous molecular pathways such as dopaminergic, adrenergic, serotonergic, and cholinergic, all of which are key players in the pathophysiology of ADHD (Mueller et al. 2017; Stevens et al. 2018).

Molecular genomic investigations have revealed that ADHD has a substantial hereditary component, with an estimated heritability rate ranging from 70 to 88% (Klein et al. 2017; Faraone and Mick 2010; Larsson et al. 2014). Nonetheless, data on the heritability of ADHD in children and adults has been inconsistent, with some studies suggesting that heritability is higher in children (75–90%) (Faraone and Mick 2010) than adults (30–50%) (Boomsma et al. 2010; Kan et al. 2013; Larsson et al. 2013), while other studies support greater heritability in adults (Biederman et al. 1995, 1996; Faraone et al. 2000a, b). The ADHD community has made considerable strides, ranging from linkage and candidate gene approach (CGA) research to genome-wide association studies (GWAS) investigations and beyond. In addition, it has been estimated that 40% of ADHD heritability is attributable to a polygenic component that includes numerous common variations single nucleotide polymorphisms (SNPs) as well as rare deletions and insertions (copy number variants—CNVs) (Lee et al. 2013; Williams et al. 2012; Martin et al. 2015).

Multiple studies on correlation, linkage, and meta-analysis exploring the genetic vulnerability of ADHD have discovered many potential genes, polymorphisms, and chromosomes linked with ADHD symptoms. In ADHD, GWAS have found genomic DNA copy number variations as well as uncommon or significant deletion/duplications (Williams et al. 2012; Jarick et al. 2014). Using fine-mapping linkage analysis, researchers found that variation in the gene encoding neural signaling might be a possible risk factor for ADHD (Acosta et al. 2016; Silva et al. 2011; Arcos-Burgos et al. 2010). In addition, multiple-gene polymorphisms were linked to alterations in neuropsychobiological processes in ADHD (Khadka et al. 2016; Kebir et al. 2009). However, the majority of individual candidate gene association studies in ADHD were performed with a limited sample size, and the identified variations were not significantly linked with ADHD pathophysiology. Furthermore, a recent genome-wide association study, GWAS-based meta-analysis research with a much larger sample size, found no relationship between previously reported candidate genes and ADHD (Demontis et al. 2019).

To develop a better understanding of the neuropsychobiology of ADHD, structural and functional imaging phenotypes might be used in conjunction with other methods, which in turn will lead to a better understanding of the clinical manifestations of ADHD phenotypes. Intermediate phenotypes are those aspects of a disease that are more closely linked to its neurobiology than its clinical symptoms and contribute to the disorder's hereditary susceptibility. Neuroimaging-derived phenotypes, such as structural, functional, and molecular features, are thought to be highly inherited and have been demonstrated to be substantially changed in ADHD and to be critical intermediary phenotypes for ADHD. Biomarker identification might aid in the differential diagnosis of ADHD. Recently, a study found several possible peripheral biomarkers in juvenile ADHD to be associated with monoaminergic pathways and the hypothalamic-pituitary-adrenal (HPA) axis (Scassellati et al. 2012).

The purpose of this chapter is to provide an overview of existing literature on genetic, pharmacogenetic, and biochemical (metabolomics) investigations in ADHD. The authors highlight ADHD neurobiology, with an emphasis on structural and functional alterations and their connections with complicated chromosomal variants using imaging genetics methods. In addition, the authors summarize the biological mechanisms involved in ADHD. Finally, the authors discuss the scope for additional research in understanding the pathophysiology of ADHD in the context of disrupted signaling pathways, which in turn could eventually lead to the discovery of possible therapeutic targets and novel treatment strategies.

5.2 Genetic Insights in ADHD

ADHD has been shown to have strong hereditary components, according to studies of the family, siblings, and adoption. Therefore, the first- and second-degree ADHD relatives are at a greater risk of developing the disorder (Chen et al. 2008). According to a recent meta-analysis of twin studies, the heritability of ADHD is estimated to be 77–88% (Faraone and Larsson 2019). Further, several multiple twin studies have reported heritability estimates reaching up to 90% (Faraone and Mick 2010; Larsson et al. 2013, 2014). The magnitude is, therefore, similar to that of autism spectrum disorder (about 80%) and schizophrenia (approximately 80%) (Styles et al. 2020; Sullivan et al. 2012; Al-Dewik and Alsharshani 2020; Al-Dewik et al. 2020; Shaltout et al. 2020).

Several molecular genetic investigations have been performed to discover risk genes for ADHD, with numerous genetic polymorphisms identified in association with ADHD (Gizer et al. 2009; Faraone et al. 2005; Li et al. 2006) (Fig. 5.2). Despite multiple investigations, identifying ADHD-risk genes continues to be a difficult task (Gizer et al. 2009; Franke et al. 2009), owing to the complicated and polygenic nature of ADHD pathogenesis. Along with genetic risk factor involvement, several other external risk variables, including the environment and potential interactions between genes and the environment, have been linked to a higher risk of ADHD (Larsson et al. 2013). The dopamine and serotonin transporter genes have been linked to attention-deficit hyperactivity disorder (ADHD) (Gizer et al. 2009; Faraone et al. 2005; Li et al. 2006).

An in-depth study of more than 50 candidate genes from European multicenter samples of about 674 families was carried out by the International Multi-site ADHD Genetics (IMAGE) project. The results revealed that many candidate genes were associated with ADHD in a substantial way (Brookes et al. 2006). Another meta-analysis based on seven linkage studies revealed that the short arm of chromosome 16 might be linked with the symptoms of ADHD (Zhou et al. 2008). Research utilizing a cadherin 13 (*CDH13*) knockout mice model discovered that *CDH13* regulates the synaptic activity of hippocampal interneurons and cognitive domains. Also, it was hypothesized that *CDH13* might be a risk gene for ADHD (Rivero et al. 2015). According to a linkage analysis-based research conducted on multigenerational families, the adhesion G protein-coupled receptor L3 (ADGRL3) gene (formerly known as *LPHN3*) variations were shown to be associated with increased vulnerability to developing ADHD (Ribases et al. 2011).

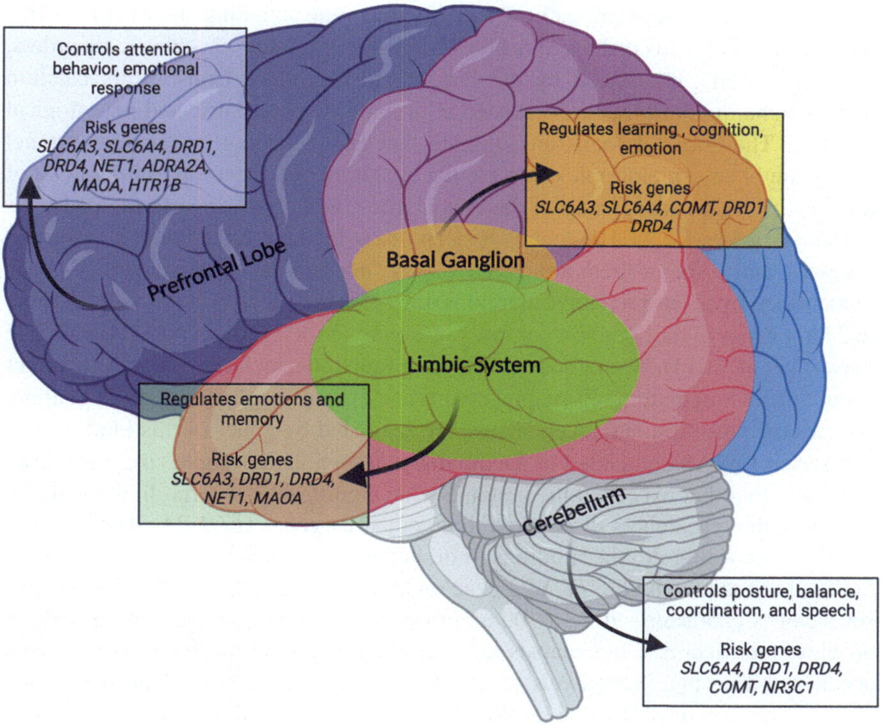

Fig. 5.2 ADHD-risk gene brain map. Multiple genes have been linked to altered structural and functional alterations in the brain, with the majority of changes occurring in the frontal lobe, basal ganglia, limbic system, and cerebellum (figure adapted from Yadav, S.K., Bhat, A.A., Hashem, S. et al., (Yadav et al. 2021))

Until recently, candidate gene research on the genetics of ADHD has largely focused on common genetic variants (CNVs) to explain the genetic etiology of the disorder. Nonetheless, the bulk of the identified genetic variants had minor effect sizes (Gizer et al. 2009; Faraone et al. 2005). With the growing evidence indicating a higher frequency of rare CNVs among individuals with psychiatric disorders, more efforts are being focused on studying rare variants' role in neurodevelopmental disorders, including ADHD. Recently, a GWAS of CNVs revealed that duplications spanning *CHRNA7* were associated with ADHD. The above findings were replicated in four independent cohorts from the USA, the UK, and Canada, yielding consistent results and thus supporting the enrichment of large, rare CNVs in ADHD (Williams et al. 2012).

ADHD-related genes are distributed across the genome, but as a result of a clustering study, it was shown that they tend to be grouped into particular functional categories. When group comparison enrichment analyses were used, it was discovered that enriched functions for the ADHD-GWAS associations were linked to neuronal projections, synaptic structures, nervous system structures, neuronal morphogenesis, cell–cell interaction, and glutamatergic signaling (Hawi et al. 2015; Yang et al. 2013). According to a GWAS on five common psychiatric disorders, including ADHD, conducted by the Psychiatric Genomics Consortium, calcium channel signaling genes were associated with multiple psychiatric and neurological disorders. The above association suggests that gene variants in calcium channel activity might have pleiotropic effects in the evolution of the neuropsychobiology of ADHD (Cross Disorder Group of the Psychiatric Genomics Consortium 2013).

The risk genes for ADHD that have been discovered are linked with a variety of processes, and some of them are referred to as "hot genes." When candidate genes are identified by at least five investigations, they are referred to as "hot genes." There are 24 hot genes in the spotlight currently, accounting for 7% of all ADHD candidate genes (Li et al. 2014). Numerous neurological functions, including neurotransmitter production and regulation of synaptic membrane dynamics, glutaminergic signaling, and various transcriptional mechanisms, are regulated by genes of this kind.

A recent GWAS conducted by Demontis et al. in a large sample size discovered 12 distinct loci that exceeded the genome-wide significance criteria. In the association areas, the researchers discovered three potential genes (*FOXP2*, *SORCS3*, and *DUSP6*), according to the findings. Forkhead/winged-helix transcription factor (*FOXP2*) is encoded by the *FOXP2* gene on chromosome 7. The *FOXP2* transcription factor is characterized by a 100-amino acid monomeric DNA-binding domain and plays a critical role in synapsis formation and neuronal mechanisms related to speech and learning (Schreiweis et al. 2014; Tsui et al. 2013). Transmembrane receptors are encoded by the *SORCS3* gene, which is found on chromosome 10 and has a high expression level in the brain. It has been shown that this gene is essential for neuronal growth and plasticity (Breiderhoff et al. 2013). The *DUSP6* gene, on the other hand, is situated on chromosome 12 and encodes for the dual-specificity phosphatase 6 enzyme, which controls neurotransmitter homeostasis by altering the amount of dopamine in synapses. DUSP6 is expressed at a low level in the brain, yet it is essential for brain development despite its low expression level (Mortensen et al. 2008; Li et al. 2007; Carithers et al. 2015).

The development of imaging and molecular genomic technologies has provided the opportunity to investigate the impact of genetic variations on the structural, functional, and neuronal connectivity of the brain, as well as investigate the dysregulated biological pathways in various neurological and psychiatric disorders among others. A vast body of studies have shown a strong connection between structural and functional brain alterations in ADHD and genetic variants in the condition using neuroimaging genomics research (Table 5.1). Recent advances in imaging and genetics have made it possible to conduct a more in-depth study of the neurological processes implicated in attention-deficit hyperactivity disorder. Such studies are necessary to better understand the connection between neurodevelopmental and neurofunctional changes associated with behavioral performance, as well as the association between genetic alterations and behavioral performance (Fig. 5.2). Recent results from numerous studies in twins indicated that global and regional brain development and functions are actively regulated by genetics, with a high heritability rate (Peper et al. 2007; McKay et al. 2014;

Table 5.1 Genetic polymorphisms associated with ADHD

Systems	Genes	Genetic variants	Results	Ref
Dopaminergic	*DRD1*	5' UTR: rs4532	DRD1: No effect of genotype on the clinical outcome or cortical development	Shaw et al. (2007)
Dopaminergic	*DRD1*	5' UTR: rs4532; rs265981	DRD1 and NET1 SNPs: No genotype effects on GM or WM volume and no group × genotype interactions	Bobb et al. (2005)
Dopaminergic	*DRD1*	3'UTR: rs4867798, rs251937, rs703748, rs835540, rs835616, rs835541, rs863126, rs265977	Absence of association	Ribasés et al. (2012)
Dopaminergic	*DRD2*	Promoter: rs1799732	Absence of association	Hasler et al. (2015)
Dopaminergic	*DRD2*	Intron: rs4630328, rs7131056, rs4245146, rs17529477, rs2002453, rs12363125, rs2283265, 3'UTR: rs2242592, rs1554929, rs2234689	Absence of association	Ribasés et al. (2012)
Dopaminergic	*DRD2*	Exon: rs6277, Intron: rs2283265	Absence of association	Mota et al. (2015)
Dopaminergic	*DRD3*	Exon: rs6280	Absence of association	Muglia et al. (2002)
Dopaminergic	*DRD3*	5'UTR: rs9825563, Exon: rs6280, Intron: rs1800828, rs10934256, rs167770, rs167771, rs324035, rs9880168, rs2134655, 3'UTR: rs3732790, rs2399496	Absence of association	Ribasés et al. (2012)
Dopaminergic	*DRD4*	Exon 3 VNTR	No volumetric differences between 7R and non-7R carriers; no group × genotype interactions	Castellanos et al. (1998)
Dopaminergic	*DRD4*	Exon 3 VNTR	*DRD4*-unaffected siblings 7R carriers: ↑ prefrontal GM volume	Durston et al. (2005)
Dopaminergic	*DRD4*	Exon 3 VNTR	*DRD4*-4R/4R carriers: No effect on WM integrity	Hong et al. (2015)

(continued)

Table 5.1 (continued)

Systems	Genes	Genetic variants	Results	Ref
Dopaminergic	*DRD4*	Exon 3 VNTR	7R carriers: ↓ volumes of the superior frontal cortex and cerebellum cortex compared to noncarriers	Monuteaux et al. (2008)
Dopaminergic	*DRD4*	Exon 3 VNTR	Putamen volume, *DRD4*-7R carriers showed opposite age relations	Richards et al. (2016)
Dopaminergic	*DRD4*	Exon 3 VNTR	*DRD4*-7R carriers: Frontal cortex volume is associated with stimulant treatment at a younger age	Schweren et al. (2016)
Dopaminergic	*DRD5*	5'Flank Dinucleotide repeat	Absence of association	Carpentier et al. (2013)
Dopaminergic	*DRD5*	5'Flank Dinucleotide repeat	Presence of association	Johansson et al. (2008)
Dopaminergic	*DRD5*	5'Flank Dinucleotide repeat	Absence of association	Squassina et al. (2008)
Dopaminergic	*DRD5*	Exon: rs2227850, 3'UTR: rs10033951	Absence of association	Ribasés et al. (2012)
Dopaminergic	*SLC6A3*	3'UTR and intron 8 VNTR haplotype	No differences in striatal activity compared with non 9–6 haplotype carriers nor 9R and 10R/10R carriers	Hoogman et al. (2013)
Dopaminergic	*SLC6A3*	3'UTR and intron 8 VNTR haplotype	*SLC6A3* 10–6 haplotype carriers: ↓ left striatal volume, irrespective of treatment	Schweren et al. (2016)
Dopaminergic	*SLC6A3*	3'UTR and intron 8 VNTR haplotype, rs37020, rs460000	No genotype × ADHD interaction effects. *SLC6A3* 10–6 "haplotype-homozygotes: ↑ activity related to successful" stop trials in pre-supplementary motor "areas, ↓ activity in superior frontal and temporal pole" areas. "rs37020 AA carriers: ↓ activity during failed stop trials in IFG," pre-supplementary motor areas, and postcentral gyrus	van Rooij et al. (2015a)

Dopaminergic	*SLC6A3*	3'UTR intron 8 VNTR haplotype	Adult ADHD 9–6 haplotype carriers ↑ 5.9% larger striatum volume relative to participants not carrying this haplotype	Onnink et al. (2016)
Dopaminergic	*SLC6A3*	3'UTR VNTR	10R/10R carriers: ↑ activity in left striatum, right dorsal premotor cortex, and temporoparietal cortical junction compared to 9R carriers	Bedard et al. (2010)
Dopaminergic	*SLC6A3*	3'UTR VNTR	10R/10R carriers: ↑ activity in frontal, medial, and parietal regions during response inhibition compared to 9R carriers; ↓ error response in the parahippocampal gyrus	Braet et al. (2011)
Dopaminergic	*SLC6A3*	3'UTR VNTR	9R carriers: ↓ activity in dorsal ACC compared to 10R/10R carriers	Brown et al. (2010)
Dopaminergic	*SLC6A3*	3'UTR VNTR	9R carriers: ↓ left medial PFC activation compared to 10R/10R carriers "Group × genotype interaction showed that 10R/10R-ADHD" "patients had ↑ activity in pre-SMA/dorsal ACC compared" to HC	Brown et al. (2011)
Dopaminergic	*SLC6A3*	3'UTR VNTR	*SLC6A3* ADHD 10/10R carriers: ↓ CN volumes	Durston et al. (2005)
Dopaminergic	*SLC6A3*	3'UTR VNTR	9R carriers: ↑ activity in CN and ↓ in cerebellar vermis compared to 10R/10R carriers Group × genotype interaction: Effect in CN is observed in ADHD and unaffected siblings, but not HC	Durston et al. (2008)
Dopaminergic	*SLC6A3*	3'UTR VNTR	10R/10R carriers: ↓ cortical thickness in right BA 46 (lateral PFC)	Fernandez-Jaen et al. (2015)
Dopaminergic	*SLC6A3*	3'UTR VNTR	10R/10R carriers: ↑ thickness in right cingulated gyrus and right BA 24	Fernandez-Jaen et al. (2018)

(continued)

Table 5.1 (continued)

Systems	Genes	Genetic variants	Results	Ref
Dopaminergic	*SLC6A3*	3'UTR VNTR	9R carriers: ↑ volumes of CN	Shook et al. (2011)
Dopaminergic	*SLC6A3*	3'UTR VNTR, exon 3 VNTR	*SLC6A3* 9R carriers: No effect on WM integrity *DRD4*-4R/4R carriers: No effect on WM integrity	Hong et al. (2015)
Dopaminergic	*SLC6A3*	3' UTR VNTR, exon 3 VNTR	*SLC6A3* 9R carriers: No effect *DRD4*-7R carriers: Thinner right orbitofrontal/ inferior prefrontal and posterior parietal cortex ADHD 7R carriers: Distinct trajectory of cortical development; normalization of the right parietal cortical region	Shaw et al. (2007)
Dopaminergic	*SLC6A3*	9–6 *SLC6A3* haplotype	ADHD: Activation in CN ↓ as a number of copies ↑, but in the control group reverse was found	Paloyelis et al. (2012)
Dopaminergic	*SLC6A3*	9–6 *SLC6A3* haplotype	Bayesian constraint-based causal discovery (BCCD) algorithm confirmed that there is no association direct link between *SLC6A3* genetic variability and brain activation, but suggested an indirect link mediated through inattention symptoms and diagnostic status of ADHD	Sokolova et al. (2015)
Dopaminergic	*SLC6A3*	*SLC6A3* 3'UTR VNTR; HTTLPR	For total GM, differential age effects were found for *SLC6A3* 9R and SLC6A4 L/L carriers, depending on the amount of positive peer affiliation	Richards et al. (2016)
Metabolic enzymes	*BCHE*	5'UTR: rs17659982, rs6802425, rs4496529, rs1837232, rs1370895, rs17659206, rs4680612, rs10936458, rs9860697, Intron: rs6445067, rs829508, 3'UTR: rs9832712	Presence of association: rs4680612, rs829508	Jacob et al. (2013)

Metabolic enzymes	*COMT*	rs4680	Met carriers: ↓ Network of WM connections linking 18 brain regions	Hong et al. (2015)
Metabolic enzymes	*COMT*	rs4680	ADHD Val/Val: ↓ FA and ↑ RD in the right cingulum (cingulated gyrus) compared to ADHD Met carriers and HC Val/Val	Kabukcu Basay et al. (2016)
Metabolic enzymes	*COMT*	rs4680	Met carriers exhibiting significantly lower functional connectivity than the Val/Val genotype	Mizuno et al. (2017)
Metabolic enzymes	*COMT*	rs4680	ADHD Met carriers: ↓ GM volume. ADHD Val/Val: ↑ GM volume in right CN compared to ADHD Met carriers and HC	Villemonteix et al. (2015)
Metabolic enzymes	*COMT*	rs4686	ADHD Met carriers: ↓ GM volume in left putamen	Shimada et al. (2017)
Metabolic enzymes	*DBH*	Promoter: rs1611115	Presence of association	
Metabolic enzymes	*DBH*	Intron: rs2007153, rs2797851, rs1548364, rs2797855, rs1541332, rs2519154, rs2797853, rs6479643, rs2097628, rs2073833, rs1611131, Exon: rs77905, 3'UTR: rs129883, rs129915	Absence of association	Ribasés et al. (2012)
Metabolic enzymes	*DBH*	Intron: rs2519152	Presence of association	Carpentier et al. (2013)
Metabolic enzymes	*DBH*	Intron: rs2519152	Absence of association	Inkster et al. (2004)
Metabolic enzymes	*DDC*	Intron: rs7803788, rs2329340, rs10499695, rs11974297, rs10499694, rs2044859, rs1982406, rs6944090, rs6592961, rs3823674, rs732215, rs11761683, rs11238131	Presence of association	Ribasés et al. (2009)
Metabolic enzymes	*MAOA*	rs1137070	ADHD TT carriers: ↑ activation in the left inferior frontal lobe, pars opercularis	Ko et al. (2018)

(continued)

Table 5.1 (continued)

Systems	Genes	Genetic variants	Results	Ref
Metabolic enzymes	*MAOA*	rs3027400, rs2072743	Absence of association	Ribasés et al. (2009)
Metabolic enzymes	*MAOB*	5'UTR: rs5906213, Intron: rs5905512, rs1799836, 3'UTR: rs3027415	Presence of association	Ribasés et al. (2009)
Metabolic enzymes	*TH*	5'UTR: rs10770140, Exon: rs6356, Intron: rs2070762	Presence of association: rs2070762	Ribasés et al. (2012)
Metabolic enzymes	*TPH1*	Intron: rs10488683, rs172423, rs1800532, rs211102	Absence of association	Ribasés et al. (2009)
Metabolic enzymes	*TPH1*	Exon 7: rs1799913, Intron: rs1800532, rs11606304, rs10488683, rs17794760, rs169806, rs10832876, rs591556, 5'UTR: rs623580	Absence of association	Johansson et al. (2010)
Metabolic enzymes	*TPH2*	Promoter: rs4570625, rs11178997	Absence of association	Baehne et al. (2009)
Metabolic enzymes	*TPH2*	5'UTR: rs7963717, Intron: rs11178999, rs4565946, rs7955501, rs1386496, Intron: rs4760818, rs4760820, rs1352250, rs17722134, rs17110690, rs12231356, rs10879354, rs1487275, rs10879357, rs10879358, rs11615016, 3'UTR: rs17110747, Exon: rs7305115	Absence of association	Johansson et al. (2010)
Neurodevelopmental network	*BDNF*	5'UTR: rs1491851, rs1491850, rs908867, rs12273363, rs10501087, rs925946, rs11030096, rs1491850, Exon: rs6265	Absence of association	Ribasés et al. (2008)
Neurodevelopmental network	*BDNF*	Exon: rs6265	Absence of association	Sánchez-Mora et al. (2010)
Neurodevelopmental network	*BDNF*	Exon: rs6265	Presence of association	Lanktree et al. (2008)

Neurodevelopmental network	*CDH13*	Intron: rs11646411, rs6565113, rs11150556	Absence of association	Salatino-Oliveira et al. (2015)
Neurodevelopmental network	*CDH13*	Exon: rs200199969, rs183971768, rs35549391, rs199759196, rs34106627, rs72807847 and R174W	Absence of association	Mavroconstanti et al. (2013)
Neurodevelopmental network	*CNTF*	5'UTR: rs17489568, 3'UTR: rs550942, rs11229549	Presence of association: rs7036351 rs3763613 rs550942	Ribasés et al. (2008)
Neurodevelopmental network	*CNTFR*	5'UTR: rs7036351, rs3763613, rs3763614, rs3763615, rs1080750, rs6476454, Intron: rs12551429, rs10758268, rs10814123, rs1124882, rs2381164, rs2381165, rs4363285, rs2274592, 3'UTR: rs4395980, rs10972144	Presence of association: rs7036351 rs3763613 rs550942	Ribasés et al. (2008)
Neurodevelopmental network	*DIRAS2*	3'UTR: rs7868194, rs7854469, rs1542479, rs16906711, Intron: rs690111, rs1017753, rs1331503, rs2297354, rs1331504, Promoter: rs7848810, rs1412005, rs689687	Presence of association	Reif et al. (2011)
Neurodevelopmental network	*LPHN3*	Intron: rs6551665, rs1947274, rs2345039	Absence of association	Arcos-Burgos et al. (2010)
Neurodevelopmental network	*NGF*	Intron: rs17540656, rs6673867, rs17033706, rs11102929, rs4332358, rs4320778, rs10776799, rs7530686, rs7513144, rs6537860, rs11102922, rs4240542, rs7520839, rs11102920, rs2856811, rs2856813, rs719452, rs910330, rs2254404, rs6327, rs6328	Presence of association: rs6327	Ribasés et al. (2008)
Neurodevelopmental network	*NGFR*	rs2537710, rs603769, rs3785931 rs657770, rs534561, rs2072445 rs741073	Absence of association	Ribasés et al. (2008)

(continued)

Table 5.1 (continued)

Systems	Genes	Genetic variants	Results	Ref
Neurodevelopmental network	*NTF3*	Exon: rs6332	Absence of association	
Neurodevelopmental network	*NTF3*	Intron: rs7484401, rs4074967, Exon: rs6332, 3'UTR: rs6489630, rs7956189, rs4073543	Absence of association	Ribasés et al. (2008)
Neurodevelopmental network	*NTF4/5*	3'UTR: rs17206784, rs4802546, 5'UTR: rs11669977	Absence of association	Ribasés et al. (2008)
Neurodevelopmental network	*NTRK1*	rs1800601, rs7522395, rs1998977 rs6674412, rs4661229, rs10908521 rs11590051, rs6334, rs2644596	Absence of association	Ribasés et al. (2008)
Neurodevelopmental network	*NTRK2*	rs1201364, rs1201363, rs1187332 rs999244, rs1147193, rs1439050 rs1187352, rs1187356, rs1187362 rs1209068, rs3780632, rs4877880 rs1662695, rs716893, rs1187274 rs7816, rs1867283, rs7855888 rs11140783, rs2120266, rs10868232 rs4877289, rs1443445, rs920776 rs995861, rs1545285, rs7045900 rs2277192, rs2277193, rs4412435 rs10868241, rs4361832, rs11795386 rs12000011, rs10512159, rs1948308 rs2378672, rs1387926, rs1387924 rs3739570, rs1490403, rs10780695rs1073049, rs1586681	Absence of association	Ribasés et al. (2008)
Neurodevelopmental network	*NTRK3*	rs7164531, rs998636, rs744993 rs744994, rs4887400, rs7176520 rs4887210, rs6496469, rs1863494 rs1346164, rs2162266, rs4887376 rs8025158, rs10520676, rs12594095	Absence of association	Ribasés et al. (2008)

		rs3784406, rs7180942, rs1834573 rs991729, rs3784416, rs1461210 rs10520673, rs3825884, rs3825885 rs12595249, rs4887351, rs1948066 rs12594283, rs12440144, rs4887348 rs1381112, rs999905, rs2009966 rs2009853, rs4887342, rs922231, rs922232, rs1461214, rs7170215, rs4887337, rs1369426, rs7164421, rs1435399, rs3784434, rs1836592, rs3903308, rs1435402, rs1435403, rs11638486, rs1369430, rs2117655, rs7176429, rs1017412, rs3743165		
Neurodevelopmental network		Intron: rs335286, rs11131328, rs10029192, rs335314, rs904243, rs1565901, rs867711, rs1565902, rs13121223, rs10011995, rs2343249, rs958862, rs2172802, rs10001410, rs1497897, rs4552500, rs11735774, rs4860429, rs2122643, rs1868790, rs10446786, rs6551665, rs1947275, rs9683662, rs6858066, rs11131347, rs1470724, rs6813183, rs12503398, rs12648453, rs7667199, rs10030755, rs1397547, rs4860106, rs17226398, rs13115125, rs997407, rs1510921, rs12509655, rs6827266, rs1397546, rs1510924, Exon: rs10434219, rs734644	Presence of association	Ribasés et al. (2011)
Neurodevelopmental network/neurite outgrowth	*NOS1*	Exon 1f-VNTR	SS carriers: ↑ activity in VS; no group × genotype interactions	Hoogman et al. (2011)

(continued)

Table 5.1 (continued)

Systems	Genes	Genetic variants	Results	Ref
Neurodevelopmental network/neurite outgrowth	*NOS1*	Exon 1f-VNTR	Female SS carriers: ↑ MD in right parietal WM tracts Males: No difference between genotype groups No genotype × diagnostic group interaction	van Ewijk et al. (2017)
Noradrenergic	*ADRA2A*	Promoter, rs1800544, rs1800545, rs553668	Absence of association	de Cerqueira et al. (2011)
Noradrenergic	*ADRA2C*	15 kb upstream of start codon: (CA/TG)n	Absence of association	De Luca et al. (2004a)
Noradrenergic	*SLC6A2*	Intron: rs2242447	Presence of association	Pazvantoğlu et al. (2013)
Noradrenergic	*SLC6A2*	Intron: rs998424	Absence of association	De Luca et al. (2004b)
Noradrenergic	*ADRA2A*	rs1800544, rs553668	rs1800544 C-allele carriers: ↓ FA in the right postcentral gyrus rs553668 T-allele carriers: ↓ FA in the right middle frontal cortex	Park et al. (2013)
Noradrenergic	*SLC6A2*	rs998424, rs3785157	NET1 SNPs: No genotype effects on GM or WM volume and no group × genotype interactions	Bobb et al. (2005)
Others	*ANK3*	Intron: rs9804190, rs10994336	Absence of association	Landaas et al. (2010)
Others	*ANKK1*	Exon 8: rs2734849, rs1800497	Absence of association	Mota et al. (2015)
Others	*CACNA1C*	Intron: rs1006737	Absence of association	Landaas et al. (2010)
Others	*CLOCK*	3'UTR: rs1801260	Presence of association	

Others	*CMTM8*	5'UTR: rs9838223, Intron: rs6550109, rs12496256, rs4955272, rs7644602, rs7632109, rs4276227, rs6803740, rs2053280	Presence of association	Weber et al. (2011)
Others	*CTNNA2*	Intron: rs13395022	Absence of association	Salatino-Oliveira et al. (2015)
Others	*DFNB31*	Intron: rs4979416	Absence of association	Weber et al. (2011)
Others	*DGKH*	Intron: rs1170191, rs1170169, rs2148004, rs994856, rs9525580, rs9525584, rs1170101, rs347412, rs347405, rs9315899	Presence of association	Weber et al. (2011)
Others	*DISC1*	Intron: rs1538979	Presence of association	Jacobsen et al. (2013)
Others	*DISC1*	Intron: rs1538979, rs17817356, rs1984895, rs1000731, rs1015101, rs999710, rs821577, rs821633, Exon: rs821616, rs6675281, rs3738401	Presence of association	Jacobsen et al. (2013)
Others	*EGFR*	Intron: rs11773818	Absence of association	Chen et al. (2008)
Others	*FOXP2*	Intron: rs12533005, rs10228350, rs10255943, rs10268637, rs10486026, rs4727799, rs17137124, rs1229761, rs7782412, rs7799652, rs12670585, rs936146, rs10953766	Presence of association	Rebases et al. (2012)
Others	*GRM7*	Intron: rs3792452, rs7623055	Absence of association	Akutagava-Martins et al. (2014)
Others	*KCNIP4*	3'UTR: rs16869915, rs7688805, rs2114474, Intron: rs10084802, rs13121188, rs10938803, rs11726872, rs2322688, rs7689421,	Presence of association	Weißflog et al. (2013)

(continued)

Table 5.1 (continued)

Systems	Genes	Genetic variants	Results	Ref
		rs2291530, rs3816874, rs990206, rs17623902, rs17624591, rs12649113, rs62410693, rs4697192, rs4697193, rs3765119, rs10516363, rs1388321, rs6447994, rs10516367, rs6850182, rs17557419, rs16870168, rs6830607, rs876477, rs1425326, rs6448034, rs17455886, rs1545914, rs1038495, rs16870748, rs16870753, rs16870755, rs983071, rs11726974, rs16870771, rs17460344, rs13110425, rs1503986, rs17547634, rs358834, rs7676963, rs17498614, rs1349383, rs1349384, rs896122, rs16871882, rs7356396, rs1513563, rs12646862, rs6448103, rs7659568, rs2323262		
Others	*MYO5B*	Intron: rs4939921	Absence of association	Landaas et al. (2010)
Others	*NCAM1*	Intron 13: rs646558	Absence of association	Mota et al. (2015)
Others	*NOS1/3*	Intron 4: rs2070744 VNTR rs1799983	Absence of association	Kittel-Schneider et al. (2015)
Others	*NOS1/3*	Exon 1c: VNTR	Presence of association	Hoogman et al. (2011)
Others	*NOS1/3*	Exon 1f: VNTR	Presence of association	
Others	*NPAS3*	Intron: rs17455703	Presence of association	

Others	*NR3C1*	5-HTTLPR, rs6189, rs6198	*NR3C1* risk haplotype carriers: ↑ positive relation between stress exposure and ADHD severity	van der Meer et al. (2016)
Others	*NR3C2*	Exon: rs5522	Absence of association	Kortmann et al. (2013)
Others	*OPRM1*	Exon: rs1799971	Presence of association	Carpentier et al. (2013)
Others	*PPP2R2C*	3'UTR: rs12511742, rs1046319, rs1046322, rs7655674, Exon: rs3796403, Intron: rs4689404, rs4689411, rs4689413, rs6856061, rs11727760, rs4689425, rs4327561, rs6851340, rs16838698, rs755403, rs872858, rs4270639, rs17721635, rs11725306, rs6831981, rs7671165, rs4689001, rs4689440, rs878283, rs16838844, rs9968250, rs4526050, rs6828090, rs4386674	Presence of association	Jacob et al. (2012)
Others	*PRKG1*	3'UTR: rs1881597	Absence of association	De Luca et al. (2002)
Others	*SLC39A3*	Intron: rs4806874	Absence of association	Weber et al. (2011)
Others	*SPOCK3*	3'UT: rs11943562, Intron: rs2318483, rs4602517, rs7689440, rs17052591, rs897511, rs17598103, rs17696409, rs17052602, rs10018183, rs9637685, rs897514, rs7660401, rs7440269, rs17702109, rs17520441, rs13113012, rs10025945, rs6553415, rs17520763, rs17052671, rs11725742, rs13114933, rs13102367, rs1346376, rs1834833, rs17702475, rs1593770, rs6857340,	Presence of association	Weber et al. (2014)

(continued)

Table 5.1 (continued)

Systems	Genes	Genetic variants	Results	Ref
		rs11722292, rs7660050, rs13128738, rs1427635, rs13117458, rs7698061, rs6822214, rs7683298, rs17053121, 5'UTR: rs1485318, rs7694278		
Others	*TSPAN8*	Intron: rs1705236	Absence of association	Landaas et al. (2010)
Others	*TTC12*	Exon: rs723077, Intron: rs2303380	Absence of association	Mota et al. (2015)
Others	*ZNF804A*	Intron: rs1344706	Absence of association	Landaas et al. (2010)
Others		5'UTR		
Serotoninergic	*5-HT1A*	3'UTR: rs878567	Absence of association	Ribasés et al. (2009)
Serotoninergic	*5-HT1D*	5'UTR: rs676643, 3'UTR: rs604030	Absence of association	Ribasés et al. (2009)
Serotoninergic	*5-HT1E*	Intron: rs828358, rs10944288, rs1581774, rs1408449, rs2209639, 3'UTR: rs9344666	Absence of association	Ribasés et al. (2009)
Serotoninergic	*5-HT1F*	5'UTR: rs1503433	Absence of association	Ribasés et al. (2009)
Serotoninergic	*5-HT2A*	Intron: rs17289394, rs2070040, rs9534511, rs2296973, rs1328684, rs2770304, rs9526246, rs1410657, rs2770296, rs4941570, rs2770293, rs6561335, rs2224721, rs7984966, rs9534495, rs2770300, rs1923887, rs6561333, rs6561332, rs7997012, rs7322347	Presence of association	Ribasés et al. (2009)
Serotoninergic	*5-HT2A*	Promoter: rs6311	Presence of association	

Serotoninergic	*5-HT2B*	Intron: rs10194776, rs16827801, rs4973377	Absence of association	Ribasés et al. (2009)
Serotoninergic	*5-HT2C*	5'UTR: rs475717, rs3813928, Exon: rs518147, rs6318, Intron: rs2428721, rs4911871, rs2497503, rs6579495	Absence of association	Ribasés et al. (2009)
Serotoninergic	*5-HT3A*	Intron: rs1150222, rs1985242, rs1176717, rs10160548	Absence of association	Ribasés et al. (2009)
Serotoninergic	*5-HT3B*	5'UTR: rs11214763, Intron: rs12270070, rs1176743, rs11214775, rs1672717, Exon: rs1176744, 3'UTR: rs7129190	Absence of association	Ribasés et al. (2009)
Serotoninergic	*5-HT4*	Intron: rs6865654, rs9686886, rs7721747, rs10223307, rs7711800, rs2910098, rs6873382, rs13166761, rs10515616, rs1368384, rs2005953, rs4597955, rs3995090, rs13156542, 3'UTR: rs4274967	Absence of association	Ribasés et al. (2009)
Serotoninergic	*5-HT5A*	Exon: rs6320, Intron: rs2241859, rs2581841, rs6597455, rs1657268	Absence of association	Ribasés et al. (2009)
Serotoninergic	*5-HT6*	Intron: rs4912138, rs9659997	Absence of association	Ribasés et al. (2009)
Serotoninergic	*5-HT7*	3'UTR: rs1298056, rs11817364, Intron: rs7904560, rs7916403, rs11186320, rs12261011, rs10785973, rs2226116, rs7084468, rs12259062	Absence of association	Ribasés et al. (2009)
Serotoninergic	*HTR1B*	5'UTR: rs130058, Exon: rs6296	Absence of association	Ribasés et al. (2009)
Serotoninergic	*HTR1B*	Exon: rs6296	Absence of association	Carpentier et al. (2013)
Serotoninergic	*SLC6A4*	Promoter: 5HTTLPR	Absence of association	
Serotoninergic	*SLC6A4*	Promoter: 5HTTLPR, Intron: rs140700	Absence of association	

(continued)

Table 5.1 (continued)

Systems	Genes	Genetic variants	Results	Ref
Serotoninergic	*SLC6A4*	Intron: rs2020942, rs140701	Absence of association	Ribasés et al. (2009)
Serotoninergic	*SLC6A4*	Promoter: VNTR	Absence of association	
Serotoninergic	*HTR1B*	5-HTTLPR, rs6296	SLC6A4 SS genotype group: ↓ activation in frontal nodes and "↑ activation in posterior nodes HTR1B genotype: Associated" with differential activation in anterior cingulate, occipital, inferior temporal, and cerebellar regions during successful stop trials No associations between SLC6A4 and HTR1B variants and ADHD or ADHD-related neural activation	van Rooij et al. (2015b)
Serotoninergic	*SLC6A4*	5-HTTLPR	For total GM, differential age effects were found for SLC6A4 L/L carriers, depending on the amount of positive peer affiliation	Richards et al. (2016)
Serotoninergic	*SLC6A4*	5-HTTLPR	S carriers: Stress exposure is associated with ↓ GM volume in the precentral gyrus, middle and superior frontal gyri, frontal pole, and cingulated gyrus Association of GxE interaction with ADHD symptom count was mediated by GM volume in frontal pole and anterior cingulated gyrus only	van der Meer et al. (2015)
Serotoninergic	*SLC6A4*	5-HTTLPR	↑ Positive relation between stress exposure and ADHD severity; interactions were reflected in GM volume of cerebellum, parahippocampal gyrus, intracalcarine cortex, and angular gyrus	van der Meer et al. (2016)

Serotoninergic	*SLC6A4*	5-HTTLPR	SLC6A4 SS genotype group: ↓ activation in frontal nodes and ↑ activation in posterior nodes No associations between SLC6A4 and HTR1B variants and ADHD or ADHD-related neural activation	van Rooij et al. (2015b)
Synaptosomal vesicle mechanism	*CACNA1A*	Intron: rs5906754	Absence of association	Sánchez-Mora et al. (2013)
Synaptosomal vesicle mechanism	*CLPX1/2/4*	3'UTR: rs6811804, rs3733358, Exon 4: rs2242237, Intron: rs4690313, rs3816676, rs2306251, rs11722977, rs10004297, rs7677766, rs9328758, rs11248042, rs11248043, rs7376690, rs6832751, rs7375209, 5UTR: rs11248047, Exon: rs2243404, 5'UTR: rs10866688, Intron: rs7718856, rs6874025, rs12520557, rs11134932, rs4242187, rs10476170, rs17065535, rs6556225, rs1560035, rs12188152, rs4867806, rs11134935, rs4868538, rs10866691, rs4868539, rs890737, rs11134938, Exon: rs1006101, rs10866692, rs930047, 3'UTR: rs13166213, rs11134942, rs2114968, Intron: rs10072860,3'UTR: rs1914321, rs499824, Intron: rs7228681, rs12456930, rs12232757, rs640401, rs509886, rs10503024 Absence of association	Absence of association	Sánchez-Mora et al. (2013)
Synaptosomal vesicle mechanism	*NSF*	Intron: rs17692129, rs7224296, rs17698176	Absence of association	Sánchez-Mora et al. (2013)

(continued)

Table 5.1 (continued)

Systems	Genes	Genetic variants	Results	Ref
Synaptosomal vesicle mechanism	*RAB3A*	3'UTR: rs874628, rs2271881, rs1044821, Intron: rs2271882, rs17683539, 5'UTR: rs2049051	Absence of association	Sánchez-Mora et al. (2013)
Synaptosomal vesicle mechanism	*SNAP-25*	5'UTR: rs1889189, rs6039769, Intron: rs363039, rs363043, rs12626080, rs6074113, rs362547, rs362570, rs6039806, rs3025873, rs362988, rs6108464, rs3787283, rs4813925, 3'UTR: rs6074121, rs6032845, rs3025879, rs6032846	Absence of association	Sánchez-Mora et al. (2013)
Synaptosomal vesicle mechanism	*SNAP-25*	3'UTR: rs3746544, rs1051312	Presence of association: rs3746544	Herken et al. (2014)
Synaptosomal vesicle mechanism	*SNAP-25*	3'UTR: rs3746544, rs1051312	Presence of association	Pazvantoğlu et al. (2013)
Synaptosomal vesicle mechanism	*SNAP-25*	3'UTR: rs3746544, rs1051312, rs8636	Absence of association	Olgiati et al. (2014)
Synaptosomal vesicle mechanism	*SNPH*	Intron: rs6109320, rs2317645, rs912105, rs7354385, rs11905792, Exon: rs6134520, rs3803947, rs3764715, 3'UTR: rs4814106, rs2281711, rs1048621	Absence of association	Sánchez-Mora et al. (2013)
Synaptosomal vesicle mechanism	*STX1A*	Intron: rs2293485, rs941298	Presence of association	Olgiati et al. (2014)
Synaptosomal vesicle mechanism	*STX1A*	Intron: rs35459363	Presence of association	Kenar et al. (2014)
Synaptosomal vesicle mechanism	*STX1A*	Intron: rs941298, rs2293485, rs4363087, Exon: rs3793243	Presence of association	Sánchez-Mora et al. (2013)
Synaptosomal vesicle mechanism	*STXBP1*	5'UTR: rs2039204, rs7852204, Intron: rs2241167	Absence of association	Sánchez-Mora et al. (2013)

Synaptosomal vesicle mechanism	*SYN3*	Promoter: rs133945, rs13394	Presence of association: rs13394	Kenar et al. (2013)
Synaptosomal vesicle mechanism	*SYP*	Intron: rs2293945, 5'UTR: rs5906754	Absence of association	Sánchez-Mora et al. (2013)
Synaptosomal vesicle mechanism	*SYT1*	Intron: rs2251214	Absence of association	Sánchez-Mora et al. (2013)
Synaptosomal vesicle mechanism	*SYT2*	Intron: rs12739678, rs907697, rs9633344, rs6427957	Presence of association	Sánchez-Mora et al. (2013)
Synaptosomal vesicle mechanism	*VAMP-1*	Exon rs12964, rs2072376, Intron: rs2534717, rs2240867, 5'UTR: rs11064213, rs10492096, rs2534724	Absence of association	Sánchez-Mora et al. (2013)
Synaptosomal vesicle mechanism	*VAMP-2*	3'UTR: rs772100314 (26 bp Ins/Del)	Presence of association	Kenar et al. (2014)
Synaptosomal vesicle mechanism	*VAMP-2*	5'UTR: rs8067606	Absence of association	Sánchez-Mora et al. (2013)

Abbreviations: *VNTR* variable number of tandem repeats, *UTR* untranslated region

Kochunov et al. 2015; Glahn et al. 2010; Blokland et al. 2014). It was also shown that genetics plays an important role in the development and function of the brain. In a review paper with an emphasis on brain circuits and genetic variations involved in the development of ADHD symptoms, the authors examined the relationship between circuitry anomalies and symptom presentation and therapy for the condition. The authors recommended that, in order to untangle ADHD causation, sophisticated research in animal models, neuromodulation, and discoveries based on pharmaco-imaging should be conducted in both basic and clinical settings (Gallo and Posner 2016).

5.2.1 Genes Linked to Anatomical Brain Alterations

Anatomical alterations in different neurological and psychiatric diseases, such as ADHD, might be studied using structural magnetic resonance imaging (MRI) for a better understanding of the impact of gene polymorphisms on anatomical changes. Global and regional brain volumes and subcortical structures that are smaller than normal, as well as multiple-gene polymorphisms, have been identified as risk factors for ADHD (Table 5.1). One such example of a gene polymorphism is the dopamine transporter solute carrier family 6-member 3 (*SLC6A3*) gene, which encodes the transmembrane proteins involved in the reuptake of dopamine from the synapse. Dopamine is a neuromodulator produced primarily in the frontal and striatal regions of the brain and is thought to have a role in cognitive performance. *SLC6A3* plays a vital role in the pathophysiology of a variety of mental conditions, including ADHD (Fig. 5.3).

SLC6A3 haplotype is linked with reduced grey matter volume in several brain regions, including the left superior occipital region, cuneus, precuneus, and right angular areas, according to research on children with ADHD. It has been proposed that anomalies in these brain areas are blamed for the visual memory deficit seen in ADHD children (Shang et al. 2018). Atomoxetine helps to increase grey matter in the prefrontal and occipital areas of ADHD children with the 10/10-Repeat allele with a variable number tandem repeat of 40 bp of the *SLC6A3* genotype associated with ADHD when compared to typically developing, long-term treatment with psychostimulant drugs like methylphenidate (MPH), which inhibits dopamine and norepinephrine reuptake.

According to a study based on cortical thickness measurements in ADHD patients, those with two copies of the 10-Repeat (10R) allele in the *SLC6A3* gene had lower cortical thickness in the right lateral prefrontal cortex than those with one or no copies of the 10R allele (Fernandez-Jaen et al. 2015). Another volumetric based research found that carriers of the *SLC6A3* haplotype 9–6 had approximately 6% larger striatum volume than noncarriers (Onnink et al. 2016), when examining the impact of three *SLC6A3* alleles (10/10 genotype, haplotypes 10–6 and 9–6) on striatum volume in ADHD patients. Fernandez-Jaen et al. (2015) discovered that ADHD children who were homozygous for *SLC6A3* with a 10R allele had substantially greater cortical thicknesses in the right ventral and cingulate gyrus than

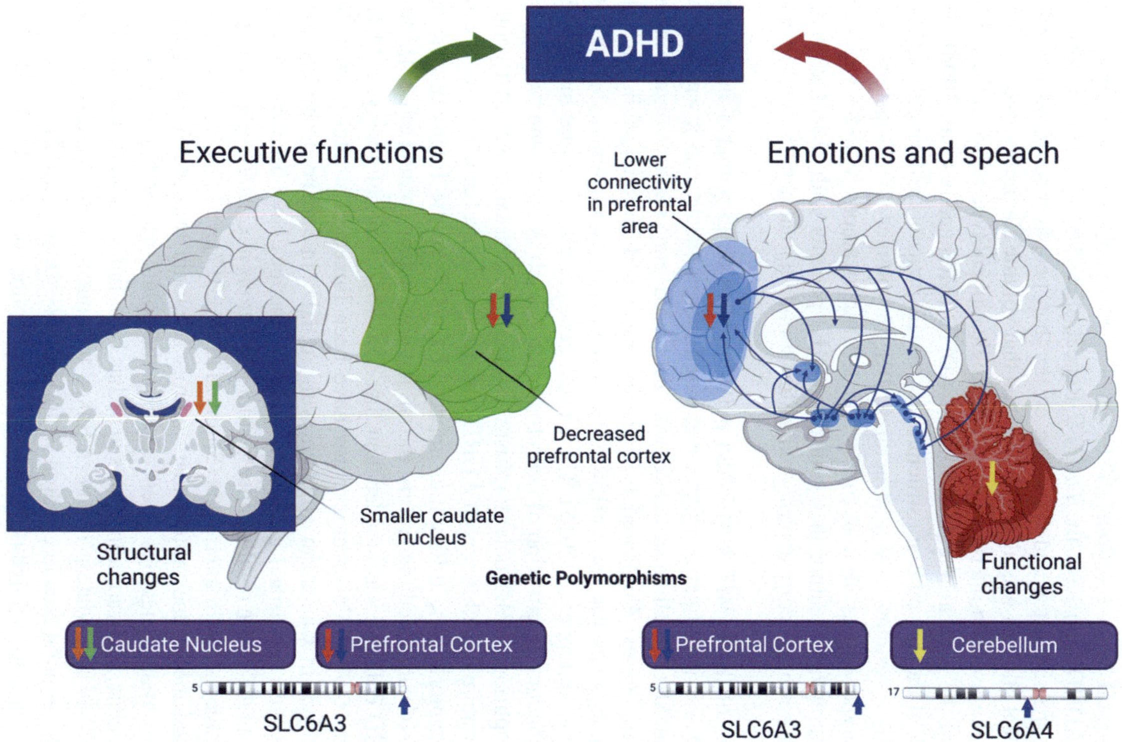

Fig. 5.3 *SLC6A3* polymorphisms are linked to a smaller caudate nucleus and prefrontal cortex in ADHD individuals. *SLC6A3* and *SLC6A4* polymorphisms in ADHD patients' brains are linked to decreased functional activity in the prefrontal cortex and cerebellum (figure adapted from Tripp G et al. (Herken et al. 2014))

9-Repeat carriers in another seminal research. The presence of the 10R allele in ADHD might also affect the cortical thickness of the cingulate gyrus (Fernandez-Jaen et al. 2018), according to the scientists.

In children with ADHD, research found a link between the *DRD4*-7-repeat allele and reduced cortical thickness in the right orbitofrontal/inferior prefrontal and posterior parietal brain locations (Shaw et al. 2007). Another study found that those with ADHD who have the *DRD4*-7-Repeat allele had a smaller volume of the superior frontal cortex and cerebellar cortex than people who do not have the *DRD4*-7-repeat allele. It was also hypothesized that volume alterations in the ADHD brain might represent an intermediate morphological phenotype between the *DRD4* genotype and the clinical manifestations of ADHD (Monuteaux et al. 2008). The impact of MPH therapy on brain structures in ADHD patients harboring the *DRD4*-7R allele was investigated using structural MRI. In younger ADHD patients with *DRD4* genotype, higher frontal cortex and left hippocampus volume was seen following therapy, indicating that younger ADHD patients with *DRD4* genotype are more susceptible to cortical remodeling after stimulant treatment (Schweren et al. 2016).

The effect of the *COMT* Val158Met polymorphism on grey matter in children with ADHD was investigated using voxel-based morphometry, and it was discovered that the Met158 allele is linked with decreased grey matter volume in the inferior frontal gyrus. Compared to typically developed children, children with ADHD homozygotes for the Val158 allele had more grey matter in the caudate nucleus (Villemonteix et al. 2015). The *COMT* gene controls enzymes involved in the production and deactivation of neurotransmitters such as dopamine, adrenaline, and norepinephrine. *COMT* polymorphism resulted in decreased grey matter volume in the left striatum in children with ADHD compared to control children. According to a study exploring the differential impact of *COMT* on the brains of children with ADHD of various ethnic groups, the striatal grey matter volume of *COMT* Met carrier ADHD children was smaller than that of *COMT* Val/Val genotype ADHD children. The *COMT* Val homozygotes have been linked to striatal grey matter volume changes in Caucasian children with ADHD. In Japanese children with ADHD, however, the striatal grey matter volume changes have been linked to the *COMT* Met allele. These results indicate ethnic variations in *COMT*'s genetic impact on ADHD patients' brain alterations (Shimada et al. 2017).

The crucial membrane protein that controls serotonin transport from synaptic gaps into presynaptic neurons is encoded by the solute carrier family 6-member 4 (*SLC6A4*) gene, and variation in this gene has been linked to an increased risk of ADHD. *SLC6A4* methylation has been linked to reduced cortical thickness in the right occipitotemporal area in children with ADHD (Park et al. 2015).

In ADHD patients, a polymorphism in the gene coding for a synaptosomal related protein (*SNAP25*) was linked to changes in grey matter volume (Soderqvist et al. 2010). The *DBH* gene, which is involved in producing the enzyme dopamine β-hydroxylase, which converts dopamine to norepinephrine, is thought to be important in the autonomic nervous system. *DBH* gene polymorphism is related to greater left insula surface area in children with ADHD with G carriers than AA homozygotes, according to a surface measurement-based study (Li et al. 2015).

5.2.2 Genes Linked to White Matter Brain Alterations

White matter is highly heritable in the human brain and is a crucial driver of interindividual variations in brain functions, including cognition, and contributing to neuropsychiatric diseases. ADHD is linked to the brain's anatomical makeup and architecture, such as white matter connections and gene polymorphism. Lesser evidence is currently available on the relationship between alterations in white matter architecture and gene polymorphisms in ADHD (Table 5.1). Diffusion tensor imaging (DTI) was utilized by Hong and colleagues to assess white matter connectivity in ADHD patients who were *COMT* Val homozygous or *COMT* Met carriers. They discovered that ADHD individuals who were *COMT* Met carriers had less white matter connections than those who were *COMT* Val homozygous (Hong et al. 2015). In another DTI research, children with ADHD who were homozygous for *COMT* Val showed substantially decreased fractional anisotropy and higher radial diffusivity in the right cingulate gyrus when compared to *COMT* Met carriers and healthy controls who were homozygous for *COMT* Val. In addition, compared to children with ADHD with homozygote *COMT* Val, children with ADHD with *COMT* Met carriers exhibited higher fractional anisotropy and axial diffusivity in the left uncinate fasciculus and reduced radial diffusivity in the left posterior corona radiata and posterior thalamic radiation. The above results indicated that the *COMT* polymorphism affected the development of white matter in Val homozygote ADHD (Kabukcu Basay et al. 2016). Another study found a substantial increase in mean diffusivity in the grey and white matter regions when the *DRD4*-5-repeat allele was expressed, which might be a risk factor for developing ADHD in children (Takeuchi et al. 2015).

5.2.3 Genes Linked to Functional Brain Alterations

The functional brain activity in individuals with ADHD has been found to be influenced by genetic differences, particularly gene polymorphisms of different genes such as *DRD4*, *SLC6A3*, *DRD1*, neuroepithelial cell transforming 1 (*NET1*), and others (Table 5.1) (Hoogman et al. 2013; Bedard et al. 2010; Paloyelis et al. 2012; Sokolova et al. 2015; Ko et al. 2018; van Rooij et al. 2015c). In comparison to children with ADHD who had the 2-repeat allele, a study utilizing resting-state functional magnetic resonance imaging (rs-fMRI) found that the lack of a 2-repeat allele of the *DRD4* gene is linked with hyperconnectivity in the default mode network and sensorimotor network and hypoconnectivity in the executive control network in children with ADHD. The finding indicates that the *DRD4*-2-repeat allele polymorphism affects the network connectivity linked to inattention activity (Qian et al. 2018a). Another study that examined the impact of the *DRD4* (4R/4R vs. 2R) gene polymorphism on regional homogeneity (ReHo) and functional connectivity in ADHD patients showed that the *DRD4*-2R allele enhanced and reduced ReHo bilaterally in the cerebellum and left angular gyrus, respectively. Patients with the *DRD4*-2R allele had decreased functional connectivity in the left striatum, right inferior frontal gyrus, bilateral lobes of the cerebellum, and higher functional connectivity in the left superior frontal, medial frontal, and rectus gyrus. Based on their results, the authors concluded that *DRD4* polymorphisms are linked to localized brain activity and particular functional connections (Qian et al. 2018b). Another study using rs-fMRI found that ADHD individuals with the *SLC6A3* polymorphism *SLC6A3* rs27048 (C)/rs429699 (T) haplotype had lower functional connectivity/ReHo in the left superior occipital gyrus, cuneus, and precuneus. In the right postcentral gyrus, there were significant interactions between ADHD disorder status (diagnosis) and CT haplotype with reduced ReHo/functional connectivity (Shang et al. 2018).

Functional connectivity in ADHD has been reported to be influenced by the *N*-methyl-D-aspartate (NMDA) and dopamine receptor genes. Using rs-fMRI, researchers looked at the impact of NMDA receptor gene glutamate ionotropic receptor NMDA type subunit 2B (*GRIN2B*) and dopamine receptor gene (*DRD4*) variations on ReHo in ADHD patients and healthy controls. According to the researchers, the ADHD group with the *GRIN2B* TC/TT genotype had lower static and dynamic ReHo in the left superior parietal surface than the healthy controls. In the right superior parietal surface, the ADHD group with the *DRD4* variable number tandem repeat (VNTR) 2R had reduced dynamic ReHo. Given the importance of the superior parietal area in the selective attention process, lower static and dynamic ReHo in the ADHD group in the superior parietal region might result in poor performance during active states and a reduced capacity to react to attention-based activities. The above research findings indicated that changes in the dopaminergic and glutamatergic systems contributed to decreased local functional connectivity in ADHD patients, resulting in attention deficits (Kim et al. 2018).

The response inhibition task was measured using a stop-signal task-based fMRI research in people with ADHD-risk alleles in the *DRD4* and *SLC6A3* genes. Carriers of the *DRD4*-7-repeat allele had lower activation in the superior frontal and middle gyrus during successful response inhibition and lower activation in the supramarginal gyrus and parietal lobule after unsuccessful response inhibition, according to the researchers. During unsuccessful trials of effective response inhibition, *SLC6A3* risk variants exhibited decreased cerebellar activity (van der Meer et al. 2017). Another study looked for relationships between potential plasticity genes (*SLC6A3*, SLC6A4, and *DRD4*) and social contexts (maternal expressed emotion and peer affiliation). The study found that in serotonin-transporter-linked polymorphic region (HTTLPR) short allele carriers, exposure to high positive peer affiliation, were associated with the least reward speeding. In contrast, exposure to low positive peer affiliation or low maternal warmth was associated with the most reward speeding. *SLC6A3* 10-repeat homozygote, on the other hand, had the longest response times when exposed to low maternal warmth. *DRD4*-7-repeat carriers, on the other hand, exhibited increased brain activity when exposed to low maternal warmth and vice versa. These results, taken together, highlight the importance of supportive social settings in sensitivity rewards and task performance, with various genotypes displaying varied responses to environmental impacts (Richards et al. 2016).

ADHD heterogeneity was discovered to be one of the major issues in a study evaluating the neurobiological pinning of a new ADHD phenotype known as ADHD-restrictive inattentive (ADHD-RI). Children with ADHD-RI had slower psychomotor speeds and more activation of the temporo-occipital regions during the Go/No-Go test when compared to typically developed children. In addition, *DRD4*-7-repeat allele was found to be more prevalent in ADHD-RI youngsters (Ercan et al. 2016). Another study examined the impact of the *DRD4*-5-repeat allele on microstructural characteristics and functional connectivity in the brain in a healthy Asian population, mainly made up of adolescents, using DTI and N-back fMRI paradigms. The presence of a 5-repeat allele was linked to poor processing speed, increased impulsivity, and a decreased ability to sustain attentional concentration in the research, indicating that the 5-repeat allele of the *DRD4* gene might contribute to the risk of developing ADHD (Takeuchi et al. 2015). Gilsbach and colleagues used a combined stimulus-response incompatibility test (IC) and a time-discrimination task (TT) to examine the impact of the *DRD4*-7-repeat allele in a healthy sample of children and adolescents. The *DRD4*-7-repeat carriers had lower neural activity of the middle and frontal gyrus in IC and lowered cerebellar activation in TT, according to the research. In addition, as compared to 7-repeat noncarriers, *DRD4*-7 carriers exhibited less coupling between frontal brain regions (Gilsbach et al. 2012). The impact of the *SLC6A3* VNTR polymorphism on the brain's activity in a working memory test in children with ADHD and typically developed children was studied using fMRI. Working memory-related activation was more significant in 9R carriers in ADHD subjects, and only 10R homozygote subjects showed higher working memory-related activation than 9R carriers in multiple brain sites in children with ADHD, including the parietal, temporal, and ventral lobes; ventral visual cortex; orbitofrontal gyrus; and head of the caudate nucleus. The results indicated that the existence of the *SLC6A3* polymorphism might have a substantial impact on the working memory of children with ADHD (Pineau et al. 2019). The impact of one (9/10) copy of the 10-repeat allele of the *SLC6A3* genotype in typically developed children was investigated using a verbal n-back task in two fMRI sessions. In a high-load run, 9/10 carriers exhibited more significant activity in the frontal–striatal–parietal areas than 10/10 carriers. Under low stress, however, subthalamic nuclei were more active in 10/10 carriers, indicating that *SLC6A3* 10R homozygosity is linked to poor performance in more demanding working memory tasks (Stollstorff et al. 2010). Another task-based research study used the Go and No-Go paradigm to see how *SLC6A3* 3'UTR genotype polymorphisms affected brain activity in adolescents with ADHD and children who were not on medication. Youth with the *SLC6A3* 3′UTR 10R/10R genotype exhibited greater activity in the left striatum, right dorsal premotor cortex, and bilaterally in the temporoparietal cortical junction when compared to ADHD patients who were heterozygous for the *SLC6A3* 3′UTR 9R allele. The above results suggested that in children with ADHD, brain activity linked to inhibitory control might vary according to the *SLC6A3* 3′UTR genotype (Bedard et al. 2010).

Dopamine and serotonin-related genes are essential in the neurobiological response to inhibition and large-scale brain activity alterations in ADHD patients.

A study reported large-scale alterations in the brain activity of response inhibition networks in prefrontal, parietal, and subcortical areas in ADHD patients in relation to *SLC6A3* and *COMT* polymorphisms. In a similar research study, numerous variations of HTR1B and solute carrier family 6 member 4 (*SLC6A4*), formerly known as 5HTT genes, revealed large-scale changes in neural activity in the frontal and parietal areas of the response inhibition network in ADHD patients (van Rooij et al. 2015a, b).

Researchers investigated whether the *COMT* polymorphism was linked to changes in cortico-cerebellar executive function in children with ADHD using rs-fMRI. ADHD *COMT* Met carriers had lower functional connectivity of right Crus I/II with the left dorsolateral prefrontal cortex when compared to ADHD children with Val/Val genotype (Mizuno et al. 2017). Brown and colleagues found that ADHD patients who were homozygous for the *SLC6A3* 10R allele had significantly lower activity in the left dorsal anterior cingulate cortex, lateral prefrontal cortex, and cerebellar vermis than *SLC6A3* 9R carriers. The *SLC6A3* 9R carriers exhibited more activity in the left dorsal anterior cingulate cortex, cerebellar vermis, and right lateral prefrontal cortex when compared to *SLC6A3* 10R carriers (Brown et al. 2010). When rs3746544 TT homozygous carriers were compared to rs3746544 G-allele carriers in a study looking at the relationship between synaptosomal associated protein (SNAP25) rs3746544 polymorphism and functional connectivity density in male children with ADHD, they found that rs3746544 TT homozygous carriers had fewer local functional connectivity hubs in the dorsal lateral prefrontal cortex and anterior cingulate cortex, as compared to rs3746544 G-allele carriers, indicating the implication of SNAP25 polymorphism in ADHD (Wang et al. 2018). Another study using a stop-signal fMRI task in adolescent boys and girls found that *MAOA* was associated with ADHD symptoms and that SNP rs12843268 "A" hemizygotes reduced *MAOA* levels and reduced ventral striatal BOLD response during the monetary incentive delay task. During the monetary incentive delay test, "G" hemizygotes of SNP rs12843268 were linked with greater *MAOA* levels and frontal gyrus and ventral striatal hyperactivation, and frontal gyrus hypoactivation during the stop-signal task (Nymberg et al. 2013). Working memory tests were performed in individuals with ADHD in another study using fMRI in order to examine how *MAOA* polymorphisms affect working memory, distraction, and dual tasking. In response to dual tasking, the authors discovered increased activation for working memory in the lower bilateral frontal lobe and pars opercularis, as well as increased activity in the lingual gyrus compared to control subjects (Ko et al. 2018).

5.3 Pharmacogenetic Insights in ADHD

Both genetic and epigenetic factors govern and regulate the amounts of brain metabolites or neurotransmitters. Changes in brain metabolite levels might be caused by gene abnormalities or polymorphisms in ADHD due to variations in the cortico-striato-thalamic-cortical networks.

Polymorphisms in *DRD4* (120 bp insertion/deletion, 48 bp VNTR) and *SLC6A2* (4 bp insertion/deletion in the promoter region) have been investigated in adults treated with immediate-release methylphenidate (IR-MPH) (Kooij et al. 2008). However, no associations were found with the outcomes of treatment response (Kooij et al. 2008) (Table 5.2). Negative associations were also found for *ADRA2A* (rs1800544, rs1800545, and rs553668) (Contini et al. 2011) (Table 5.2).

Research using proton magnetic resonance spectroscopy (^{1}H MRS) to examine the connection between *SLC6A3* gene polymorphisms and brain metabolite responses in individuals with ADHD after taking the MPH medication showed no significant changes in *N*-acetylaspartate (NAA), total creatine tCr, or choline-containing compound (Cho) levels (Table 5.2). Only high amounts of tCr were seen in the cerebellum of ADHD patients with the *SLC6A3* 10R genotype following MPH administration (Inci Kenar et al. 2016) (Table 5.2). The restoration of cerebral blood flow and glucose metabolism after psychostimulant treatment in ADHD might have contributed to the rise in Cr levels (hypermetabolic state) after MPH delivery. Recent studies showed that the metabolism of both Methylphenidate and Atomoxetine is influenced by Cytochrome P450 (CYP)2D6 polymorphisms (Brown et al. 2019; Bishop et al. 2021). Atomoxetine is a selective norepinephrine reuptake inhibitor (SNRI). Variant-specific prescribing guidance is reported in clinical pharmacogenetic implementation consortium (CPIC). The CPIC guideline described the dosing recommendations for atomoxetine based on CYP2D6 genotype for children (Brown et al. 2019; Bishop et al. 2021).

Another study utilizing MRS research looked at brain metabolite responses to MPH therapy in ADHD individuals with the *SNAP25* gene polymorphism. After MPH treatment, individuals with SNAP25 DdeI (rs1051312) and SNAP-25 MnlI (rs3746544) polymorphisms in the anterior cingulate brain area had high levels of NAA, indicating that alterations linked to MPH might affect NAA levels in ADHD (Unal et al. 2016).

Another MRS research explored how MPH affected the levels of neurometabolites in ADHD individuals with synapsin III (*SYN3*) gene polymorphisms. The researchers discovered greater Cho levels in the striatum of ADHD patients with the synapsin III rs133945 polymorphism, and a higher NAA level in those with the synapsin III rs133945 polymorphism (Basay et al. 2016). A similar research study used MRS to examine how MPH medication affected brain metabolite levels in ADHD patients who had a *COMT* gene polymorphism. After treatment with MPH, increased levels of NAA were found in the anterior cingulate cortex and prefrontal dorsolateral cortex regions of Val/Val and Val/Met genotype (rs4680) carriers. Also, elevated levels of Cho in the striatum of Val/Val and Val/Met genotype (rs4680) carriers were seen indicating that MPH improved neuronal function and activity (Ozturk et al. 2016).

5.4 Metabolomic Insights in ADHD

There have been relatively few studies on the biochemical levels of proteins implicated in ADHD. Some studies looked at circadian rhythms by examining melatonin levels (Baird et al. 2012; Van Veen et al. 2010), while others investigated the oxidative stress pathway (Bulut et al. 2013; Karababa et al. 2017). A summary of metabolomic studies in ADHD is presented in Table 5.3.

Table 5.2 Summary of pharmacogenetic studies on ADHD

Systems	Genes	Location	Polymorphisms	Populations	Sample size		Dosages	Duration times	Results	Ref
					R	*NR/PL*				
Dopaminergic	*SLC6A3*	3′UTR	40 bp VNTR	Netherlands	16	26	0.5 mg kg^{-1} per day; 1 mg kg^{-1}per day	3 weeks	Absence of association	Kooij et al. (2008)
Dopaminergic	*SLC6A3*	3′UTR	40 bp VNTR	Brazil	136	35	0.3 mg kg^{-1} per day	4 weeks	Absence of association	Contini et al. (2010)
Dopaminergic	*SLC6A3*	3′UTR	40 bp VNTR	Boston	66	40	0.5 mg kg^{-1} per day; 1 mg kg^{-1}per day	6 weeks	Absence of association	Mick et al. (2006)
Dopaminergic	*SLC6A3*	5′UTR	rs2652511	Brazil	136	35	0.3 mg kg^{-1} per day	4 weeks	Absence of association	Contini et al. (2010)
Dopaminergic	*SLC6A3*	Intron 8	30 bp VNTR	Brazil	136	35	0.3 mg kg^{-1} per day	4 weeks	Absence of association	Contini et al. (2010)
Dopaminergic	*DRD4*	Exon 3	48 bp VNTR	Netherlands	16	26	0.5 mg/kg/day	3 weeks	Absence of association	Kooij et al. (2008)
Dopaminergic	*DRD4*	Exon 3	48 bp VNTR	Netherlands	16	26	1 mg/kg/day	3 weeks	Absence of association	Kooij et al. (2008)
Dopaminergic	*DRD4*	Exon 3	48 bp VNTR	Brazil	128	27	0.3 mg/kg/day	4 weeks	Absence of association	Contini et al. (2012)
Dopaminergic	*DRD4*	Promoter	dup120bp	Netherlands	16	26	0.5 mg/kg/day-1 mg/kg/day	3 weeks	Absence of association	Kooij et al. (2008)

Serotoninergic	*HTR1B*	Exon 1	rs11568817, rs6296, rs13212041	Brazil	136	28	0.3 mg/kg/day	4 weeks	Absence of association	Contini et al. (2012)
Serotoninergic	*SLC6A4*	Promoter	5-HTTLPR	Brazil	136	28	0.3 mg/kg/day	4 weeks	Absence of association	Contini et al. (2012)
Noradrenergic	*SLC6A2*	Promoter	4 bp ins/del	Netherlands	16	26	0.5 mg/kg/day	3 weeks	Absence of association	Kooij et al. (2008)
Noradrenergic	*SLC6A2*	Promoter	4 bp ins/del	Netherlands	16	26	1 mg/kg/day	3 weeks	Absence of association	Kooij et al. (2008)
Noradrenergic	*ADRA2A*[e]	Promoter, promoter, 3UTR	rs1800544, rs1800545, rs553668	Brazil	137	28	0.3 mg/kg/day	4 weeks	Absence of association	Contini et al. (2011)
Metabolism enzymes	*TPH2*	Intron, promoter	rs1843809, rs4570625	Brazil	136	28	0.3 mg/kg/day	4 weeks	Absence of association	Contini et al. (2012)
Metabolism enzymes	*DBH*	5'UTR	rs1611115	Brazil	136	28	0.3 mg/kg/day	4 weeks	Absence of association	Contini et al. (2012)
Metabolism enzymes	*COMT*	Exon 4	rs4680	Brazil	136	28	0.3 mg/kg/day	4 weeks	Absence of association	Contini et al. (2012)
Synaptosomal vesicle mechanism	*SNAP25*	3'UTR, intron	rs3746544, rs363020	Brazil	136	28	0.3 mg/kg/day	4 weeks	Absence of association	Contini et al. (2012)

Abbreviations: *NR* no responders, *PL* placebo, *R* responders

Table 5.3 Summary of metabolomic studies on ADHD

Systems	Biomarkers	Biological fluids	Populations	Sample size ADHD (N)	Sample size CTRL (N)	Results	Ref
Circadian rhythms	Cortisol	Saliva	Swedish	28	28	No alterations	Hirvikoski et al. (2009)
Circadian rhythms	Cortisol	Saliva	Germany	18	18	No alterations	Lackschewitz et al. (2008)
Circadian rhythms	Cortisol	Saliva	UK	13	19	Alterations	Baird et al. (2012)
Circadian rhythms	Cortisol	Saliva	Israel	24	25	No alterations	Raz and Leykin (2015)
Polyunsaturated fatty acids	AA	RBC	Canada	36	35	No alterations	Young et al. (2004)
Polyunsaturated fatty acids	AA	Serum	Canada	36	35	No alterations	Young et al. (2004)
Polyunsaturated fatty acids	AA	Serum	Finland	26	36	No alterations	Laasonen et al. (2009)
Polyunsaturated fatty acids	AA	Serum	Germany	15	15	Alterations	Irmisch et al. (2013)
Polyunsaturated fatty acids	DHA	RBC	Canada	36	35	Alterations	Young et al. (2004)
Polyunsaturated fatty acids	DHA	Serum	Canada	36	35	Alterations	Young et al. (2004)
Polyunsaturated fatty acids	DHA	–	Finland	26	36	No alterations	Laasonen et al. (2009)
Polyunsaturated fatty acids	DHA	–	Germany	15	15	Alterations	Irmisch et al. (2013)
Polyunsaturated fatty acids	EPA	RBC	Canada	36	35	No alterations	Young et al. (2004)
Polyunsaturated fatty acids	EPA	Serum	Canada	36	35	No alterations	Young et al. (2004)
Polyunsaturated fatty acids	EPA		Finland	26	36	No alterations	Laasonen et al. (2009)

Polyunsaturated fatty acids	EPA	–	Germany	15	15	Alterations	Irmisch et al. (2013)
Polyunsaturated fatty acids	DHLA	RBC	Canada	36	35	No alterations	Young et al. (2004)
Polyunsaturated fatty acids	DHLA	Serum	Canada	36	35	No alterations	Young et al. (2004)
Polyunsaturated fatty acids	DHLA	–	Finland	26	36	No alterations	Laasonen et al. (2009)
Polyunsaturated fatty acids	DHLA	–	Germany	15	15	Alterations	Irmisch et al. (2013)
Circadian rhythms	Melatonin	Saliva	Netherlands	34	24	Alterations	
Circadian rhythms	Melatonin	Saliva	UK	13	19	Alterations	Baird et al. (2012)
Oxidative pathway	AREase	Serum	Turkey	35	29	Alterations	
Oxidative pathway	Folate	Plasma	Turkey	32	32	Alterations	Karababa et al. (2017)
Oxidative pathway	Hcy	Plasma	Turkey	32	32	Alterations	Karababa et al. (2017)
Oxidative pathway	MDA	Plasma	Turkey	20	21	Alterations	Bulut et al. (2007)
Oxidative pathway	MDA	Serum	Turkey	35	29	Alterations	Bulut et al. (2007)
Oxidative pathway	NO	Serum	Turkey	20	21	Alterations	Selek et al. (2008)
Oxidative pathway	NO	Serum	Germany	14	33	No alterations	Kittel-Schneider et al. (2015)
Oxidative pathway	OS	Plasma	Turkey	50	31	Alterations	Selek et al. (2012)
Oxidative pathway	OSI	Plasma	Turkey	32	32	No alterations	Karababa et al. (2017)
Oxidative pathway	PON	Serum	Turkey	35	29	Alterations	Bulut et al. (2013)
Oxidative pathway	SOD	Serum	Turkey	20	21	Alterations	Selek et al. (2008)
Oxidative pathway	TAS	Plasma	Turkey	50	31	Alterations	Selek et al. (2012)
Oxidative pathway	TAS	Plasma	Turkey	32	32	No alterations	Karababa et al. (2017)
Oxidative pathway	TOS	Plasma	Turkey	50	31	Alterations	Selek et al. (2012)
Oxidative pathway	TOS	Plasma	Turkey	32	32	No alterations	Karababa et al. (2017)

(continued)

Table 5.3 (continued)

Systems	Biomarkers	Biological fluids	Populations	Sample size ADHD (N)	Sample size CTRL (N)	Results	Ref
Oxidative pathway	Vitamin B12	Plasma	Turkey	32	32	No alterations	Karababa et al. (2017)
Polyunsaturated fatty acidsa	Fatty acids	Serum	Canada	26	–	FA treatment	Young et al. (2005)
Polyunsaturated fatty acidsa	OL	Serum	Finland	26	36	No alterations	Laasonen et al. (2009)
Polyunsaturated fatty acidsa	OL	Serum	Germany	15	15	Alterations	Irmisch et al. (2013)
Polyunsaturated fatty acidsa	PAOL	Serum	Finland	26	36	No alterations	Laasonen et al. (2009)
Polyunsaturated fatty acidsa	PAOL	Serum	Germany	15	15	Alterations	Irmisch et al. (2013)
Others	Adiponectin	Serum	Norway	44	29	Alterations	Mavroconstanti et al. (2014)
Others	Albumin	Serum	Germany	20	20	Alterations	Grabemann et al. (2014)
Others	*BDNF*[q]	Serum	Spain	54	59	Alterations	Corominas-Roso et al. (2013)
Others	*BDNF*[q]	Serum	Brazil	54	–	Atomoxetine treatment	Ramos-Quiroga et al. (2014b)
Others	HVA and 5-HIAA	CSF[t]	USA	36	30	No alterations	Reimherr et al. (1984)
Others	Monoamines	Plasma	USA	36	–	Selegiline treatment	Ernst et al. (1997)
Others	Monoamine metabolites	Plasma	USA	36	–	Selegiline treatment	Ernst et al. (1997)

Abbreviations: *AA* Arachidonic acid, *DHA* docosahexaenoic acid, *EPA* eicosapentaenoic acid, *DHLA* dihomogammalinolenic acid, *MDA* malondialdehyde, *NO* nitric oxide, *OS* oxidative status, *PON* paraoxonase, *SOD* superoxide dismutase, *TAS* total antioxidative status, *TOS* total oxidative status, *PAOL* palmitoleic acid, *BDNF* brain-derived neurotrophic factor, *HVA* homovanillic acid, *5-HIAAs* 5-hydroxyindoleacetic acid

A groundbreaking study showed that plasma homocysteine levels were lower, and serum folic acid levels were greater in adult ADHD patients compared to controls, but no difference was observed in blood vitamin B12, total antioxidant status, or oxidative stress index (Karababa et al. 2017) (Table 5.3). By contrast, another study found that people with ADHD had substantially greater total antioxidant status, total oxidant status, and oxidative stress index than controls (Selek et al. 2012) (Table 5.3). Bulut and colleagues (Bulut et al. 2007) found differences in malondialdehyde levels between ADHD cases and controls. In subsequent research, the same authors found that adults with ADHD had greater malondialdehyde levels and lower paraoxonase and arylesterase levels when compared to controls (Bulut et al. 2013). Finally, a study found that patients' mean nitric oxide metabolite levels were considerably greater than controls, but their superoxide dismutase activity was much lower (Selek et al. 2008). In another study, adults with ADHD did not demonstrate any significant changes in the levels of nitrogen oxides (NOx) when compared to controls (Kittel-Schneider et al. 2015). In addition, serum *BDNF* (Corominas-Roso et al. 2013), adiponectin (Mavroconstanti et al. 2014), albumin (Grabemann et al. 2014), and cerebrospinal fluid metabolites homovanillic acid and 5-hydroxyindoleacetic acid are further important peripheral indicators (Reimherr et al. 1984). A recent meta-analysis reported no significant changes in salivary cortisol levels between adults with ADHD and controls, although individuals with ADHD had lower serum DHA levels. Even after Bonferroni adjustment, the latter relationship remained significant, and heterogeneity was not significant. Other polyunsaturated fatty acids (PUFAs), such as arachidonic acid (AA), eicosapentaenoic acid (EPA), and dihomogammalinolenic acid (DHLA), had no difference in blood levels between patients and controls (Table 5.3).

5.5 Current Challenges and Future Perspective

The accurate and exact diagnosis or profile of ADHD symptoms is indeed a dilemma. The significant variability in clinical presentation, symptom trajectory, and treatment response is a barrier for patients with ADHD and their healthcare providers (Mamiya et al. 2021). It is particularly difficult to determine whether the symptoms are signs of ADHD or just those of a hyperactive young child or whether they are indicators of other neurodevelopmental or neuropsychiatric disorders (Bruchmuller et al. 2012; Fresson et al. 2019; Thomas et al. 2015).

The main challenge in diagnosing ADHD is that its symptoms are common in the general population, with a range of factors contributing to that (Hamed et al. 2015). Because there are numerous subtypes of ADHD (inattention, hyperactivity, and mixed), as well as comorbidity with mental disorders, establishing the basis of these phenotypes would need a collaborative research strategy that includes ADHD and other neuropsychiatric diseases.

Establishing causal connections between brain oscillations and ADHD is considered to be one of the most challenging tasks ahead. Improved neuroimaging methods, in conjunction with experimental manipulations like sophisticated neuromodulation and pharmacological therapies, would almost definitely need a greater understanding of brain networks and their functions. Researchers might be able to alter the brain circuitry involved in ADHD-related behaviors selectively using well-studied animal models and cutting-edge technologies such as in vivo optogenetics. Combining human and animal neuroimaging genetic investigations to explore how risk genes related to ADHD neurobiology affect brain alterations in knockout mice models is a promising future direction for ADHD research. This will help identify the abnormal biological processes involved in the pathophysiology of ADHD.

The intermediate or endophenotype approach might be used to map the effects of individual risk genes on neurobiological features such as brain structure, chemistry, and ultimately function. Furthermore, integrating neuroimaging-related endophenotypes with genetic networks is now regarded as an explanatory combinatorial model for fully understanding ADHD etiology, including the development of polygenic risk scores.

The use of a mix of neuroimaging, mental genetics, and behavioral genetics might not only help in the diagnosis of ADHD, but also prove to be a useful tool for tailoring medicine. In addition, future research integrating pharmacogenetics and metabolomics would broaden the diagnostic and treatment options for ADHD.

Despite the large body of research on neurobiological correlates of ADHD that might eventually guide a precision medicine approach, surprising heterogeneity of findings, especially in child studies, remains a huge concern (Yadav et al. 2021). This is because child samples include both those who will have ADHD in adulthood and those who will not. Differences in the developmental stage might also explain the higher variability in childhood since biological and psychological development occurs dramatically throughout infancy. Another hypothesis is that the child samples have more phenotypic variability.

As a result, more thorough designs of investigations are required. Although potential genes and neurotransmitter systems are implicated in ADHD, genome-wide correlations between ADHD symptoms and specific genetic variants are yet to be discovered. Targeted next-generation sequencing of coding and noncoding regions might reveal different genes and pathways linked to ADHD, paving the way for more accurate diagnostic tools and treatment outcomes (Faraone and Larsson 2019).

The epigenetic processes involved in ADHD and their mechanism of interaction with inherited factors or risks need to be investigated. Due to the wide-ranging consequences of the heterogenic nature of ADHD, it might be challenging to unravel the entire genetic profile. It is also essential to link the increasing evidence of genetic anomalies in ADHD with measures of brain functioning in longitudinal research to determine whether the brain abnormalities develop over time. In addition it may be helpful to conduct studies that use the same measurements to assess neurobiological continuity in children and adults with ADHD.

Data specificity for ADHD is another issue. *SLC6A3*, *DRD4*, and *COMT* have all been linked to mental illnesses, including major depressive disorder, schizophrenia, anxiety disorders, bipolar disorder, and obsessive-compulsive disorder (Kabukcu

Basay et al. 2016). Comorbidity is a critical problem for adults with ADHD since it is so common (Takeuchi et al. 2015). Despite this, many studies on Axis I illnesses in adults ignore comorbidity, perhaps because ADHD was formerly thought to be a childhood disease.

The Structured Clinical Interview for DSM (SCID), for example, was utilized as a diagnostic interview in the majority of research into adults. ADHD would have gone unnoticed since it is not listed in the SCID. As a result, the lack of specificity for genetic connections might be related to the fact that studies of these other comorbidities did not include ADHD. Non-specificity also indicates that similar biological processes related to dopaminergic and noradrenergic systems might be shared across mental illnesses, as recently shown (Kabukcu Basay et al. 2016). Future research should include ADHD in the evaluation of comorbidities and also address the concomitant treatment of ADHD and comorbid Axis I disorders, which is a new and vital issue.

5.6 Conclusions

The neurobiology of ADHD is complex and includes many brain pathways, but the neurotransmitters dopamine, noradrenaline, and serotonin are the most important in the pathogenesis of ADHD. A better knowledge of ADHD etiology needs the joint work of psychologists, psychiatrists, geneticists, and neuroscientists. Given the complexity in symptomatology and multiple origins of ADHD, considerable research efforts are required to investigate ADHD-related genetic and neurological changes. Larger scale, multicenter neuroimaging genetic methods are currently being developed, providing one promising path for translating the genetic architecture of this polygenic disease. With rapid advances in research, it is believed that researchers will acquire a better knowledge of ADHD in the coming years, enabling the development of new medicines that are more effective than those presently available.

References

Acosta MT et al (2016) ADGRL3 (LPHN3) variants are associated with a refined phenotype of ADHD in the M.T.A. study. Mol Genet Genomic Med 4:540–547

Akutagava-Martins GC, Salatino-Oliveira A, Genro JP, Contini V, Polanczyk G, Zeni C et al (2014) Glutamatergic copy number variants and their role in attention-deficit/hyperactivity disorder. Am J Med Genet B Neuropsychiatr Genet 165B:502–509

Al-Dewik N, Alsharshani M (2020) New horizons for molecular genetics diagnostic and research in autism spectrum disorder. Adv Neurobiol 24:43–81

Al-Dewik N et al (2020) Overview and introduction to autism spectrum disorder (A.S.D.). Adv Neurobiol 24:3–42

Arcos-Burgos M, Jain M, Acosta MT, Shively S, Stanescu H, Wallis D et al (2010) A common variant of the latrophilin 3 gene, LPHN3, confers susceptibility to ADHD and predicts the effectiveness of the stimulant medication. Mol Psychiatry 15:1053–1066

Baehne CG, Ehlis AC, Plichta MM, Conzelmann A, Pauli P, Jacob C et al (2009) Tph2 gene variants modulate response control processes in adult ADHD patients and healthy individuals. Mol Psychiatry 14:1032–1039

Baird AL, Coogan AN, Siddiqui A, Donev RM, Thome J (2012) Adult attention-deficit hyperactivity disorder is associated with alterations in circadian rhythms at the behavioural, endocrine and molecular levels. Mol Psychiatry 17:988–995

Basay O et al (2016) The impact of synapsin III gene on the neurometabolite level alterations after single-dose methylphenidate in attention-deficit hyperactivity disorder patients. Neuropsychiatr Dis Treat 12:1141–1149

Bedard AC et al (2010) Dopamine transporter gene variation modulates activation of striatum in youth with ADHD. NeuroImage 53:935–942

Biederman J, Faraone SV, Mick E, Spencer T, Wilens T, Kiely K et al (1995) High risk for attention deficit hyperactivity disorder among children of parents with childhood onset of the disorder: a pilot study. Am J Psychiatry 152:431–435

Biederman J, Faraone S, Milberger S, Curtis S, Chen L, Marrs A et al (1996) Predictors of persistence and remission of ADHD into adolescence: results from a four-year prospective follow-up study. J Am Acad Child Adolesc Psychiatry 35:343–351

Bishop J et al (2021) Dopamine transporter and CYP2D6 genotype associations with ADHD treatment response: results from the methylphenidate-atomoxetine crossover study (MACRO). Biol Psychiatry 89(9):S233

Blokland GA et al (2014) Genetic effects on the cerebellar role in working memory: same brain, different genes? NeuroImage 86:392–403

Bobb AJ et al (2005) Support for association between ADHD and two candidate genes: NET1 and DRD1. Am J Med Genet B Neuropsychiatr Genet 134B:67–72

Boomsma DI, Saviouk V, Hottenga JJ, Distel MA, de Moor MH, Vink JM et al (2010) Genetic epidemiology of attention deficit hyperactivity disorder (ADHD index) in adults. PLoS One 5: e10621

Braet W et al (2011) fMRI activation during response inhibition and error processing: the role of the DAT1 gene in typically developing adolescents and those diagnosed with ADHD. Neuropsychologia 49:1641–1650

Breiderhoff T et al (2013) Sortilin-related receptor SORCS3 is a postsynaptic modulator of synaptic depression and fear extinction. PLoS One 8:e75006

Brookes KJ et al (2006) A common haplotype of the dopamine transporter gene associated with attention-deficit/hyperactivity disorder and interacting with maternal use of alcohol during pregnancy. Arch Gen Psychiatry 63:74–81

Brown AB et al (2010) Effect of dopamine transporter gene (SLC6A3) variation on dorsal anterior cingulate function in attention-deficit/hyperactivity disorder. Am J Med Genet B Neuropsychiatr Genet 153B:365–375

Brown AB et al (2011) Relationship of DAT1 and adult ADHD to task-positive and task-negative working memory networks. Psychiatry Res 193:7–16

Brown JT, Bishop JR, Sangkuhl K, Nurmi EL, Mueller DJ, Dinh JC, Gaedigk A, Klein TE, Caudle KE, McCracken JT, de Leon J, Leeder JS (2019) Clinical pharmacogenetics implementation consortium guideline for cytochrome P450 (CYP)2D6 genotype and atomoxetine therapy. Clin Pharmacol Ther 106(1):94–102. https://doi.org/10.1002/cpt.1409

Bruchmuller K, Margraf J, Schneider S (2012) Is ADHD diagnosed in accord with diagnostic criteria? Overdiagnosis and influence of client gender on diagnosis. J Consult Clin Psychol 80: 128–138

Buitelaar JK et al (2012) Long-term efficacy and safety outcomes with OROS-MPH in adults with ADHD. Int J Neuropsychopharmacol 15:1–13

Bulut M, Selek S, Gergerlioglu HS, Savas HA, Yilmaz HR, Yuce M et al (2007) Malondialdehyde levels in adult attention-deficit hyperactivity disorder. J Psychiatry Neurosci 32:435–438

Bulut M, Selek S, Bez Y, Kaya MC, Gunes M, Karababa F et al (2013) Lipid peroxidation markers in adult attention deficit hyperactivity disorder: new findings for oxidative stress. Psychiatry Res 209:638–642

Carithers LJ et al (2015) A novel approach to high-quality postmortem tissue procurement: the GTEx project. Biopreserv Biobank 13:311–319

Carpentier PJ, Arias Vasquez A, Hoogman M, Onnink M, Kan CC, Kooij JJ et al (2013) Shared and unique genetic contributions to attention-deficit/hyperactivity disorder and substance use disorders: a pilot study of six candidate genes. Eur Neuropsychopharmacol 23:448–457

Castellanos FX et al (1998) Lack of an association between a dopamine-4 receptor polymorphism and attention deficit/hyperactivity disorder: genetic and brain morphometric analyses. Mol Psychiatry 3:431–434

Chen W et al (2008) DSM-IV combined type ADHD show familial association with sibling trait scores: a sampling strategy for QTL linkage. Am J Med Genet B Neuropsychiatr Genet 147B:1450–1460

Contini V, Victor MM, Marques FZ, Bertuzzi GP, Salgado CA, Silva KL et al (2010) Response to methylphenidate is not influenced by DAT1 polymorphisms in a sample of Brazilian adult patients with ADHD. J Neural Transm 117:269–276

Contini V, Victor MM, Cerqueira CC, Polina ER, Grevet EH, Salgado CA et al (2011) Adrenergic alpha 2 A receptor gene is not associated with methylphenidate response in adults with ADHD. Eur Arch Psychiatry Clin Neurosci 261:205–211

Contini V, Victor MM, Bertuzzi GP, Salgado CA, Picon FA, Grevet EH et al (2012) No significant association between genetic variants in seven candidate genes and response to methylphenidate treatment in adult patients with ADHD. J Clin Psychopharmacol 32:820–823

Corominas-Roso M, Ramos-Quiroga JA, Ribases M, Sanchez-Mora C, Palomar G, Valero S et al (2013) Decreased serum levels of brain-derived neurotrophic factor in adults with attention-deficit hyperactivity disorder. Int J Neuropsychopharmacol 30:1–9

Cortese S, Castellanos FX (2012) Neuroimaging of attention-deficit/hyperactivity disorder: current neuroscience-informed perspectives for clinicians. Curr Psychiatry Rep 14:568–578

Cross Disorder Group of the Psychiatric Genomics Consortium (2013) Identification of risk loci with shared effects on five major psychiatric disorders: a genome-wide analysis. Lancet 381: 1371–1379. https://doi.org/10.1016/S0140-6736(12)62129-1

de Cerqueira CC, Polina ER, Contini V, Marques FZ, Grevet EH, Salgado CA et al (2011) ADRA2A polymorphisms and ADHD in adults: possible mediating effect of personality. Psychiatry Res 186:345–350

De Luca V, Muglia P, Jain U, Basile VS, Sokolowski MB, Kennedy JL (2002) A drosophila model for attention deficit hyperactivity disorder (ADHD): no evidence of association with PRKG1 gene. Neuromolecular Med 2:281–287

De Luca V, Muglia P, Vincent JB, Lanktree M, Jain U, Kennedy JL (2004a) Adrenergic alpha 2C receptor genomic organization: association study in adult ADHD. Am J Med Genet B Neuropsychiatr Genet 127B:65–67

De Luca V, Muglia P, Jain U, Kennedy JL (2004b) No evidence of linkage or association between the norepinephrine transporter (N.E.T.) gene MnlI polymorphism and adult ADHD. Am J Med Genet B Neuropsychiatr Genet 124B:38–40

Demontis D et al (2019) Discovery of the first genome-wide significant risk loci for attention-deficit/hyperactivity disorder. Nat Genet 51:63–75

Durston S et al (2005) Differential effects of DRD4 and DAT1 genotype on fronto-striatal grey matter volumes in a sample of subjects with attention deficit hyperactivity disorder, their unaffected siblings, and controls. Mol Psychiatry 10:678–689

Durston S et al (2008) Dopamine transporter genotype convey familial risk of attention-deficit/hyperactivity disorder through striatal activation. J Am Acad Child Adolesc Psychiatry 47:61–67

Ercan ES et al (2016) Decreasing ADHD phenotypic heterogeneity: searching for neurobiological underpinnings of the restrictive inattentive phenotype. Eur Child Adolesc Psychiatry 25:273–282

Ernst M, Liebenauer LL, Tebeka D, Jons PH, Eisenhofer G, Murphy DL et al (1997) Selegiline in ADHD adults: plasma monoamines and monoamine metabolites. Neuropsychopharmacology 16:276–284

Faraone SV, Larsson H (2019) Genetics of attention deficit hyperactivity disorder. Mol Psychiatry 24(4):562–575

Faraone SV, Mick E (2010) Molecular genetics of attention deficit hyperactivity disorder. Psychiatr Clin North Am 33:159–180

Faraone SV, Biederman J, Monuteaux MC (2000a) Toward guidelines for pedigree selection in genetic studies of attention deficit hyperactivity disorder. Genet Epidemiol 18:1–16

Faraone SV, Biederman J, Feighner JA, Monuteaux MC (2000b) Assessing symptoms of attention deficit hyperactivity disorder in children and adults: which is more valid? J Consult Clin Psychol 68:830–842

Faraone SV, Perlis RH, Doyle AE, Smoller JW, Goralnick JJ, Holmgren MA, Sklar P (2005) Molecular genetics of attention-deficit/hyperactivity disorder. Biol Psychiatry 57:1313–1323

Faraone SV, Biederman J, Mick E (2006) The age-dependent decline of attention deficit hyperactivity disorder: a meta-analysis of follow-up studies. Psychol Med 36(2):159–165

Faraone SV, Asherson P, Banaschewski T, Biederman J, Buitelaar JK, Ramos-Quiroga JA, Rohde LA, Sonuga-Barke EJ, Tannock R, Franke B (2015) Attention-deficit/hyperactivity disorder. Nat Rev Dis Primers 1:15020

Fernandez-Jaen A et al (2015) Cortical thickness differences in the prefrontal cortex in children and adolescents with ADHD in relation to dopamine transporter (DAT1) genotype. Psychiatry Res 233:409–417

Fernandez-Jaen A et al (2018) Cingulate cortical thickness and dopamine transporter (DAT1) genotype in children and adolescents with ADHD. J Atten Disord 22:651–660

Franke B, Neale BM, Faraone SV (2009) Genome-wide association studies in ADHD. Hum Genet 126:13–50

Fresson M, Meulemans T, Dardenne B, Geurten M (2019) Overdiagnosis of ADHD in boys: stereotype impact on neuropsychological assessment. Appl Neuropsychol Child 8:231–245

Gallo EF, Posner J (2016) Moving towards causality in attention-deficit hyperactivity disorder: an overview of neural and genetic mechanisms. Lancet Psychiatry 3:555–567

Gilsbach S et al (2012) Effects of the DRD4 genotype on neural networks associated with executive functions in children and adolescents. Dev Cogn Neurosci 2:417–427

Gizer IR, Ficks C, Waldman ID (2009) Candidate gene studies of ADHD: a meta-analytic review. Hum Genet 126:51–90

Glahn DC et al (2010) Genetic control over the resting brain. Proc Natl Acad Sci U S A 107:1223–1228

Grabemann M, Mette C, Zimmermann M, Wiltfang J, Kis B (2014) Serum albumin correlates with affective prosody in adult males with attention-deficit hyperactivity disorder. Psychiatry Res 217:198–201

Haavik J, Halmoy A, Lundervold AJ, Fasmer OB (2010) Clinical assessment and diagnosis of adults with attention-deficit/hyperactivity disorder. Expert Rev Neurother 10:1569–1580

Hamed AM, Kauer AJ, Stevens HE (2015) Why the diagnosis of attention deficit hyperactivity disorder matters. Front Psychiatry 6:168–168

Hasler R, Salzmann A, Bolzan T, Zimmermann J, Baud P, Giannakopoulos P et al (2015) DAT1 and DRD4 genes involved in key dimensions of adult ADHD. Neurol Sci 36:861–869

Hawi Z et al (2015) The molecular genetic architecture of attention deficit hyperactivity disorder. Mol Psychiatry 20:289–297

Herken H, Erdal ME, Kenar AN, Unal GA, Cakaloz B, Ay ME et al (2014) Association of SNAP-25 gene Ddel and Mnll polymorphisms with adult attention deficit hyperactivity disorder. Psychiatry Investig 11:476–480

Hirvikoski T, Lindholm T, Nordenstrom A, Nordstrom AL, Lajic S (2009) High self-perceived stress and many stressors, but normal diurnal cortisol rhythm, in adults with ADHD (attention-deficit/hyperactivity disorder). Horm Behav 55:418–424

Hong SB et al (2015) COMT genotype affects brain white matter pathways in attention-deficit/ hyperactivity disorder. Hum Brain Mapp 36:367–377

Hoogman M, Aarts E, Zwiers M, Slaats-Willemse D, Naber M, Onnink M et al (2011) Nitric oxide synthase genotype modulation of impulsivity and ventral striatal activity in adult ADHD patients and healthy comparison subjects. Am J Psychiatry 168:1099–1106

Hoogman M et al (2013) The dopamine transporter haplotype and reward-related striatal responses in adult ADHD. Eur Neuropsychopharmacol 23:469–478

Inci Kenar AN et al (2016) Relationship between the DAT1 gene and the effects of methylphenidate administration in adult attention deficit hyperactivity disorder: a magnetic resonance spectroscopy study. Eur Rev Med Pharmacol Sci 20:1373–1378

Inkster B, Muglia P, Jain U, Kennedy JL (2004) Linkage disequilibrium analysis of the dopamine beta-hydroxylase gene in persistent attention deficit hyperactivity disorder. Psychiatr Genet 14: 117–120

Irmisch G, Richter J, Thome J, Sheldrick AJ, Wandschneider R (2013) Altered serum mono- and polyunsaturated fatty acid levels in adults with ADHD. Atten Defic Hyperact Disord 5:303–311

Jacob C, Nguyen TT, Weißflog L, Herrmann M, Liedel S, Zamzow K et al (2012) PPP2R2C as a candidate gene of a temperament and character trait-based endophenotype of ADHD. Atten Defic Hyperact Disord 4:145–152

Jacob CP, Weber H, Retz W, Kittel-Schneider S, Heupel J, Renner T et al (2013) Acetylcholine-metabolizing butyrylcholinesterase (BCHE) copy number and single nucleotide polymorphisms and their role in attention-deficit/hyperactivity syndrome. J Psychiatr Res 47:1902–1908

Jacobsen KK, Halmøy A, Sánchez-Mora C, Ramos-Quiroga JA, Cormand B, Haavik J et al (2013) DISC1 in adult ADHD patients: an association study in two European samples. Am J Med Genet B Neuropsychiatr Genet 162B:227–234

Jarick I et al (2014) Genome-wide analysis of rare copy number variations reveals PARK2 as a candidate gene for attention-deficit/hyperactivity disorder. Mol Psychiatry 19:115–121

Johansson S, Halleland H, Halmøy A, Jacobsen KK, Landaas ET, Dramsdahl M et al (2008) Genetic analyses of dopamine-related genes in adult ADHD patients suggest an association with the DRD5-microsatellite repeat, but not with DRD4 or SLC6A3 VNTRs. Am J Med Genet B Neuropsychiatr Genet 147B:1470–1475

Johansson S, Halmøy A, Mavroconstanti T, Jacobsen KK, Landaas ET, Reif A et al (2010) Common variants in the TPH1 and TPH2 regions are not associated with persistent ADHD in a combined sample of 1,636 adult cases and 1,923 controls from four European populations. Am J Med Genet B Neuropsychiatr Genet 153B:1008–1015

Kabukcu Basay B et al (2016) White matter alterations related to attention-deficit hyperactivity disorder and COMT val(158)met polymorphism: children with valine homozygote attention-deficit hyperactivity disorder have altered white matter connectivity in the right cingulum (cingulate gyrus). Neuropsychiatr Dis Treat 12:969–981

Kan KJ, Dolan CV, Nivard MG, Middeldorp CM, van Beijsterveldt CE, Willemsen G et al (2013) Genetic and environmental stability in attention problems across the lifespan: evidence from the Netherlands twin register. J Am Acad Child Adolesc Psychiatry 52:12–25

Karababa IF, Savas SN, Selek S, Cicek E, Cicek EI, Asoglu M et al (2017) Homocysteine levels and oxidative stress parameters in patients with adult ADHD. J Atten Disord 21(6):487–493. https://doi.org/10.1177/1087054714538657

Kebir O, Tabbane K, Sengupta S, Joober R (2009) Candidate genes and neuropsychological phenotypes in children with ADHD: a review of association studies. J Psychiatry Neurosci 34:88–101

Kenar AN, Edgünlü T, Herken H, Erdal ME (2013) Association of synapsin III gene with adult attention deficit hyperactivity disorder. DNA Cell Biol 32:430–434

Kenar AN, Ay Oİ, Herken H, Erdal ME (2014) Association of VAMP-2 and Syntaxin 1A genes with adult attention deficit hyperactivity disorder. Psychiatry Investig 11:76–83

Khadka S et al (2016) Multivariate imaging genetics study of M.R.I. grey matter volume and SNPs reveals biological pathways correlated with brain structural differences in attention deficit hyperactivity disorder. Front Psychiatry 7:128

Kim JI, Yoo JH, Kim D, Jeong B, Kim BN (2018) The effects of GRIN2B and DRD4 gene variants on local functional connectivity in attention-deficit/hyperactivity disorder. Brain Imaging Behav 12:247–257

Kittel-Schneider S, Reuß M, Meyer A, Weber H, Gessner A, Leistner C et al (2015) Multi-level biomarker analysis of nitric oxide synthase isoforms in bipolar disorder and adult ADHD. J Psychopharmacol 29:31–38

Klein M et al (2017) Brain imaging genetics in ADHD and beyond—mapping pathways from gene to disorder at different levels of complexity. Neurosci Biobehav Rev 80:115–155

Ko CH et al (2018) The altered brain activation of phonological working memory, dual tasking, and distraction among participants with adult ADHD and the effect of the MAOA polymorphism. J Atten Disord 22:240–249

Kochunov P et al (2015) Heritability of fractional anisotropy in human white matter: a comparison of Human Connectome Project and ENIGMA-DTI data. NeuroImage 111:300–311

Kooij JS, Boonstra AM, Vermeulen SH, Heister AG, Burger H, Buitelaar JK et al (2008) Response to methylphenidate in adults with ADHD is associated with a polymorphism in SLC6A3 (DAT1). Am J Med Genet B Neuropsychiatr Genet 147B:201–208

Kortmann GL, Contini V, Bertuzzi GP, Mota NR, Rovaris DL, Paixão-Côrtes VR et al (2013) The role of a mineralocorticoid receptor gene functional polymorphism in the symptom dimensions of persistent ADHD. Eur Arch Psychiatry Clin Neurosci 263:181–188

Laasonen M, Hokkanen L, Leppamaki S, Tani P, Erkkila AT (2009) Project DyAdd: fatty acids in adult dyslexia, ADHD, and their co-morbid combination. Prostaglandins Leukot Essent Fatty Acids 81:89–96

Lackschewitz H, Huther G, Kroner-Herwig B (2008) Physiological and psychological stress responses in adults with attention-deficit/hyperactivity disorder (ADHD). Psychoneuroendocrinology 33:612–624

Landaas ET, Johansson S, Jacobsen KK, Ribasés M, Bosch R, Sánchez-Mora C et al (2010) An international multicenter association study of the serotonin transporter gene in persistent ADHD. Genes Brain Behav 9:449–458

Lanktree M, Squassina A, Krinsky M, Strauss J, Jain U, Macciardi F et al (2008) Association study of brain-derived neurotrophic factor (BDNF) and LIN-7 homolog (LIN-7) genes with adult attention-deficit/hyperactivity disorder. Am J Med Genet B Neuropsychiatr Genet 147B:945–951

Larsson H, Asherson P, Chang Z, Ljung T, Friedrichs B, Larsson JO et al (2013) Genetic and environmental influences on adult attention deficit hyperactivity disorder symptoms: a large Swedish population-based study of twins. Psychol Med 43:197–207

Larsson H, Chang Z, D'Onofrio BM, Lichtenstein P (2014) The heritability of clinically diagnosed attention deficit hyperactivity disorder across the lifespan. Psychol Med 44:2223–2229

Lee SH, Ripke S, Neale BM, Faraone SV, Purcell SM, Perlis RH, Mowry BJ, Thapar A, Goddard ME, Witte JS et al (2013) Genetic relationship between five psychiatric disorders estimated from genome-wide SNPs. Nat Genet 45(9):984–994

Li D, Sham PC, Owen MJ, He L (2006) Meta-analysis shows a significant association between dopamine system genes and attention deficit hyperactivity disorder (ADHD). Hum Mol Genet 15:2276–2284

Li C, Scott DA, Hatch E, Tian X, Mansour SL (2007) Dusp6 (Mkp3) is a negative feedback regulator of FGF-stimulated ERK signalling during mouse development. Development 134: 167–176

Li Z, Chang SH, Zhang LY, Gao L, Wang J (2014) Molecular genetic studies of ADHD and its candidate genes: a review. Psychiatry Res 219:10–24

Li J et al (2015) The cortical surface area of the insula mediates the effect of DBH rs7040170 on novelty seeking. NeuroImage 117:184–190

Mamiya PC, Arnett AB, Stein MA (2021) Precision medicine care in ADHD: the case for neural excitation and inhibition. Brain Sci 11(1):91

Martin J, O'Donovan MC, Thapar A, Langley K, Williams N (2015) The relative contribution of common and rare genetic variants to ADHD. Transl Psychiatry 5:e506

Martinez-Raga J, Ferreros A, Knecht C, de Alvaro R, Carabal E (2017) Attention-deficit hyperactivity disorder medication use: factors involved in prescribing, safety aspects and outcomes. Ther Adv Drug Saf 8:87–99

Mavroconstanti T, Johansson S, Winge I, Knappskog PM, Haavik J (2013) Functional properties of rare missense variants of human CDH13 found in adult attention-deficit/hyperactivity disorder (ADHD) patients. PLoS One 8:e71445

Mavroconstanti T, Halmøy A, Haavik J (2014) Decreased serum levels of adiponectin in adult attention deficit hyperactivity disorder. Psychiatry Res 216:123–130

McKay DR et al (2014) Influence of age, sex and genetic factors on the human brain. Brain Imaging Behav 8:143–152

Mick E, Biederman J, Spencer T, Faraone SV, Sklar P (2006) Absence of association with DAT1 polymorphism and response to methylphenidate in a sample of adults with ADHD. Am J Med Genet B Neuropsychiatr Genet 141B:890–894

Mizuno Y et al (2017) Catechol-O-methyltransferase polymorphism is associated with the cortico-cerebellar functional connectivity of executive function in children with attention-deficit/hyperactivity disorder. Sci Rep 7:4850

Monuteaux MC et al (2008) A preliminary study of dopamine D4 receptor genotype and structural brain alterations in adults with ADHD. Am J Med Genet B Neuropsychiatr Genet 147B:1436–1441

Mortensen OV, Larsen MB, Prasad BM, Amara SG (2008) Genetic complementation screen identifies a mitogen-activated protein kinase phosphatase, MKP3, as a regulator of dopamine transporter trafficking. Mol Biol Cell 19:2818–2829

Mota NR, Rovaris DL, Kappel DB, Picon FA, Vitola ES, Salgado CA et al (2015) NCAM1-TTC12-ANKK1-DRD2 gene cluster and the clinical and genetic heterogeneity of adults with ADHD. Am J Med Genet B Neuropsychiatr Genet 168(6):433–444. https://doi.org/10.1002/ajmg.b.32317

Mueller A, Hong DS, Shepard S, Moore T (2017) Linking ADHD to the neural circuitry of attention. Trends Cogn Sci 21:474–488

Muglia P, Jain U, Kennedy JL (2002) A transmission disequilibrium test of the Ser9/Gly dopamine D3 receptor gene polymorphism in adult attention-deficit hyperactivity disorder. Behav Brain Res 130:91–95

Nigg JT (2013) Attention deficits and hyperactivity-impulsivity: what have we learned, what next? Dev Psychopathol 25:1489–1503

Nymberg C et al (2013) Neural mechanisms of attention-deficit/hyperactivity disorder symptoms are stratified by MAOA genotype. Biol Psychiatry 74:607–614

Ogundele MO (2018) Behavioural and emotional disorders in childhood: a brief overview for paediatricians. World J Clin Pediatr 7:9–26

Olgiati P, Mandelli L, Alberti S, Lia L, Serretti A, Tiwari AK et al (2014) Role of synaptosome-related (SNARE) genes in adults with attention deficit hyperactivity disorder. Psychiatry Res 215:799–800

Onnink AM et al (2016) Enlarged striatal volume in adults with ADHD carrying the 9-6 haplotype of the dopamine transporter gene DAT1. J Neural Transm 123:905–915

Ozturk O et al (2016) The effect of single-dose methylphenidate on neurometabolites according to COMT gene Val158Met polymorphism in the patient with attention deficit hyperactivity disorder: a study using magnetic resonance spectroscopy. Clin Psychopharmacol Neurosci 14:184–193

Paloyelis Y, Mehta MA, Faraone SV, Asherson P, Kuntsi J (2012) Striatal sensitivity during reward processing in attention-deficit/hyperactivity disorder. J Am Acad Child Adolesc Psychiatry 51:722–732.e9

Park S et al (2013) White-matter connectivity and methylphenidate-induced changes in attentional performance according to alpha2A-adrenergic receptor gene polymorphisms in Korean children with attention-deficit hyperactivity disorder. J Neuropsychiatry Clin Neurosci 25:222–228

Park S et al (2015) Associations between serotonin transporter gene (SLC6A4) methylation and clinical characteristics and cortical thickness in children with ADHD. Psychol Med 45:3009–3017

Pazvantoğlu O, Güneş S, Karabekiroğlu K, Yeğin Z, Erenkuş Z, Akbaş S et al (2013) The relationship between the presence of ADHD and certain candidate gene polymorphisms in a Turkish sample. Gene 528:320–327

Peper JS, Brouwer RM, Boomsma DI, Kahn RS, Hulshoff Pol HE (2007) Genetic influences on human brain structure: a review of brain imaging studies in twins. Hum Brain Mapp 28:464–473

Pineau G et al (2019) Dopamine transporter genotype modulates brain activity during a working memory task in children with ADHD. Res Dev Disabil 92:103430

Polanczyk G, Rohde LA (2007) Epidemiology of attention-deficit/hyperactivity disorder across the lifespan. Curr Opin Psychiatry 20:386–392

Polanczyk GV, Willcutt EG, Salum GA, Kieling C, Rohde LA (2014) ADHD prevalence estimates across three decades: an updated systematic review and meta-regression analysis. Int J Epidemiol 43:434–442

Posner J, Polanczyk GV, Sonuga-Barke E (2020) Attention-deficit hyperactivity disorder. Lancet 395:450–462

Qian A et al (2018a) Effects of the 2-repeat allele of the DRD4 gene on neural networks associated with the prefrontal cortex in children with ADHD. Front Hum Neurosci 12:279

Qian A et al (2018b) Dopamine D4 receptor gene associated with the frontal-striatal-cerebellar loop in children with ADHD: a resting-state fMRI study. Neurosci Bull 34:497–506

Quintero J et al (2018) Health care and societal costs of the management of children and adolescents with attention-deficit/hyperactivity disorder in Spain: a descriptive analysis. BMC Psychiatry 18:40–40

Ramos-Quiroga JA, Nasillo V, Fernández-Aranda F, Casas M (2014a) Addressing the lack of studies in attention-deficit/hyperactivity disorder in adults. Expert Rev Neurother 14:553–567

Ramos-Quiroga JA, Corominas-Roso M, Palomar G, Gomez-Barros N, Ribases M, Sanchez-Mora C et al (2014b) Changes in the serum levels of brain-derived neurotrophic factor in adults with attention deficit hyperactivity disorder after treatment with atomoxetine. Psychopharmacology (Berl) 231:1389–1395

Raz S, Leykin D (2015) Psychological and cortisol reactivity to experimentally induced stress in adults with ADHD. Psychoneuroendocrinology 60:7–17

Rebases M, Sánchez-Mora C, Ramos-Quiroga JA, Bosch R, Gómez N, Nogueira M et al (2012) An association study of sequence variants in the forkhead box P2 (FOXP2) gene and adulthood attention-deficit/hyperactivity disorder in two European samples. Psychiatr Genet 22:155–160

Reif A, Nguyen TT, Weissflog L, Jacob CP, Romanos M, Renner TJ et al (2011) DIRAS2 is associated with adult ADHD, related traits, and co-morbid disorders. Neuropsychopharmacology 36:2318–2327

Reimherr FW, Wender PH, Ebert MH, Wood DR (1984) Cerebrospinal fluid homovanillic acid and 5-hydroxy-indoleacetic acid in adults with attention deficit disorder, residual type. Psychiatry Res 11:71–78

Ribasés M, Hervás A, Ramos-Quiroga JA, Bosch R, Bielsa A, Gastaminza X et al (2008) Association study of 10 genes encoding neurotrophic factors and their receptors in adult and child attention-deficit/hyperactivity disorder. Biol Psychiatry 63:935–945

Ribasés M, Ramos-Quiroga JA, Hervás A, Bosch R, Bielsa A, Gastaminza X et al (2009) Exploration of 19 serotoninergic candidate genes in adults and children with attention-deficit/hyperactivity disorder identifies the association for 5HT2A, D.D.C. and MAOB. Mol Psychiatry 14:71–85

Ribases M et al (2011) Contribution of LPHN3 to the genetic susceptibility to ADHD in adulthood: a replication study. Genes Brain Behav 10:149–157

Ribasés M, Ramos-Quiroga JA, Sánchez-Mora C, Bosch R, Richarte V, Palomar G et al (2011) Contribution of LPHN3 to the genetic susceptibility to ADHD in adulthood: a replication study. Genes Brain Behav 10:149–157

Ribasés M, Ramos-Quiroga JA, Hervás A, Sánchez-Mora C, Bosch R, Bielsa A et al (2012) Candidate system analysis in ADHD: evaluation of nine genes involved in dopaminergic neurotransmission identifies association with DRD1. World J Biol Psychiatry 13:281–292

Richards JS et al (2016) Adolescent behavioural and neural reward sensitivity: a test of the differential susceptibility theory. Transl Psychiatry 6:e771

Rivero O et al (2015) Cadherin-13, a risk gene for ADHD and co-morbid disorders, impacts GABAergic function in the hippocampus and cognition. Transl Psychiatry 5:e655

Salatino-Oliveira A, Genro JP, Polanczyk G, Zeni C, Schmitz M, Kieling C et al (2015) Cadherin-13 gene is associated with hyperactive/impulsive symptoms in attention/deficit hyperactivity disorder. Am J Med Genet B Neuropsychiatr Genet 168B:162–169

Samea F et al (2019) Brain alterations in children/adolescents with ADHD revisited: a neuroimaging meta-analysis of 96 structural and functional studies. Neurosci Biobehav Rev 100:1–8

Sánchez-Mora C, Ribasés M, Ramos-Quiroga JA, Casas M, Bosch R, Boreatti-Hümmer A et al (2010) Meta-analysis of brain-derived neurotrophic factor p.Val66Met in adult ADHD in four European populations. Am J Med Genet B Neuropsychiatr Genet 153B:512–523

Sánchez-Mora C, Cormand B, Ramos-Quiroga JA, Hervás A, Bosch R, Palomar G et al (2013) Evaluation of common variants in 16 genes involved in the regulation of neurotransmitter release in ADHD. Eur Neuropsychopharmacol 23:426–435

Scassellati C et al (2012) Biomarkers and attention-deficit/hyperactivity disorder: a systematic review and meta-analyses. J Am Acad Child Adolesc Psychiatry 51(10):1003–1019.e20

Schreiweis C et al (2014) Humanized Foxp2 accelerates learning by enhancing transitions from declarative to procedural performance. Proc Natl Acad Sci U S A 111:14253–14258

Schweren LJ et al (2016) Age and DRD4 genotype moderate associations between stimulant treatment history and cortex structure in attention-deficit/hyperactivity disorder. J Am Acad Child Adolesc Psychiatry 55:877–885.e3

Scott JG, Mihalopoulos C, Erskine HE, Roberts J, Rahman A (2016) Childhood mental and developmental disorders. In: Patel V et al (eds) Mental, neurological, and substance use disorders: disease control priorities, vol 4, 3rd edn, pp 145–161

Selek S, Savas H, Gergerlioglu S, Bulut M, Yilmaz HR (2008) Oxidative imbalance in adult attention deficit/hyperactivity disorder. Biol Psychol 79:256–259

Selek S, Bulut M, Ocak AR, Kalenderoglu A, Savas H (2012) Evaluation of total oxidative status in adult attention deficit hyperactivity disorder and its diagnostic implications. J Psychiatr Res 46: 451–455

Shaltout E, Al-Dewik N, Samara M, Morsi H, Khattab A (2020) Psychological comorbidities in autism spectrum disorder. In: Essa M, Qoronfleh M (eds) Personalized food intervention and therapy for autism spectrum disorder management. Advances in neurobiology, vol 24. Springer, Cham. https://doi.org/10.1007/978-3-030-30402-7_6

Shang CY, Lin HY, Tseng WY, Gau SS (2018) A haplotype of the dopamine transporter gene modulates regional homogeneity, gray matter volume, and visual memory in children with attention-deficit/hyperactivity disorder. Psychol Med 48:2530–2540

Shaw P et al (2007) Polymorphisms of the dopamine D4 receptor, clinical outcome, and cortical structure in attention-deficit/hyperactivity disorder. Arch Gen Psychiatry 64:921–931

Shaw P, Malek M, Watson B, Greenstein D, de Rossi P, Sharp W (2013) Trajectories of cerebral cortical development in childhood and adolescence and adult attention-deficit/hyperactivity disorder. Biol Psychiatry 74(8):599–606

Shimada K et al (2017) Ethnic differences in COMT genetic effects on striatal grey matter alterations associated with childhood ADHD: a voxel-based morphometry study in a Japanese sample. World J Biol Psychiatry 18:322–328

Shook D et al (2011) Effect of dopamine transporter genotype on caudate volume in childhood ADHD and controls. Am J Med Genet B Neuropsychiatr Genet 156B:28–35

Silva JP et al (2011) Latrophilin 1 and its endogenous ligand Lasso/teneurin-2 form a high-affinity transsynaptic receptor pair with signalling capabilities. Proc Natl Acad Sci U S A 108:12113–12118

Simon V, Czobor P, Balint S, Meszaros A, Bitter I (2009) Prevalence and correlates of adult attention-deficit hyperactivity disorder: meta-analysis. Br J Psychiatry 194:204–211

Soderqvist S et al (2010) The SNAP25 gene is linked to working memory capacity and maturation of the posterior cingulate cortex during childhood. Biol Psychiatry 68:1120–1125

Sokolova E et al (2015) Causal discovery in an adult ADHD data set suggests an indirect link between DAT1 genetic variants and striatal brain activation during reward processing. Am J Med Genet B Neuropsychiatr Genet 168:508–515

Squassina A, Lanktree M, De Luca V, Jain U, Krinsky M, Kennedy JL et al (2008) Investigation of the dopamine D5 receptor gene (DRD5) in adult attention deficit hyperactivity disorder. Neurosci Lett 432:50–53

Stevens MC, Pearlson GD, Calhoun VD, Bessette KL (2018) Functional neuroimaging evidence for distinct neurobiological pathways in attention-deficit/hyperactivity disorder. Biol Psychiatry Cogn Neurosci Neuroimaging 3:675–685

Stollstorff M et al (2010) Neural response to working memory load varies by dopamine transporter genotype in children. NeuroImage 53:970–977

Styles M, Alsharshani D, Samara M, Alsharshani M, Khattab A, Qoronfleh MW, Al-Dewik NI (2020) Risk factors, diagnosis, prognosis and treatment of autism. Front Biosci (Landmark Ed) 25:1682–1717. https://doi.org/10.2741/4873

Sullivan PF, Daly MJ, O'Donovan M (2012) Genetic architectures of psychiatric disorders: the emerging picture and its implications. Nat Rev Genet 13(8):537–551

Takeuchi H et al (2015) Cognitive and neural correlates of the 5-repeat allele of the dopamine D4 receptor gene in a population lacking the 7-repeat allele. NeuroImage 110:124–135

Thomas R, Sanders S, Doust J, Beller E, Glasziou P (2015) Prevalence of attention-deficit/hyperactivity disorder: a systematic review and meta-analysis. Pediatrics 135:e994–e1001

Tsui D, Vessey JP, Tomita H, Kaplan DR, Miller FD (2013) FoxP2 regulates neurogenesis during embryonic cortical development. J Neurosci 33:244–258

Unal GA, Inci Kenar AN, Tepeli E, Kiroglu Y, Herken H (2016) Relationship between the SNAP-25 gene and the effects of methylphenidate on the anterior cingulate cortex of patients with adult attention deficit hyperactivity disorder: a magnetic resonance spectroscopy study. Eur Rev Med Pharmacol Sci 20:2443–2449

van der Meer D et al (2015) Brain correlates of the interaction between 5-HTTLPR and psychosocial stress mediating attention deficit hyperactivity disorder severity. Am J Psychiatry 172:768–775

van der Meer D et al (2016) Interplay between stress response genes associated with attention deficit hyperactivity disorder and brain volume. Genes Brain Behav 15:627–636

van der Meer D et al (2017) Effects of dopaminergic genes, prenatal adversities, and their interaction on attention-deficit/hyperactivity disorder and neural correlates of response inhibition. J Psychiatry Neurosci 42:113–121

van Ewijk H et al (2017) Female-specific association of NOS1 genotype with white matter microstructure in ADHD patients and controls. J Child Psychol Psychiatry 58:958–966

van Rooij D et al (2015a) Influence of DAT1 and COMT variants on neural activation during response inhibition in adolescents with attention-deficit/hyperactivity disorder and healthy controls. Psychol Med 45:3159–3170

van Rooij D et al (2015b) Variation in serotonin neurotransmission genes affects neural activation during response inhibition in adolescents and young adults with ADHD and healthy controls. World J Biol Psychiatry 16:625–634

van Rooij D et al (2015c) Altered neural connectivity during response inhibition in adolescents with attention-deficit/hyperactivity disorder and their unaffected siblings. Neuroimage Clin 7:325–335

Van Veen MM, Kooij JJ, Boonstra AM, Gordijn MC, Van Someren EJ (2010) Delayed circadian rhythm in adults with attention-deficit/hyperactivity disorder and chronic sleep-onset insomnia. Biol Psychiatry 67:1091–1096

Villemonteix T et al (2015) Structural correlates of COMT Val158Met polymorphism in childhood ADHD: a voxel-based morphometry study. World J Biol Psychiatry 16:190–199

Volkow ND, Swanson JM (2013) Clinical practice: adult attention deficit-hyperactivity disorder. N Engl J Med 369:1935–1944

Wang C et al (2018) The impact of SNAP25 on brain functional connectivity density and working memory in ADHD. Biol Psychol 138:35–40

Weber H, Kittel-Schneider S, Gessner A, Domschke K, Neuner M, Jacob CP et al (2011) Cross-disorder analysis of bipolar risk genes: further evidence of DGKH as a risk gene for bipolar disorder, but also unipolar depression and adult ADHD. Neuropsychopharmacology 36:2076–2085

Weber H, Scholz CJ, Jacob CP, Heupel J, Kittel-Schneider S, Erhardt A et al (2014) SPOCK3, a risk gene for adult ADHD and personality disorders. Eur Arch Psychiatry Clin Neurosci 64: 409–421

Weißflog L, Scholz CJ, Jacob CP, Nguyen TT, Zamzow K, Groß-Lesch S et al (2013) KCNIP4 as a candidate gene for personality disorders and adult ADHD. Eur Neuropsychopharmacol 23:436–447

Weyandt L, Swentosky A, Gudmundsdottir BG (2013) Neuroimaging and ADHD: fMRI, P.E.T., D.T.I. findings, and methodological limitations. Dev Neuropsychol 38:211–225

Wilens TE, Spencer TJ (2010) Understanding attention-deficit/hyperactivity disorder from childhood to adulthood. Postgrad Med 122:97–109

Willcutt EG (2012) The prevalence of DSM-IV attention-deficit/hyperactivity disorder: a meta-analytic review. Neurotherapeutics 9:490–499

Williams NM, Franke B, Mick E, Anney RJ, Freitag CM, Gill M, Thapar A, O'Donovan MC, Owen MJ, Holmans P et al (2012) Genome-wide analysis of copy number variants in attention deficit hyperactivity disorder: the role of rare variants and duplications at 15q13.3. Am J Psychiatry 169(2):195–204

Yadav SK et al (2021) Genetic variations influence brain changes in patients with attention-deficit hyperactivity disorder. Transl Psychiatry 11(1):349

Yang L et al (2013) Polygenic transmission and complex neuro developmental network for attention deficit hyperactivity disorder: genome-wide association study of common and rare variants. Am J Med Genet B Neuropsychiatr Genet 162B:419–430

Young GS, Maharaj NJ, Conquer JA (2004) Blood phospholipid fatty acid analysis of adults with and without attention-deficit/hyperactivity disorder. Lipids 39:117–123

Young GS, Conquer JA, Thomas R (2005) Effect of randomized supplementation with high dose olive, flax or fish oil on serum phospholipid fatty acid levels in adults with attention deficit hyperactivity disorder. Reprod Nutr Dev 45:549–558

Yurtbasi P et al (2018) Comparison of neurological and cognitive deficits in children with ADHD and anxiety disorders. J Atten Disord 22:472–485

Zhou K et al (2008) Meta-analysis of genome-wide linkage scans of attention deficit hyperactivity disorder. Am J Med Genet B Neuropsychiatr Genet 147B:1392–1398

Genomic Profiling of ADHD

6

Arokiasamy Justin Thenmozhi, Chinnasamy Dhanalakshmi, and Thamilarasan Manivasagam

Abstract

The inheritance of attention-deficit hyperactivity disorder (ADHD) is more common in children and adults and therefore more research in the field of genetics was carried out. The experiments indicated that the genetic factors played a crucial role in the etiology and course of the disease. Numerous studies initially focused on the candidate genes for ADHD particularly those genes involved in the dopaminergic, noradrenergic, and serotonergic neurotransmission systems. In the recent past, the association of ADHD with the candidate genes linked to neuronal growth and plasticity, and the glutaminergic system, has been published. This chapter reviews the single-nucleotide polymorphisms found in the candidate genes and recaps the results of genome-wide association studies (GWAS). GWAS helps in the discovery of new ADHD genes in a hypothesis-free manner. The GWAS findings are redirecting the future of the ADHD research towards novel gene systems and processes. The association between genetic experts (researchers), clinicians, and statisticians is needed in the future to identify more novel ADHD genes.

A. Justin Thenmozhi
Department of Biochemistry, School of Biological Sciences, Madurai Kamaraj University, Madurai, Tamil Nadu, India

Department of Biochemistry and Biotechnology, Faculty of Science, Annamalai University, Chidambaram, Tamil Nadu, India

C. Dhanalakshmi
Department of Pathology, University of Arizona College of Medicine and College of Pharmacy, Tucson, AZ, USA

T. Manivasagam (✉)
Department of Biochemistry and Biotechnology, Faculty of Science, Annamalai University, Chidambaram, Tamil Nadu, India

M. W. Qoronfleh et al. (eds.), *Proteins Associated with Neurodevelopmental Disorders*, Nutritional Neurosciences, https://doi.org/10.1007/978-981-15-9781-7_6

Keywords

Attention-deficit hyperactivity disorder · ADHD · Genome-wide association study (GWAS)

6.1 Introduction

Attention-deficit/hyperactivity disorder (ADHD) is a common neurodevelopmental disease that affects up to 8–12% of children globally. About 65% of them have ADHD symptoms and neuropsychological problems even in adulthood. ADHD symptoms reversely affect various aspects in the child's or adult's academic success, health, and social relationship with their families, friends, and society. Academic and communal outputs, stressed child-parent associations, and enhanced consumption and expenses on healthcare services are notable outcomes. In the early years of the twentieth century, people believed that children with hyperactive symptoms suffered from a gloomy fault of decent control. During the period of 1930s, theories indicating the involvement of slight brain damage and/or brain dysfunction originated based on the resemblance of behavioral disturbances seen in encephalitis or traumatic birth. The therapeutic role of amphetamines for the management of ADHD symptoms was also demonstrated at the same time. Initially, the disease was termed as the "hyperactive child syndrome" and renamed to "hyperactive reaction of childhood." In the 1980s, the term "attention-deficit disorder" was introduced by DSM-III, and finally in 1994 it was coined as "attention-deficit hyperactivity disorder" in DSM-IV (Polanczyk et al. 2015). The causes of ADHD are proved to be more complex.

Some reports indicated that there are differences in the ADHD inheritance of children (75–90%) and adults (30–50%) (Faraone and Mick 2010) whereas other studies found higher prevalence in adults (Faraone et al. 2000). The candidate gene, linkage, and genome-wide association studies (GWAS) demonstrated that the occurrence of 40% of ADHD inheritance was accounted by the polygenic responsibility containing single-nucleotide polymorphisms (SNPs) (more common variants) and copy number variants (insertions/deletions) (Lee et al. 2013; Martin et al. 2015). The GWAS (Demontis et al. 2017; Grove et al. 2019; Pardiñas et al. 2018) carried out in the children with ADHD, schizophrenia, and autism found 12, 145, and 5 autonomous linked loci, respectively. Although most cases of ADHD occur due to genetic disturbances, the exposure to environmental toxins and their interactions also contribute to the risk of ADHD (Banerjee et al. 2007). Previous studies indicated that the risk of ADHD increased due to exposure to environmental contaminants like polychlorinated biphenyls and lead (Eubig et al. 2010), and biological factors including very low birth weight of babies (Hack et al. 2009), prenatal exposure to nicotine (Ernst et al. 2001), stress (Rodriguez and Bohlin 2005) and alcohol (Han et al. 2015).

6.2 Genetic Overlap

Various twin and adoption experiments demonstrated that the inheritance of ADHD was found between 70 and 90% (Kotte et al. 2013; Thapar et al. 2013), which is as high as other psychiatric disorders like schizophrenia, autism, and bipolar disorder (~75–80%) (Sullivan et al. 2012). Moreover, as ADHD arises due to polygenic genetic background, in which multiple genetic variants contributed, detection of risk genes is challenging (Franke et al. 2009; Gizer et al. 2009).

For the detection of risk genes, genetic studies involved two approaches: (1) hypothesis-driven and (2) hypothesis-free approaches (Table 6.1).

6.3 Psychiatric Comorbidity

Various twin and sibling experiments indicated that about 45% of covariance in genetic factors was found across externalizing, internalizing, and phobia symptoms; 31% in neurodevelopmental symptoms; and 10–36% in psychiatric symptoms (Pettersson et al. 2016; Waldman et al. 2006). The results of two studies demonstrated that 18 and 38% of the SNP heritability of the mother was responsible for internalizing, externalizing, and attention problems (Pappa et al. 2015; Neumann et al. 2016). Few studies found the genetic relations between ADHD and antisocial behavior, substance-abuse, oppositional defiant, and conduct disorders (Nadder et al. 2002; Kuja-Halkola et al. 2015; Capusan et al. 2015). Studies from the USA (Ronald et al. 2010), the UK (Ronald et al. 2008), and Sweden (Ronald et al. 2014) confirmed the genetic overlap in children with ADHD and autism. Genetic overlaps were responsible for the coincidence of internalizing disorders such as attempted and completed suicide with ADHD (Ljung et al. 2014). Experiments indicated the relationship between ADHD and depression, and the coincidence was triggered by

Table 6.1 Detection of risk genes by hypothesis-driven and hypothesis-free approaches (Klein et al. 2017)

Risk genes detected by hypothesis-driven approach (candidate genes and their associates)	Risk genes detected by hypothesis-free approach
Genes coding for dopamine and serotonin transporters (*SLC6A3/DAT1* and *SLC6A4/5HTT*)	A locus on the short arm of chromosome 16—cadherin 13 (*CDH13*)
Genes encoding D4 and D5 dopamine receptors (*DRD4* and *DRD5*)	Latrophilin 3 (*LPHN3*) gene on chromosome 4
Gene for a serotonin receptor (*HTR1B*)	
Gene encoding synaptosomal-associated protein 25, *SNAP25*	
Genes encoding tryptophan hydroxylase 2 (*TPH2*), adrenoceptor alpha 2A (*ADRA2A*), dopamine beta-hydroxylase (*DBH*), and monoamine oxidase A (*MAOA*)	
Genes for *ADRAB2*, *DAT1*, *DRD4*, *TPH2*, and *MAOA*	

shared genetic factors (Faraone and Biederman 1997, 1998). Only a few studies showed the familial link of ADHD to intellectual impairment. A report demonstrated that the intelligence quotient of average individual was nine points greater as compared to ADHD patients (Frazier et al. 2004), and another indicated that the individuals with ID and their relatives were prone to ADHD as compared to peoples without ID and their relatives (Antshel et al. 2006). The involvement of genetic factors in nonpsychiatric comorbidity such as asthma, obesity, and epilepsy was explored (Mogensen et al. 2011; Chen et al. 2017; Brikell et al. 2018).

6.4 Genetic Linkage Studies

Being the earliest genome-wide method, the genetic linkage studies involved searching the DNA segment transmitted within families of ADHD. By involving the Genome Scan Meta-Analysis, Zhou et al. (2008) indicated a significant genome-wide linkage on particular loci (64–83 Mb) of chromosome 16. Most ADHD linkage studies involve the offspring or parents of different people. Arcos-Burgos et al. (2004) studied about 16 multigenerational Colombian families and found the link to chromosomes 4 (4q13.2), 5 (5q33.3), 8 (8q11.23), 11 (11q22), and 17 (17p11) and another region (*LPHN3*). In a study by the International Multisite ADHD Gene project, the analysis of 51 genes from 674 European ADHD families exhibited the overlapping for DAT1, DRD4, ADRAB2, TPH2, and MAOA genes (Brookes et al. 2006).

6.5 Candidate Gene Association Studies

In the beginning, genetic studies of ADHD were associated with the search of genes linked to the cause of ADHD. As ADHD drugs target monoaminergic transmission (dopamine and noradrenaline), many experiments observed "candidate genes" in the pathways. Gizer et al. (2009) indicated that the candidate genes for ADHD included *DAT1*, *DRD4 and DRD5*, *5HTT and HTR1B*, and *SNAP25* and *BAIAP2*I (brain-specific angiogenesis inhibitor 1-associated protein 2 gene). As 3′-untranslated region of *SLC6A3* contains 40 bp variable number of tandem repeats, two variants are formed with 9- (9R) and 10-repeats (10R) due to polymorphism. The 9R allele is connected to adults with ADHD (Faraone and Mick 2010), while 10R allele is for children (Franke et al. 2010) (Table 6.2).

6.6 Genome-Wide Significant Common Variants

Genome-wide association studies (GWAS) examine the whole genome to identify common (greater than 1% of the population) DNA variants having minor etiologic effects. Initial studies on ADHD (Neale et al. 2010; Yang et al. 2013) did not show any genome-wide DNA variant, although about 5000 samples were collected from

Table 6.2 ADHD candidate genes

Name of the gene and protein	Chromosome position	SNP in marker gene
ADRA1B (adrenoceptor alpha 1B)	5q33.3	Six SNPs (rs2030373, rs6884105, rs756275, rs6892282, rs6888306, and rs13162302)
ADRA2A (adrenoceptor alpha 2A)	10q25.2	rs1800544 SNP in the promoter region—G-allele rs553668 SNP in the promoter region—T-allele
ADRA2C (adrenoceptor alpha 2C)	4p16.3	ADRA2C (GT)n repeat polymorphism (STR marker adra2c1)
ADRB1 (adrenoceptor beta 1)	10q25.3	rs10885531
ADRB2 (adrenoceptor beta 2)	5q31-q32	rs17108817
ASTN2 (astrotactin 2)	9q33	C-allele of rs12376789
BCHE (butyryl cholinesterase)	3q26.1–q26.2	rs4680612 and rs829508
BDNF (brain-derived neurotrophic factor)	11p14.1	rs6265
CALY (calcyon neuron-specific vesicular protein)	10q26.3	rs4838721A and rs2275723C
CCSER1/FAM190A (coiled serine-rich protein I)	4q22.1	rs12505502
CDH13 (cadherin 13)	16q23.3	rs11150556
CHRNA3 (cholinergic receptor, nicotinic alpha 3)	15q25.1	rs578776 and rs3743078
CHRNA4 (cholinergic receptor, nicotinic alpha 4)	20q13.33	rs3787138
CHRNA7 (cholinergic receptor, nicotinic alpha 7)	5q13.3	D15S165 and D15S1360 (microsatellite markers)
CNTF (ciliary neurotrophic factor)	11q12	rs550942
COMT (catechol-*O*-methyl transferase)	22q11.21	rs4680
CPLX2 (complexin 2)	5Q35.2	rs7448069
DBH (dopamine beta hydroxylase)	9q34	rs1076150, rs2873804, rs1548364, rs2519154, and rs1108580
DDC (dopamine decarboxylase)	7p12.1	rs3887825, rs3807566, rs7786398, rs10499695, and rs6969081
DIRAS2 (DIRAS family, GTP-binding RAS-like 2)	9q22.32	rs1331503, rs1412005, rs1331503, rs2297354, rs1331504, rs7848810, rs1412005, and rs689687
DRD1 (dopamine receptor D1)	5q34.q35	rs10039221, rs11747728
DRD2/ANNK1 (dopamine receptor D2)	11q22.q23	rs1800496, rs1801028, and rs1799732
DRD3 (dopamine receptor D3)	3q13.3	rs747302, rs1800955
DRD4 (dopamine receptor D4)	11p15	rs4646984 and rs4646983
DRD5 (dopamine receptor D5)	4p16.1	rs6283
FADS2 (fatty acid desaturase 2)	11q12.2	rs498793
FTO (fat mass and obesity associated)	16q12.2	rs8050136

(continued)

Table 6.2 (continued)

Name of the gene and protein		Chromosome position	SNP in marker gene
GDNF (glial cell-derived neurotrophic factor)		5p13.1-p12	rs2910710, rs11111, rs3749692, rs2910797
GPRC5B (G-protein-coupled receptor, class C, group 5, member B)		16p12	rs6497416
GRIN2A (glutamate receptor, ionotropic *N*-methyl-D-aspartate 2A)		16p13.2	rs8049651 polymorphism
GRM5 (glutamate receptor, metabotropic 5)		11q14.3	*rs*7341475
GRM7 (glutamate receptor, metabotropic 7)		3p26-p25	rs3792452
HES1 (Hes family bHLH transcription factor 1)		3q28-q29	rs11689432
HTR1A (5-hydroxytryptamine (serotonin) receptor 1A, G protein coupled)		5q11.2-q13	rs10042486, rs1423691, rs878567
HTR1B (5-hydroxytryptamine (serotonin) receptor 1B, G protein coupled)		6q13	rs6296 and rs6298
HTR1E (5-hydroxytryptamine (serotonin) receptor 1E, G protein coupled)		6q14-q15	rs11962946, rs722763
HTR2A (5-hydroxytryptamine (serotonin) receptor 2A, G protein coupled)		13q14-q21	rs3125, rs7330636
HTR2C (5-hydroxytryptamine (serotonin) receptor 2C, G protein coupled)		Xq23	rs3813929, rs518147
HTR3A (5-hydroxytryptamine (serotonin) receptor 3A, G protein coupled)		11q23.1-q23.2	rs1062613, rs1176744,
HTR3B (5-hydroxytryptamine (serotonin) receptor 3B, G protein coupled)		11q23.1	rs3891484, rs3758987, rs11606194, rs1176746 rs1176744, rs2276307
LPHN3 (Latrophilin 3)		4q13.1	rs6813183, rs1355368, and rs734644
MAOA (Monoamine oxidase A)		Xp11.4-p11.3	rs6323, rs1137070, rs3027407
MAOB	Monoamine oxidase B	Xp11.4-p11.3	rs4824562, rs56220155, rs2283728, rs2283727, rs3027441, rs6324, rs3027440
NOS1	Nitric oxide synthase 1	12q24.22	SNP in exon 1f-VNTR
PNMT	Phenyl ethanolamine *N*-methyl transferase	17q12	rs3764351
PRKG1	Protein kinase, cGMP-dependent, type I	10q11.2	
SLC1A3	Solute carrier family 1, member 3	5p13	rs2269272

(continued)

Table 6.2 (continued)

Name of the gene and protein		Chromosome position	SNP in marker gene
SLC6A2/ NET1	Solute carrier family 6, member 2	16q12.2	rs28386840
SLC6A3/ DAT1	Solute carrier family 6, member 3	5p15.3	rs2937639
SLC6A4/ 5HTT	Solute carrier family 6, member 4	17q11.2	rs140701
SLC9A9/ NHE9	Solute carrier family 9, member 9	3q24	rs13058809, rs1992426, rs4330252, rs6414353, rs6770565, rs7613679
SLC18A2/ VMAT2	Solute carrier family 18, member 2	10q25	rs363256, rs363279
SNAP25	Synaptosomal-associated protein	20p12-p11.2	rs363040, rs363043, rs362584, rs6108463
SPOCK3	Sparc/osteonectin, cwcv and kazal-like domains proteoglycan	4q32.3	rs7689440, rs897511
STX1A	Syntaxin 1A	7q11.2	rs2228607
SYP	Synaptophysin	Xp11.23-p11.22	rs10861968, rs1465044, rs12581451, rs7315638, rs2251214
SYT1	Synaptotagmin 1	12q21.22	rs35459363
TH	Tyrosine hydroxylase	11p15.5	rs3842727
TPH1	Tryptophan hydroxylase 1	11p15.3-p14	rs211102
TPH2 (tryptophan hydroxylase 2)		12q15	rs2129575
VAMP2	Vesicle-associated membrane protein 2	17p13.1	Insertion/deletion polymorphism of 26 bp, referred to as 26 bp Ins/Del

trios (parents and ADHD child), ADHD, and normal children. The molecular landscape obtained from these experiments and with others indicated that genes controlling neurite outgrowth were significantly involved in the etiology of ADHD (Poelmans et al. 2011). Studies conducted later indicated that the pathways controlling the synthesis and release of neurotransmitter, neuronal growth, and formation of axons were responsible for ADHD (Mooney et al. 2016; Aebi et al. 2016).

A cluster of ADHD researchers completed a GWAS meta-analysis involving about 20,000 ADHD patients and about 35,000 controls (Demontis et al. 2017). Among the 12 genes, *FOXP2* (controls dopamine levels in ADHD-linked brain regions) was specifically distinguished as earlier experiments indicted their involvement in adult ADHD. In addition, Demontis et al. (2017) indicated various genome-wide significant loci such as *DUSP6* as a regulator of dopamine levels in the synapses, *ST3GAL3 and MEF2C* as the mutant forms found in ID and other psychiatric disorders, *SEMA6D* as a regulator of neuronal wiring, and *LINC00461* to be responsible for educational attainment.

6.7 Common Variant ADHD as a Polygenic Disorder

The GWAS indicated that the inheritance of ADHD could be due to the polygenic role of numerous variants having low effects (Faraone et al. 2005). The polygenic score of ADHD was established by quantification of ADHD risk scores in a single-sample subset and viewing that in a dose-dependent manner of validation subsets of ADHD. Martin et al. (2014) reported the genetic overlap between ADHD and ASDs, which was confirmed by twin study data (Ronald et al. 2014; Polderman et al. 2014) and gene set analyses (Bralten et al. 2018). Other polygenic studies also confirmed the genetic overlap between ADHD and conduct disorder (Faraone et al. 1991, 1997), schizophrenia and bipolar disorder Larsson et al. (2013), and depression (Faraone et al. 1991). The polygenic risk of ADHD (Demontis et al. 2017) was highly correlated with about 220 disorders and traits, including IQ, lung cancer, coronary artery disease, neuroticism, obesity, depression, smoking, school achievement, and cross-disorder GWAS. The GWAS by Cross-Disorder Group of the Psychiatric Genomics Consortium (2013), analyzed children with psychiatric disorders (bipolar disorder, autism, schizophrenia, and major depressive disorder) with ADHD and indicated the presence of genetic overlap among the ITIH3, CACNA1C, AS3MT, and CACNB2 with ADHD children.

6.8 Rare Variants and Genetic Syndromes

Numerous chromosomal aberrations were present in ADHD children and other developmental diseases (Williams et al. 2012). FMR1 is a gene encoding an RNA-binding protein, and its diminished function leads to mental retardation (fragile X syndrome). The patients suffered from enhanced glutamatergic transmission and diminished GABA signaling. A study by Lo-Castro et al. (2011) indicated that about 31.5%, 7.4%, and 14.8% belonged to inattentive, hyperactive, and combined type, respectively, and were affected by FXS. The locus of neurofibromin 1 (NF1) is found in chromosome 17q11.2, whose mutation leads to the skin, CNS, and eye tumors. About 33% of children who are affected by nonfunctional NF1 presented with ADHD symptoms (Kayl et al. 2000). The pathological connections between ADHD and NF1 might arise due to the damage in basal ganglia.

In children affected by tuberous sclerosis complex (genetic disease linked to brain tumor, reduced development, skin abrasions, and benign tumors of other organ systems, having epileptic seizures and cognitive impairment), Turner syndrome and Klinefelter syndrome, Williams-Beuren syndrome (microdeletion on chromosome 7 and linked with symptoms such as elf-like facial expression, pulmonary and cardiovascular abnormalities), and DiGeorge syndrome (22q11 deletion), the prevalence of ADHD is as high as about 60% (Leyfer et al. 2006; de Vries et al. 2006; Bruining et al. 2010; Hoeffding et al. 2017).

The sentence is corrected as Martin et al. (2015) indicated that the presence of rare single-nucleotide variants (SNVs) (0.3–1% of the entire human DNA) containing polymorphisms of a base pair and copy number (CNVs) was responsible for the role of heredity in ADHD. An experiment involving about 2800 ADHD children found an increase in CNVs in locus 15q13.3 (Williams et al. 2012) and another in 16p13.11 (Williams et al. 2010). Few genome-wide screening studies demonstrated that the deletion in the gene coding for neuropeptide Y on chromosome 7p15.2–15.3 (Lesch et al. 2011) and GRM5, 7, and 8 (coding for glutamate receptor, metabotropic 5, 7, and 8) (Elia et al. 2012) was found in ADHD children.

6.9 Diagnosis and Therapeutic Approaches to ADHD

Lesions in the dopaminergic and noradrenergic neuronal pathways (reduced volume and activity of related brain areas) and their dysfunction have been reported to underlie ADHD behavior such as attention, emotion, and behavior (Heyer and Meredith 2017). Genetic studies indicated that ADHD is not only a simple "catecholaminergic" disease but a multifactorial disease which involved the abnormality of various processes including "neurite growth," "synaptic plasticity," and/or "glutamatergic signal transmission" (Demontis et al. 2018). Another new approach in the application of genetics is prediction, also known as pharmacogenomics. Presently, therapy is primarily dependent upon the enhancement of the dopaminergic neurotransmission. Therefore, more effects of genes concerned in the arbitration of dopaminergic effects could be anticipated. Few studies indicated that polymorphisms in genes of dopaminergic neurotransmission or synapse could lead to stimulant therapy.

6.10 Future Directions in Genetics Research

Until recently, conflicting and unsatisfactory results were obtained from the genetic studies. Although the role of inheritance is considered in ADHD, most of the linkage studies did not indicate wide-ranging overlaps, except few meta-analyses. As ADHD is regarded as a multi-genetic disorder, less knowledge about the genetic component of ADHD was shown by candidate gene-based experiments. GWAS carried out initially did not show any significant changes; however, the findings of studies done later redirected future ADHD research to the novel gene systems and processes. In general, GWAS in psychiatric disorders, including ADHD, were found to be poor (demonstrated less than 10% of variants) as compared to other multifactorial disorders because (1) these are complex and multi-genetic diseases that can be diagnosed with studies involving a large population; (2) the interaction of gene with other genes and environment plays a heavy role in the inheritance; (3) apart from variations in the SNPs found in the majority of experiments, commonly occurring changes in DNA structure (insertions, deletions, and duplications) were less studied; (4) the effect of rare genetic variants in the cause of ADHD is more than the expected; and (5) it is difficult to correlate the clinical diagnosis of psychiatric disorders with the genetic studies.

6.11 Conclusions

Association between the genetic experts (researchers), clinicians, and statisticians is needed in future for the identification of more novel ADHD genes. As the low-frequency gene variants were linked with the inheritance of individual patients, their detection might pave the way for accepting their functions and finding the relationship between each gene to symptoms and pathology. It will lead to the development of prediction and prevention strategies for diagnostic purposes or therapeutic strategies.

References

Aebi M, van Donkelaar MM, Poelmans G, Buitelaar JK, Sonuga-Barke EJ, Stringaris A et al (2016) Gene-set and multivariate genome-wide association analysis of oppositional defiant behavior subtypes in attention-deficit/hyperactivity disorder. Am J Med Genet B Neuropsychiatr Genet 171:573–588

Antshel KM, Phillips MH, Gordon M, Barkley R, Faraone SV (2006) Is ADHD a valid disorder in children with intellectual delays? Clin Psychol Rev 26:555–572

Arcos-Burgos M, Castellanos FX, Pineda D, Lopera F, David Palacio J, Guillermo Palacio L et al (2004) Attention-deficit/hyperactivity disorder in a population isolate: linkage to Loci at 4q13.2, 5q33.3, 11q22, and 17p11. Am J Hum Genet 75:998–1014

Banerjee TD, Middleton F, Faraone SV (2007) Environmental risk factors for attention-deficit hyperactivity disorder. Acta Paediatr 96:1269–1274

Bralten J, van Hulzen KJ, Martens MB, Galesloot TE, Arias Vasquez A, Kiemeney LA et al (2018) Autism spectrum disorders and autistic traits share genetics and biology. Mol Psychiatry 23(5): 1205–1212

Brikell I, Ghirardi L, D'Onofrio BM, Dunn DW, Almqvist C, Dalsgaard S et al (2018) Familial liability to epilepsy and attention-deficit/hyperactivity disorder: a nationwide cohort study. Biol Psychiatry 83:173–180

Brookes K, Xu X, Chen W, Zhou K, Neale B, Lowe N et al (2006) The analysis of 51 genes in DSM-IV combined type attention deficit hyperactivity disorder: association signals in DRD4, DAT1 and 16 other genes. Mol Psychiatry 11:934–953

Bruining H, de Sonneville L, Swaab H, de Jonge M, Kas M, van Engeland H, Vorstman J (2010) Dissecting the clinical heterogeneity of autism spectrum disorders through defined genotypes. PLoS One 5:e10887

Capusan AJ, Bendtsen P, Marteinsdottir I, Kuja-Halkola R, Larsson H (2015) Genetic and environmental contributions to the association between attention deficit hyperactivity disorder and alcohol dependence in adulthood: a large population-based twin study. Am J Med Genet B Neuropsychiatr Genet 168:414–422

Chen Q, Kuja-Halkola R, Sjolander A, Serlachius E, Cortese S, Faraone SV et al (2017) Shared familial risk factors between attention-deficit/hyperactivity disorder and overweight/obesity—a population-based familial coaggregation study in Sweden. J Child Psychol Psychiatry 58:711–718

Cross-Disorder Group of the Psychiatric Genomics Consortium (2013) Identification of risk loci with shared effects on five major psychiatric disorders: a genome-wide analysis. Lancet 381: 1371–1379

de Vries PJ, Gardiner J, Bolton PF (2006) Neuropsychological attention deficits in tuberous sclerosis complex (TSC). Am J Med Genet A 149(3):387–395

Demontis D, Walters RK, Martin J, Mattheisen M, Als TD, Agerbo E et al (2017) Discovery of the first genome-wide significant risk loci for ADHD. BioRxiv 14558:1–43

Demontis D, Walters RK, Martin J, Mattheisen M, Als TD, Agerbo E et al (2018) Discovery of the first genome-wide significant risk loci for attention deficit/hyperactivity disorder. Nat Genet 51: 63–75

Elia J, Glessner JT, Wang K, Takahashi N, Shtir CJ, Hadley D, Sleiman PMA et al (2012) Genome-wide copy number variation study associates metabotropic glutamate receptor gene networks with attention deficit hyperactivity disorder. Nat Genet 44(1):78–84

Ernst M, Moolchan ET, Robinson ML (2001) Behavioral and neural consequences of prenatal exposure to nicotine. J Am Acad Child Adolesc Psychiatry 40:630–641

Eubig PA, Aguiar A, Schantz SL (2010) Lead and PCBs as risk factors for attention deficit/ hyperactivity disorder. Environ Health Perspect 118:1654–1667

Faraone SV, Biederman J (1997) Do attention deficit hyperactivity disorder and major depression share familial risk factors? J Nerv Ment Dis 185:533–541

Faraone SV, Biederman J (1998) Depression: a family affair. Lancet 351:158

Faraone SV, Mick E (2010) Molecular genetics of attention deficit hyperactivity disorder. Psychiatr Clin North Am 33(1):159–180

Faraone SV, Biederman J, Keenan K, Tsuang MT (1991) Separation of DSM-III attention deficit disorder and conduct disorder: evidence from a family-genetic study of American child psychiatric patients. Psychol Med 21(1):109–121

Faraone SV, Biederman J, Mennin D, Wozniak J, Spencer T (1997) Attention-deficit/hyperactivity disorder with bipolar disorder: a familial subtype? J Am Acad Child Adolesc Psychiatry 36: 1378–1387

Faraone SV, Biederman J, Monuteaux MC (2000) Toward guidelines for pedigree selection in genetic studies of attention deficit hyperactivity disorder. Genet Epidemiol 18(1):1–16

Faraone SV, Perlis RH, Doyle AE, Smoller JW, Goralnick JJ, Holmgren MA et al (2005) Molecular genetics of attention-deficit/hyperactivity disorder. Biol Psychiatry 57:1313–1323

Franke B, Neale BM, Faraone SV (2009) Genome-wide association studies in ADHD. Hum Genet 126:13–50

Franke B, Vasquez AA, Johansson S, Hoogman M, Romanos J, Boreatti-Hummer A, Heine M et al (2010) Multicenter analysis of the SLC6A3/DAT1 VNTR haplotype in persistent ADHD suggests differential involvement of the gene in childhood and persistent ADHD. Neuropsychopharmacology 35:656–664

Frazier TW, Demaree HA, Youngstrom EA (2004) Meta-analysis of intellectual and neuropsychological test performance in attention-deficit/hyperactivity disorder. Neuropsychology 18:543–555

Gizer IR, Ficks C, Waldman ID (2009) Candidate gene studies of ADHD: a meta-analytic review. Hum Genet 126:51–90

Grove J, Ripke S, Als TD, Mattheisen M, Walters R, Won H, Pallesen J et al (2019) Identification of common genetic risk variants for autism spectrum disorder. Nat Genet 51:431–444

Hack M, Taylor HG, Schluchter M, Andreias L, Drotar D, Klein N (2009) Behavioral outcomes of extremely low birth weight children at age 8 years. J Dev Behav Pediatr 30:122–130

Han JY, Kwon HJ, Ha M, Paik KC, Lim MH, Gyu Lee S et al (2015) The effects of prenatal exposure to alcohol and environmental tobacco smoke on risk for ADHD: a large population-based study. Psychiatry Res 225:164–168

Heyer DB, Meredith RM (2017) Environmental toxicology: sensitive periods of development and neurodevelopmental disorders. Neurotoxicology 58:23–41

Hoeffding LK, Trabjerg BB, Olsen L, Mazin W, Sparso T, Vangkilde A et al (2017) Risk of psychiatric disorders among individuals with the 22q11.2 Deletion or Duplication. A Danish Nationwide, Register-Based Study. JAMA Psychiatry 74(3):282–290

Kayl E, Moore BD, Slopis JM, Jackson EF, Leeds NE (2000) Quantitative morphology of the corpus callosum in children with neurofibromatosis and attention-deficit hyperactivity disorder. J Child Neurol 15(2):90–96

Klein M, Onnink M, Donkelaar MV, Wolfers T, Harich B, Shi Y, Dammers J, Arias-Vásquez A et al (2017) Brain imaging genetics in ADHD and beyond—mapping pathways from gene to disorder at different levels of complexity. Neurosci Biobehav Rev 80:115–155

Kotte A, Faraone VV, Biederman J (2013) Association of genetic risk severity with ADHD clinical characteristics. Am J Med Genet B Neuropsychiatr Genet 162B:718–733

Kuja-Halkola R, Lichtenstein P, D'Onofrio BM, Larsson H (2015) Codevelopment of ADHD and externalizing behavior from childhood to adulthood. J Child Psychol Psychiatry 56:640–647

Larsson H, Ryden E, Boman M, Langstrom N, Lichtenstein P, Landen M (2013) Risk of bipolar disorder and schizophrenia in relatives of people with attention-deficit hyperactivity disorder. Br J Psychiatry 203:103–106

Lee SH, Ripke S, Neale BM, Faraone SV, Purcell SM, Perlis RH, Mowry BJ, Thapar A, Goddard ME, Witte JS et al (2013) Genetic relationship between five psychiatric disorders estimated from genome-wide SNPs. Nat Genet 45(9):984–994

Lesch KP, Selch S, Renner TJ, Jacob C, Nguyen TT, Hahn T et al (2011) Genome-wide copy number variation analysis in attention-deficit/hyperactivity disorder: association with neuropeptide Y gene dosage in an extended pedigree. Mol Psychiatry 16(5):491–503

Leyfer OT, Folstein SE, Bacalman S, Davis NO, Dinh E, Morgan J, Tager-Flushberg H, Lainhart JE (2006) Comorbid psychiatric disorders in children with autism: interview development and rates of disorders. J Autism Dev Disord 36(7):849–861

Ljung T, Chen Q, Lichtenstein P, Larsson H (2014) Common etiological factors of attention-deficit/hyperactivity disorder and suicidal behavior: a population-based study in Sweden. JAMA Psychiatry 71:958–964

Lo-Castro A, D'Agati E, Curatolo P (2011) ADHD and genetic syndromes. Brain Dev 33(6): 456–461

Martin J, Hamshere ML, Stergiakouli E, O'Donovan MC, Thapar A (2014) Genetic risk for attention-deficit/hyperactivity disorder contributes to neurodevelopmental traits in the general population. Biol Psychiatry 76:664–671

Martin J, O'Donovan MC, Thapar A, Langley K, Williams N (2015) The relative contribution of common and rare genetic variants to ADHD. Transl Psychiatry 5:e506

Mogensen N, Larsson H, Lundholm C, Almqvist C (2011) Association between childhood asthma and ADHD symptoms in adolescence—a prospective population-based twin study. Allergy 66: 1224–1230

Mooney MA, McWeeney SK, Faraone SV, Hinney A, Hebebrand J, Consortium I et al (2016) Pathway analysis in attention deficit hyperactivity disorder: an ensemble approach. Am J Med Genet B Neuropsychiatr Genet 171:815–826

Nadder TS, Rutter M, Silberg J, Maes H, Eaves L (2002) Genetic effects on the variation and covariation of attention deficit-hyperactivity disorder (ADHD) and oppositional-defiant disorder/conduct disorder (ODD/CD) symptomatologies across informant and occasion of measurement. Psychol Med 32:39–53

Neale BM, Medland S, Ripke S, Anney RJ, Asherson P, Buitelaar J et al (2010) Case-control genome-wide association study of attention-deficit/hyperactivity disorder. J Am Acad Child Adolesc Psychiatry 49:906–920

Neumann A, Pappa I, Lahey BB, Verhulst FC, Medina-Gomez C, Jaddoe VW et al (2016) Single nucleotide polymorphism heritability of a general psychopathology factor in children. J Am Acad Child Adolesc Psychiatry 55:1038–1045

Pappa I, Mileva-seitz VR, Bakermans-kranenburg MJ, Tiemeier H, Van Ijzendoorn MH (2015) The magnificent seven: a quantitative review of dopamine receptor d4 and its association with child behavior. Neurosci Biobehav Rev 57:175–186

Pardiñas AF, Holmans P, Pocklington AJ, Escott-Price V, Ripke S, Carrera N et al (2018) Common schizophrenia alleles are enriched in mutation-intolerant genes and in regions under strong background selection. Nat Genet 50(3):381–389

Pettersson E, Larsson H, Lichtenstein P (2016) Common psychiatric disorders share the same genetic origin: a multivariate sibling study of the Swedish population. Mol Psychiatry 21:717–721

Poelmans G, Pauls DL, Buitelaar JK, Franke B (2011) Integrated genome-wide association study findings: identification of a neurodevelopmental network for attention deficit hyperactivity disorder. Am J Psychiatry 168:365–377

Polanczyk GV, Salum GA, Sugaya LS, Caye A, Rohde LA (2015) Annual research review: a meta-analysis of the worldwide prevalence of mental disorders in children and adolescents. J Child Psychol Psychiatry 56:345–365

Polderman TJC, Hoekstra RA, Posthuma D, Larsson H (2014) The co-occurrence of autistic and ADHD dimensions in adults: an etiological study in 17,770 twins. Transl Psychiatry 4:ed435

Rodriguez A, Bohlin G (2005) Are maternal smoking and stress during pregnancy related to ADHD symptoms in children? J Child Psychol Psychiatry 46:246–254

Ronald A, Simonoff E, Kuntsi J, Asherson P, Plomin R (2008) Evidence for overlapping genetic influences on autistic and ADHD behaviours in a community twin sample. J Child Psychol Psychiatry 49:535–542

Ronald A, Edelson LR, Asherson P, Saudino KJ (2010) Exploring the relationship between autistic-like traits and ADHD behaviors in early childhood: findings from a community twin study of 2-year-olds. J Abnorm Child Psychol 38:185–196

Ronald A, Larsson H, Anckarsater H, Lichtenstein P (2014) Symptoms of autism and ADHD: a Swedish twin study examining their overlap. J Abnorm Psychol 123:440–451

Sullivan PF, Magnusson C, Reichenberg A, Boman M, Dalman C et al (2012) Family history of schizophrenia and bipolar disorder as risk factors for autism. Arch Gen Psychiatry 69:1099–1103

Thapar A, Cooper M, Eyre O, Langley K (2013) What have we learnt about the causes of ADHD? J Child Psychol Psychiatry 54:3–16

Waldman ID, Nigg JT, Gizer IR, Park L, Rappley MD, Friderici K (2006) The adrenergic receptor α-2A gene (ADRA2A) and neuropsychological executive functions as putative endophenotypes for childhood ADHD. Cogn Affect Behav Neurosci 6(1):18–30

Williams NM, Zaharieva I, Martin A, Langley K, Mantripragada K, Fossdal R, Stefansson H et al (2010) Rare chromosomal deletions and duplications in attention-deficit hyperactivity disorder: a genome-wide analysis. Lancet 376(9750):1401–1408

Williams NM, Franke B, Mick E, Anney RJ, Freitag CM, Gill M, Thapar A, O'Donovan MC, Owen MJ, Holmans P et al (2012) Genome-wide analysis of copy number variants in attention deficit hyperactivity disorder: the role of rare variants and duplications at 15q13.3. Am J Psychiatry 169(2):195–204

Yang L, Neale BM, Liu L, Lee SH, Wray NR, Ji N et al (2013) Polygenic transmission and complex neuro developmental network for attention deficit hyperactivity disorder: genome-wide association study of both common and rare variants. Am J Med Genet B Neuropsychiatr Genet 162: 419–430

Zhou K, Dempfle A, Arcos-Burgos M, Bakker SC, Banaschewski T, Biederman J et al (2008) Meta-analysis of genome-wide linkage scans of attention deficit hyperactivity disorder. Am J Med Genet B Neuropsychiatr Genet 147B:1392–1398

The Role of Protein Kinases in the Cause and Progression of Attention-Deficit Hyperactivity Disorder

7

Thamilarasan Manivasagam, Arokiasamy Justin-Thenmozhi, M. Walid Qoronfleh, and Asokan Prema

Abstract

Protein phosphorylation is reported to participate in numerous cellular processes like cell division, metabolism, viability, and cell death. It is determined by the group of enzymes called protein kinases. Changes in the functions of these enzymes might result in pathological alterations in the cell and have been implicated in the cause of numerous diseases, including vascular and inflammatory diseases, neurological disorders, and cancer. Various protein kinases such as A, G, C, CaMKII, casein kinase 1, cyclin-dependent kinase 5, MAPK, GSK-3β/Akt/mTOR/Wnt, ERK, TGF, and tyrosine kinases were reported to be involved in the cause and progression of ADHD. As alterations in the mRNA expression and level of various protein kinases were found in ADHD children, the protein kinases could act as biomarkers in ADHD. Furthermore, various protein kinases serve as promising targets for numerous neurological diseases that require

T. Manivasagam
Department of Biochemistry and Biotechnology, Faculty of Science, Annamalai University, Chidambaram, Tamil Nadu, India

A. Justin-Thenmozhi (✉)
Department of Biochemistry and Biotechnology, Faculty of Science, Annamalai University, Chidambaram, Tamil Nadu, India

Department of Biochemistry, School of Biological Sciences, Madurai Kamaraj University, Madurai, Tamil Nadu, India

M. Walid Qoronfleh
Q3CG Research Institute (QRI), Research and Policy Division, Ypsilanti, MI, USA

21HealthStreet, Consulting Services, London, UK

Applied Biomedicine, Inc., Doha, Qatar

A. Prema
Department of Chemistry, Government College of Arts and Science, Tittagudi, Tamil Nadu, India

M. W. Qoronfleh et al. (eds.), *Proteins Associated with Neurodevelopmental Disorders*, Nutritional Neurosciences, https://doi.org/10.1007/978-981-15-9781-7_7

effective therapy. The modulators hold remarkable guarantees as therapeutic agents for various CNS disorders, including ADHD.

Keywords

Attention-deficit/hyperactivity disorder · ADHD · Protein kinases · Secondary messenger · CNS disorders · Neurodevelopmental disorders

7.1 Introduction

Protein kinases are enzymes predominantly involved in the addition of a γ-phosphate group of adenosine triphosphate (ATP) to the R group of several amino acids of a protein. In contrast, phosphatases eliminate the added phosphate group from them. In combination, the above two processes are vital in posttranslational modifications that regulate various cellular functions, including metabolism, transport, division, existence, and death, mainly through numerous external stimuli. Transfer of PO_4 converts the hydrophobic apolar protein into hydrophilic polar, thus permitting them to change their conformation, hence allowing interaction with other proteins and thereby inducing to form or detach protein complexes (Alberts et al. 2007).

Human genome sequencing studies indicated that around 2% of the genome codes for protein kinase. Additionally, the structure of about 500 protein kinases was determined by X-ray protein crystallography. The website of Cell Signaling Technology PhosphoSitePlus (www.phosphosite.org) indicated that about 200,000 phosphosites exist in human proteins. Another website, KinexusPhosphoNET (www.phosphonet.ca), lists almost the same number of phosphosites and predicts the possible existence of an additional 760,000 phosphosites in humans. Out of 21,000 proteins encoded by the human genome, more than 14,000 proteins (two-thirds) were reported to be phosphorylated. More than 33% of the phosphorylation process occurred by *O*-phosphorylation of serine (Ser), threonine (Thr), and tyrosine residues (Tyr). The majority of phosphorylated residues are Ser (86.4%), followed by Thr (11.8%) and Tyr (1.8%) (Schwartz and Murray 2011).

7.2 Classification of Protein Kinases

Although there are various protein kinases, the main classification is based on the specificity of substrate and amino acid arrangements of catalytic sites (Fig. 7.1).

Subfamilies of protein kinases are classified based on the mechanism of action as AGC, CaMK, CK1, CMGC, STE, TK, and TKL (Table 7.1).

7.3 Protein Kinases A, G, C

Protein kinase A (PKA), a key regulator of various cellular functions, is the essential upstream kinase of CREB (cAMP response element-binding protein), which phosphorylates CREB in response to various effector molecules, including the

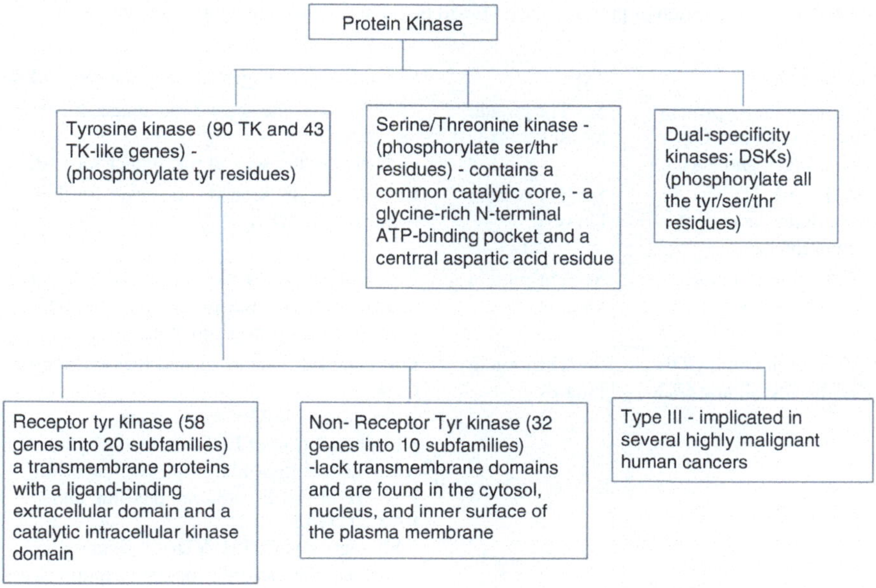

Fig. 7.1 Classification of protein kinases based on the phosphorylation site (Shchemelinin et al. 2006)

brain-derived neurotrophic factor (BDNF) gene (Leal et al. 2017). The PKA/CREB/BDNF pathway is liable for the neuron survival, synaptic morphology, and efficacy of synaptic transmission (Zhong et al. 2018). PKA signaling pathway is also called "central hub" that interrelates with other signaling pathways (Robinson-White and Stratakis 2002) such as MAPK, PKC, and PKB pathways. Few studies interlink the role of PKA signaling with ADHD. Children having a mutation in *PRKAR1A* (that encodes the RIα regulatory subunit of PKA) lead to an enhanced occurrence of learning disorders, ADHD, anxiety, and depression (Keil et al. 2014).

Alterations in dopamine signaling are primary risk factors for ADHD. Dopamine transporter (DAT) mediates the reuptake of dopamine from the synaptic cleft to cytosol. It controls the concentration and the extent of its action at synaptic receptors, which offer serious regulatory effect on overattention and behaviors (Robbins 2003). The DAT might be critically involved in the dopaminergic dysfunction associated with ADHD. The intracellular signaling of DAT occurred through the PKA and AKT pathways. ADHD is associated with increased DAT expression in the striatum (Hall et al. 2014) and with specific polymorphisms in the *DAT* gene (Greenwood et al. 2013).

Neurobeachin (NBEA), which belongs to the family of BEACH domain protein, is a large cytosolic protein (3000–4000 a.a.) attached peripherally with membranes and playing a vital role in the intracellular targeting of membrane proteins. It is found in the central synapses, which helps in its formation (Su et al. 2004). Neurobeachin is a member of the A-kinase anchoring protein (AKAP) family. Its amino-terminal has an AKAP domain, which recruits cAMP-dependent protein kinase A (PKA) by

Table 7.1 Classification of protein kinases based on the mechanism of action (Shchemelinin et al. 2006)

Protein kinase	Types	Secondary messenger/downstream molecules
AGC containing protein kinases A, G, and C (PKA, PKG, and PKC)	Ser/Thr protein kinases	cAMP acts as the second messenger for PKA and PKG Ca^{2+} acts as the second messenger for PKC
CaMK—Ca^{2+}/calmodulin-dependent protein kinases	Ser/Thr protein kinases	Ca^{2+} acts as the second messenger for their members (CaMK I, II, III, IV, and V)
CK1—casein kinase 1 or cell kinase 1	Ser/Thr protein kinases	CK1 isoforms (α, α-like, γ1, 2, and 3, δ and ε) activate other kinases/pathways (Wnt, p53, hippo, and hedgehog signaling)
CMGC containing CDK, MAPK, GSK3, and CLK	Ser/Thr protein kinases	DAG and Ca^{2+} act as second messengers for CDK Activation of MAPK leads to the phosphorylation of ERK1/2, resulting in regulation of light lilac oval, such as c-Jun, Fos, cyclin D1, E-cadherin, paxillin, and PAX2/ITGA8 GSK regulates CREB, VDAC, signal transducer and activator of transcription-3, MCL-1, heat-shock factor 1 (HSF1), Myc, NF-κB, c/EBP, p53, β-catenin, and nuclear factor of activated T cells
STE—sterile kinase	Ser/Thr protein kinases	FOXO1, FOXO3, MAPK, caspase, IκB, YAP, TAZ, Akt, H2B, P13K, PEST, NDRs (nuclear Dbf2 related), etc.
TK—tyrosine kinase	Tyr protein kinases	IP3 and DAG act as second messengers Activated by FGFR, PDGFR, and VEGFR, act by Tie, Trk, Eph, DDR, CCK4, Alk, Met, TAM, Musk, Lmr, Kin-9 and -16, VKR, Src subgroup, JAK, FAK, and FER
TKL—tyrosine kinase-like	Ser/Thr protein kinases similar to TK	MLK (acts by MAPK cascade), MLKL (TNF receptors and NF-kB), RAF (ErK-MAPK), STKR (TGFb), LRRK, LISK (LIMK and TESK), IRAK (NF-kB and JNK), RIPK (coupling TNF receptors with NF-kB)

binding to its regulatory RIIα subunit. Changes in the functions of NBEA and DAT occurred due to SNP and/or genetic deletion, leading to the alteration in the selection and/or phosphorylation of substrates by kinases, resulting in psychological disorders including ADHD (Kitagishi et al. 2015).

In the presence of few nutrients such as ω-3-polyunsaturated fatty acids (PUFA), the PKA signaling pathway controls cell growth and division (Enns et al. 2009). Administration of PUFAs such as eicosapentaenoic acid (EPA) (Szentandrássy et al. 2007) and bitter melon seed oil containing isomers of conjugated linolenic acid (Chen et al. 2012) enhanced the phosphorylation and activation of PKA, whereas genistein, a soy-derived isoflavone, activates the cAMP/PKA signaling cascade and might improve the symptoms of neuronal disorders (Liu et al. 2006). Several clinical studies indicated the diminished level of ω-3-PUFA in the blood of child and adolescent ADHD patients, and their decreased prenatal intake was linked with ADHD risk, whereas their supplement improved the symptoms of ADHD (Fuentes-Albero et al. 2019; Yonezawa et al. 2018; Chang et al. 2018; Lopez-Vicente et al. 2019). Administration of EPA induced less but significant enhancement of inattention and hyperactivity in ADHD children (Chang et al. 2018; Bloch and Qawasmi 2011). Both the in vitro and in vivo experiments indicated that the n-3-PUFA supplement to PC12 cells enhanced their viability (Bartl et al. 2014), and n-3-PUFA along with MPH treatment synergistically improved the MPH therapeutic response in ADHD children (Firouzkouhi Moghaddam et al. 2017). Liu et al. (2015a, b) demonstrated that PUFA might alter various signaling pathways by enhancing the activity of adenylate cyclase and PKA, thereby influencing serotonin, beta-adrenergic, and dopamine receptors. Diet low in protein and energy restriction leads to reduced cAMP signaling, which is distinguished by a lowered PKA activity (O'Brien et al. 1998), suggesting that dietary protein might modulate PKA activity.

7.4 CaMKII

Ca^{2+}/calmodulin–dependent protein kinase II (CaMKII), a protein present in the postsynaptic and presynaptic neurons of all regions of brain, was shown to be involved in the induction of long-term potentiation, neuronal plasticity, and learning and memory processes. CaMKII is able to regulate the receptors of the ionotropic glutamate type, adenylate cyclase type III, assembly of cytoskeletal components, and subunits of the GABA receptors (Papa et al. 1998).

CaMKII signaling cascade is reported to be involved in the regulation of dopaminergic neurotransmission after the chronic treatment with amphetamine (AMPH), a central nervous system stimulant and methamphetamine (METH), a first-choice psychostimulant for ADHD in the animal model of ADHD. Another experiment indicated that the administration of AMPH enhanced the extracellular levels of dopamine in the nucleus accumbens shell of the animal model of ADHD, which was eliminated by co-administration of CaMKII inhibitor, indicating that induction of a CaMKII-dependent mechanism is needed (Pierce and Kalivas 1997). After a single injection of METH, diminished activity of CaMKII was found in five different regions (hippocampus, nucleus accumbens, parietal cortex, frontal cortex, and striatum) of the rat brain (Papa et al. 1998).

The stroke-prone spontaneously hypertensive rats (SHRSP) exhibited ADHD-like behaviors, including cognitive deficits. Yabuki et al. (2014) demonstrated that an increase in CaMKII activity (needed for memory and learning process) was found in the medial prefrontal cortex but not in the hippocampus in SHRSP rats. It is also indicated that the increased CaMKII autophosphorylation in the mPFC induced the phosphorylation of the CaMKII substrate α-amino-3-hydroxy-5-methyl-4-isoxazolepropionic acid-type glutamate receptor subunit 1 (GluR1) (Ser-831). The

Ca^{2+}-dependent phosphorylation levels of factors such as extracellular signal-regulated kinase (ERK) and protein kinase C (PKC) were unchanged in the SHRSP mPFC. Also, protein levels of the dopamine D2 receptor (D2R) but not the dopamine D1 receptor (D1R) were increased in the SHRSP mPFC. Acute MPH (1 mg/kg, p.o.) administration attenuated aberrant CaMKII in the mice model of ADHD.

7.5 Casein Kinase 1

Casein kinase 1 (CK1), a member of the eukaryotic protein kinase family, is a critical player in various physiological functions such as cell signaling, cellular trafficking, and circadian rhythm (Panek et al. 1997; Murakami et al. 1999; Lowrey et al. 2000); behavioral traits (Lee et al. 2001; Xu et al. 2005); cell cycle regulation (Behrend et al. 2000); and pathological (amyloid-β formation and tauopathies) processes (Li et al. 2004; Flajolet et al. 2007). Synaptic signals from the dopaminergic system are integrated by a protein called dopamine and cAMP-regulated phosphoprotein (DARPP-32) (Desdouits et al. 1995). Impairment in the nigrostriatal pathway, particularly dopaminergic neurotransmission, has been related to ADHD (Tripp and Wickens 2008; Luman et al. 2010). The CK1 controls the site-specific phosphorylation of DARPP-32, thereby influencing the function of neurons. There are seven isoforms of CK1 (α, β, γ1, γ2, γ3, δ, and ϵ) found in vertebrates, of which CK1δ is predominantly present in the brain.

Transgenic mice with CK1δ overexpression (CK1δ OE) in the forebrain exhibited a reduction in the levels of D1R and D2R dopamine receptors, hyperactivity, lowered anxiety, and deficiencies in nesting behavior. D-amphetamine- or methylphenidate-injected CK1δ OE mice exhibited hypoactivity, indicating the key role of CK1 in the dopamine signaling pathway (Zhou et al. 2010). Another study by the same group (Zhou et al. 2020) demonstrated the lowered visual attention, abnormal fronto-striatal and synaptic connections, diminished glutamatergic and GABAergic transmission, changes in Drd1a medium spiny neurons, and reduced D-amphetamine-mediated place preference, representing a disturbance in the dopamine-dependent reward pathway in CK1δ OE mice. Zhou et al. (2020) concluded that more studies are needed in future to examine the effect of CK1δ in the cause and pathology of ADHD.

7.6 Cyclin-Dependent Kinase 5

Cdk5, a neuronal protein kinase, plays a key role in the early development of neurons such as migration, growth of axons and dendrite, formation of the synapse, remodeling of microtubules, and corticogenesis, which are required for learning and memory (McLinden et al. 2012; Shah and Lahiri 2017). Out of the eight gene variants analyzed, three variants of CDK5 (rs2069454, rs2069456, and rs2069459) were linked with the cause of ADHD (Maitra et al. 2017). Transgenic mice deficient in CDK-like 5 (CDKL5) genes exhibited primary symptoms of ADHD such as enhanced aggressiveness, locomotion, and impulsivity along with impairment in learning. CDKL5 deficiency interrupts dopamine synthesis and reduces the expression of forkhead box protein P2 and μ-opioid receptor (social communication-related key essential genes) in the corticostriatal circuit (Jhang et al. 2017). Jhang et al. (2020) showed that mice lacking CDKL5 diminished the phosphorylation of striatal dopamine transporter resulting in an enhanced level of extracellular dopamine and reduced locomotion. Moreover, oral administration of methylphenidate, a DAT inhibitor, diminished the hyper-locomotive symptoms of ADHD in Cdkl5null mice, which indicated that the CDKL5 might have a primary role in the control of movement and the therapeutic progress for hyperactivity diseases.

Association of CDK5 with the homologous cofactors p35 or p39 determined its activity by CDK5R1 and CDK5R2, respectively. Transgenic mice lacking p35 demonstrated enhanced cerebral glucose uptake and hyperactivity. Knockout mice displayed a reduction in locomotion and enhancement in prefrontal cortex neurotransmitter levels and dopaminergic activity in striatal and PFC slices after the treatment with methylphenidate (Drerup et al. 2010). Bouchard et al. (2010) indicated that exposure to organophosphates resulting in the dysregulation of Cdk5 (Wang et al. 2006) has a connection with the incidence of ADHD in humans.

7.7 MAPK

MAPKs are the key players reported to be involved in the maintenance of neuroplasticity and regulation of inflammatory processes. p38 MAPK and JNK are the vital members of MAPK family that are reported as targets for anti-inflammation (Feng et al. 2016). Various in vivo models of numerous diseases indicated that the MAPK signaling pathway inhibition leads to a reduction in neuroinflammation (Crown et al. 2008; Lee et al. 2011). The nuclear transcription factor (NF-κB) signaling pathway, a downstream pathway of MAPK, is a classic pathway related to inflammation and activation of immune cells in the brain. Its activation promotes the production of pro-inflammatory factors such as TNF-α, IL-1β, and IL-6 (Hayden and Ghosh 2011). Some drugs reduce neuroinflammatory responses by inhibiting MAPK and NF-κB pathways to play a therapeutic role in brain disorders (Qin et al. 2018). A combination of GWAS, transcriptome analysis of mice models, and candidate gene studies indicated that the aggressive behaviors of ADHD were linked with enriched common pathways such as GPCR signaling pathway. Both ERK/MAPK and Rho-GTPase signaling molecules coordinated internally through the receptors of serotonin, glutamate, dopamine, and GABA signaling pathways to GPCR signaling (Zhang-James et al. 2019).

7.8 GSK-3β/Akt/mTOR/Wnt Signaling Pathways

GSK-3β is reported to be involved in the pathogenesis of neuropsychiatric disorders, including ADHD, and acted as a therapeutic target in DA dysfunction-linked diseases. Association of two SNPs at positions −1727A/T (rs3755557) and −50C/T (rs334558) of GSK-3β is present nearer to the transcriptional start site and upstream of the coding sequence with ADHD (Shim et al. 2012). The physiological role of DA is mediated by two different types of G protein-coupled receptors. The D1-like receptors (D1 and D5) are mostly coupled to Gs_{α} (stimulatory), and the D2-like receptors (D2, D3, and D4) are coupled to $G_i/G_{o\alpha}$ (inhibitory) receptors (Missale et al. 1998). The D1-like receptor is the mediator of cAMP, which phosphorylates PKA, thereby activating DARPP-32, resulting in the expression of DA-induced behavior (Greengard 2001). The D2-type receptors regulate inositol signaling, Akt-GSK-3 signaling, ion-channel permeability, and phosphatase activity (Ralph et al. 2001) independent of cAMP signaling pathway, leading to regulation of DA-linked behaviors. Inhibition of GSK-3 by lithium or other inhibitors totally eliminates DA-associated behavior in transgenic mice model of ADHD and enhances the chances of targeting the Akt-GSK-3 signaling pathway as a therapeutic tool (Fig. 7.2).

Grünblatt et al. (2019) found a link with gene variants of LRP5 and LRP6 (receptor of Wnt signaling pathway) in Caucasian European child and adolescent ADHD populations. Wang et al. (2020) found that the ADHD patients had little grey matter and miR-126-5p, miR-30e-5p, and miR-140-3p, which correlated with Wnt signaling pathways. Moreover, there is a negative relationship with ADHD to the genetic loci corresponding to the surface area and thickness of the cerebral cortex, which is clustered with genes responsible for the Wnt pathway (Wang et al. 2020).

The mammalian target of rapamycin (mTOR) signaling pathway plays a crucial role in various processes of neuronal development and maintenance of synaptic plasticity (Jaworski and Sheng 2006). It is a constituent of the phosphatidylinositol 3-kinase (PI3K) and functions at the main junction of that cell survival pathway, which acts both up- and downstream of Akt. The protein that is required for both cell division and growth is regulated by this pathway through the formation of two protein complexes: mTOR complex I (mTORC1) and mTOR complex II (mTORC2) (Zarogoulidis et al. 2014).

The genome-wide association studies by the Psychiatric Genomic Consortium of ADHD identified that the Wnt/catenin signaling pathway is linked with brain-related pathways, including the formation of the synapse, development of neurons, and dopaminergic transmission (Demontis et al. 2019). Various studies demonstrated a strong correlation between ADHD and components of Wnt signaling pathway, including the α-catenin (involved in the formation of catenin–cadherin cell–adhesion complexes that is required for synaptic transmission) and Kv channel-interacting protein 4 (negative feedback loop) of this pathway (Lesch et al. 2008; Weissflog et al. 2013). Further GWAS studies (Aebi et al. 2016) revealed a significant link between Wnt/β-catenin signaling in a subpopulation of ADHD patients and childhood ADHD.

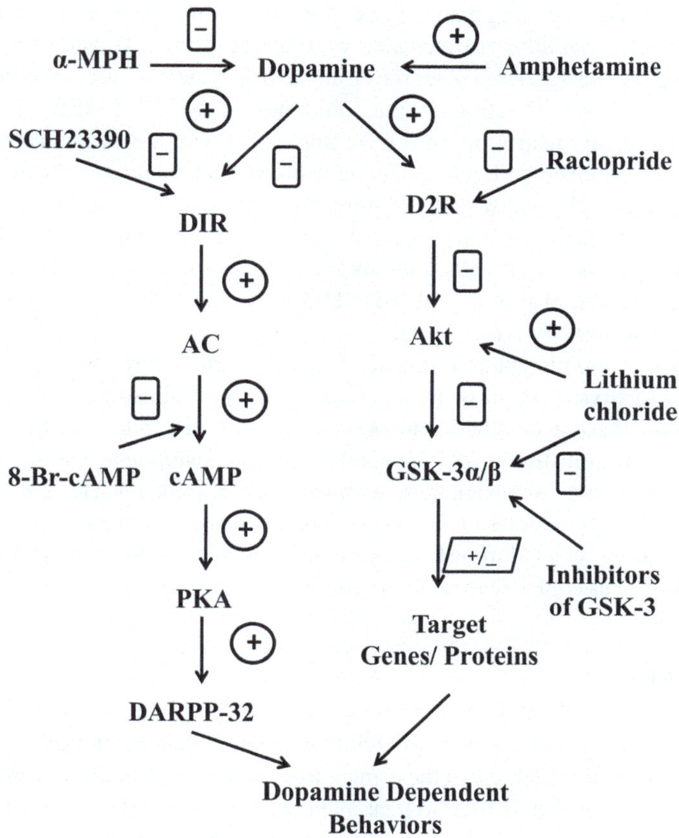

Fig. 7.2 Dopamine (DA) receptor signaling

MPH activates Wnt signaling in cellular models of ADHD (SH-SY5Y and PC12 cells) without inhibiting DAT, which was confirmed by the administration of GBR-12909, a selective dopamine transporter inhibitor showing the opposite effects (Grünblatt et al. 2018). MPH-injected rodents showed the changes in the mRNA expression related to proteins involved in the development of neurons, synapse, and axons, and the Wnt-signaling pathways (Sadasivan et al. 2012; Dela Pena et al. 2013). Transcriptome analysis indicated that the treatment of MPH leads to the origination of enriched transcripts of Wnt pathway in various regions of rodent's brain including substantia nigra (Sadasivan et al. 2012), striatum, and frontal cortex (Dela Pena et al. 2013) and lymphocytes of adult ADHD patients (Schwarz et al. 2015). Oakes et al. (2019) indicated that the levels of β-catenin, vascular endothelial growth factor, and tropomyosin receptor kinase B were enhanced after a low dose (1 mg/kg) of chronic MPH treatment (indicating enhanced cell division and viability of hippocampal cells). In contrast, their high dose (10 mg/kg) of chronic treatment reduced the levels of these growth factors.

The Akt/mTOR signaling pathway, a connected pathway to Wnt, is reported to be involved in cell viability, maintenance of synaptic plasticity, neurogenesis, and memory processing (Sánchez-Alegría et al. 2018), which is also involved in the pathology of ADHD. Mutations in the components (TSC1/2, RHEB, phosphatase tensin homolog, and neurofibromin) were implicated with the pathology of ADHD (Lee 2015). Schmitz et al. (2019) demonstrated that acute treatment of MPH modified the mTOR pathway by diminishing the activity of Akt and mTOR substrates, while their chronic treatment enhanced the activity of the substrates. The above experiments strengthen the hypothesis of Wnt and Akt/mTOR signaling involvement in the pathology of ADHD and confirm the notation that these pathways might directly affect MPH.

Low socioeconomic status might lead to enhanced cortisol levels in pregnant women due to anxiety, depression, and trauma, or by administration of dexamethasone or betamethasone (synthetic glucocorticoids) to avoid preterm birth (Alexander et al. 2012). Administration of glucocorticoids can manipulate various signaling pathways, including Wnt signaling pathways, or diminish ERK and PI3K/Akt signaling pathways (Odaka et al. 2017). Treatment of dexamethasone to human neural progenitor cells diminished expression of Wnt genes and enhanced the expression of Dickkopf 1 (Moors et al. 2012).

7.9 ERK

ERK1 null mice were also noted to exhibit behavioral excitement profiles similar to ADHD patients. However, when these mice were treated with lithium, olanzapine, or valproate, drugs commonly used to treat manic phases of ADHD or bipolar disorder, hyperactive behaviors were reversed (Engel et al. 2009).

7.10 TGF

Mice devoid of TGF-β signaling showed hyperactivity and inflexibility in learning and memory behavior. This experiment supported the impact of TGF-β signaling pathway in the regulation of excitatory/inhibitory synaptic input of DA and GABAergic neurons and its role in the cause and progression of neuropsychiatric disorders (Luo et al. 2016).

7.11 Tyrosine Kinase

Brain-derived neurotrophic factor (BDNF) is mainly involved in the production, liberation, and uptake of dopamine in dopaminergic neurons (Castellanos et al. 2002; Plessen et al. 2006) and participated in the differentiation, survival, and plasticity of neurons (Corominas-Roso et al. 2013). Changes in the activity of BDNF/tyrosine kinase B are found in the midbrain of ADHD children that might lead to the development of hyperactivity (Li et al. 2014). Diminished circulatory levels of BDNF were found in both ADHD children and adolescence (Corominas-Roso et al. 2013). Polymorphisms in the genes of BDNF (rs11030101 and rs10835210) were reported to be linked with the risk of ADHD (Cho et al. 2010; Kwon et al. 2015). Polymorphism in BDNF (rs6265/Val) is more common in ADHD females than their male counterparts (Li et al. 2014). This polymorphism was linked with the vulnerability to neuronal disorders and anxiety (Gadow et al. 2009), which might elucidate some psychiatric comorbid disorders. Mice with lowered BDNF expression in various brain regions (hippocampus, cortex, and hypothalamus) exhibited hyperactivity and aggression (Rios et al. 2001), which indicated the relationship between activities of BDNF and the control of motor functions, leading to hyperactivity symptoms. Reduced hippocampal BDNF and TrkB were found in SHRSP rats, which is responsible for cognitive and memory impairment (Jeong et al. 2014). In DAT knockout mice, reduced mRNA expression of BDNF and TrkB receptors was reported in the frontal cortex (Fumagalli et al. 2003). About 5% decrease in the volume of various regions of the brain, including the amygdala, prefrontal cortex, cerebellum, hippocampus, corpus callosum, basal ganglia, and temporal lobe, was found in ADHD children (Amico et al. 2011), similar to BDNF knockout mice (Ouchi et al. 2013). BDNF-deficient mice exhibited diminished hippocampal and cortical lipids and proteins of the myelin sheath (Tsai 2003). Pharmacological agents like antidepressants, methylphenidate, and serotonin reuptake inhibitors enhance the levels of BDNF (Tsai 2003; Simchon-Tenenbaum et al. 2015).

7.12 Conclusions

Alterations in the mRNA expression and level of various protein kinases were found in ADHD. Therefore, recognizing all these features will lead to setting up of the clinical use of these PK modulators as biomarkers in ADHD. As numerous protein kinases act as promising targets for various neurological diseases that require effective therapy, they hold remarkable guarantee as therapeutic agents for CNS disorders, including ADHD. However, the expansion of kinase-targeted therapeutic agents for CNS diseases must overcome various challenges like blood-brain barrier penetrance, selectivity and affinity of the target, and therapeutic impact. Although various candidate molecules that target CNS protein kinases are now in various phases of preclinical and clinical studies, none are being carried out in ADHD.

References

Aebi M et al (2016) Gene-set and multivariate genome-wide association analysis of oppositional defiant behavior subtypes in attention-deficit/hyperactivity disorder. Am J Med Genet B Neuropsychiatr Genet 171:573–588

Alberts B, Johnson A, Lewis J, Raff M, Roberts K, Walter P (2007) In: Anderson M, Granum S (eds) Molecular biology of the cell, 5th edn. Garland Science, New York, p 175
Alexander N et al (2012) Impact of antenatal synthetic glucocorticoid exposure on endocrine stress reactivity in term-born children. J Clin Endocrinol Metab 97:3538–3544
Amico F, Stauber J, Koutsouleris N, Frodl T (2011) Anterior cingulate cortex gray matter abnormalities in adults with attention deficit hyperactivity disorder: a voxel-based morphometry study. Psychiatry Res 191:31–35
Bartl J, Walitza S, Grünblatt E (2014) Enhancement of cell viability after treatment with polyunsaturated fatty acids. Neurosci Lett 559:56–60
Behrend L et al (2000) IC261, a specific inhibitor of the protein kinases casein kinase 1-delta and -epsilon, triggers the mitotic checkpoint and induces p53-dependent postmitotic effects. Oncogene 19:5303–5313
Bloch MH, Qawasmi A (2011) Omega-3 fatty acid supplementation for the treatment of children with attention-deficit/hyperactivity disorder symptomatology: systematic review and meta-analysis. J Am Acad Child Adolesc Psychiatry 50:991–1000
Bouchard MF, Bellinger DC, Wright RO, Weisskopf MG (2010) Attention-Deficit/Hyperactivity disorder and urinary metabolites of organophosphate pesticides. Pediatrics 125:1270–1277
Castellanos F, Lee PP, Sharp W, Jeffries NO, Greenstein DK, Clasen LS et al (2002) Developmental trajectories of brain volume abnormalities in children and adolescents with attention-deficit/hyperactivity disorder. JAMA 288:1740–1748
Chang JP, Su KP, Mondelli V, Pariante CM (2018) Omega-3 polyunsaturated fatty acids in youths with attention deficit hyperactivity disorder: a systematic review and meta-analysis of clinical trials and biological studies. Neuropsychopharmacology 43:534–545
Chen PH, Chen GC, Yang MF, Hsieh CH, Chuang SH, Yang HL, Kuo YH, Chyuan JH, Chao PM (2012) Bitter melon seed oil-attenuated body fat accumulation in diet-induced obese mice is associated with cAMP-dependent protein kinase activation and cell death in white adipose tissue. J Nutr 142:1197–1204
Cho SC, Kim HW, Kim BN, Kim JW, Shin MS, Chung S et al (2010) Gender-specific association of the brain-derived neurotrophic factor gene with attention-deficit/hyperactivity disorder. Psychiatry Investig 7:285–290
Corominas-Roso M, Ramos-Quiroga JA, Ribases M, Sanchez-Mora C, Palomar G, Valero S et al (2013) Decreased serum levels of brain-derived neurotrophic factor in adults with attention-deficit hyperactivity disorder. Int J Neuropsychopharmacol 16:1267–1275
Crown ED, Gwak YS, Ye Z, Johnson KM, Hulsebosch CE (2008) Activation of p38MAP kinase is involved in central neuropathic pain following spinal cord injury. Exp Neurol 213:257–267
Dela Pena I et al (2013) Neuronal development genes are key elements mediating the reinforcing effects of methamphetamine, amphetamine, and methylphenidate. Psychopharmacology 230: 399–413
Demontis D et al (2019) Discovery of the first genome-wide significant risk loci for attention deficit/hyperactivity disorder. Nat Genet 51:63–75
Desdouits F, Siciliano JC, Greengard P, Girault JA (1995) Dopamine- and cAMP regulated phosphoprotein DARPP-32: phosphorylation of Ser-137 by casein kinase I inhibits dephosphorylation of Thr-34 by calcineurin. Proc Natl Acad Sci U S A 92:2682–2685
Drerup JM, Hayashi K, Cui H, Mettlach GL, Long MA, Marvin M, Sun X, Goldberg MS, Lutter M, Bibb JA (2010) Attention-deficit/hyperactivity phenotype in mice lacking the cyclin-dependent kinase 5 cofactor p35. Biol Psychiatry 68:1163–1171
Engel SR, Creson TK, Hao Y, Shen Y, Maeng S, Nekrasova T, Landreth GE, Manji HK, Chen G (2009) The extracellular signal-regulated kinase pathway contributes to the control of behavioral excitement. Mol Psychiatry 14(4):448–461
Enns LC, Morton JF, Mangalindan RS, McKnight GS, Schwartz MW, Kaeberlein MR, Kennedy BK, Rabinovitch PS, Ladiges WC (2009) Attenuation of age-related metabolic dysfunction in mice with a targeted disruption of the Cβ subunit of protein kinase A. J Gerontol A Biol Sci Med Sci 64:1221–1231

Feng YJ, Li YY, Lin XH, Li K, Cao MH (2016) Anti-inflammatory effect of cannabinoid agonist WIN55, 212 on mouse experimental colitis is related to inhibition of p38MAPK. World J Gastroenterol 22:9515–9524

Firouzkouhi Moghaddam M, Shamekhi M, Rakhshani T (2017) Effectiveness of methylphenidate and PUFA for the treatment of patients with ADHD: a double-blinded randomized clinical trial. Electron Physician 9:4412–4418

Flajolet M et al (2007) Regulation of Alzheimer's disease amyloid-beta formation by casein kinase I. Proc Natl Acad Sci U S A 104:4159–4164

Fuentes-Albero M, Martínez-Martínez MI, Cauli O (2019) Omega-3 long-chain polyunsaturated fatty acids intake in children with attention deficit and hyperactivity disorder. Brain Sci 9:120–120

Fumagalli F, Racagni G, Colombo E, Riva MA (2003) BDNF gene expression is reduced in the frontal cortex of dopamine transporter knockout mice. Mol Psychiatry 8:898–899

Gadow KD, Roohi J, DeVincent CJ, Kirsch S, Hatchwell E (2009) Association of COMT (Val158Met) and BDNF (Val66Met) gene polymorphisms with anxiety, ADHD and tics in children with autism spectrum disorder. J Autism Dev Disord 39(11):1542–1551

Greengard P (2001) Science 294:1024–1030

Greenwood TA, Joo EJ, Shekhtman T, Sadovnick AD, Remick RA, Keck PE, McElroy SL, Kelsoe JR (2013) Association of dopamine transporter gene variants with childhood ADHD features in bipolar disorder. Am J Med Genet B 162B:137–145

Grünblatt E, Bartl J, Walitza S (2018) Methylphenidate enhances neuronal differentiation and reduces proliferation concomitant to activation of Wnt signal transduction pathways. Transl Psychiatry 8:51

Grünblatt E et al (2019) The involvement of the canonical Wnt-signaling receptor LRP5 and LRP6 gene variants with ADHD and sexual dimorphism: Association Study and Meta-Analysis. Am J Med Genet 180:365–376

Hall FS, Itokawa K, Schmitt A, Moessner R, Sora I, Lesch KP, Uhl GR (2014) Decreased vesicular monoamine transporter 2 (VMAT2) and dopamine transporter (DAT) function in knockout mice affects the aging of dopaminergic systems. Neuropharmacology 76:146–155

Hayden MS, Ghosh S (2011) NF-kappaB in immunobiology. Cell Res 21:223–244

Jaworski J, Sheng M (2006) The growing role of mTOR in neuronal development and plasticity. Mol Neurobiol 34:205–219

Jeong HI, Ji ES, Kim SH, Kim TW, Baek SB, Choi SW (2014) Treadmill exercise improves spatial learning ability by enhancing brain-derived neurotrophic factor expression in the attention-deficit/hyperactivity disorder rats. J Exerc Rehabil 10:162–167

Jhang CL, Huang TN, Hsueh YP, Liao W (2017) Mice lacking cyclin-dependent kinase-like 5 manifest autistic and ADHD-like behaviours. Hum Mol Genet 26:3922–3934

Jhang CL, Lee HY, Chen JC, Liao W (2020) Dopaminergic loss of cyclin-dependent kinase-like 5 recapitulates methylphenidate-remediable hyperlocomotion in mouse model of CDKL5 deficiency disorder. Hum Mol Genet 29(14):2408–2419

Keil MF, Lyssikatos C, Shaikh M, Belyavskaya E, Elliott B, Batista D et al (2014) Effects of PRKAR1A mutations in behavior and brain function. European Congress of Endocrinology. Wroclaw: Endocrine Abstracts

Kitagishi Y, Minami A, Nakanishi A, Ogura Y, Matsuda S (2015) Neuron membrane trafficking and protein kinases involved in autism and ADHD. Int J Mol Sci 16:3095–3115

Kwon HJ, Ha M, Jin HJ, Hyun JK, Shim SH, Paik KC et al (2015) Association between BDNF gene polymorphisms and attention deficit hyperactivity disorder in Korean children. Genet Test Mol Biomarkers 19:366–371

Leal G, Bramham CR, Duarte CB (2017) BDNF and hippocampal synaptic plasticity. Vitam Horm 104:153–195

Lee DY (2015) Roles of mTOR signaling in brain development. Exp Neurobiol 24:177–187

Lee C, Etchegaray JP, Cagampang FR, Loudon AS, Reppert SM (2001) Post-translational mechanisms regulate the mammalian circadian clock. Cell 107:855–867

Lee MK, Han SR, Park MK, Kim MJ, Bae YC, Kim SK et al (2011) Behavioral evidence for the differential regulation of p-p38 MAPK and p-NF-kappa B in rats with trigeminal neuropathic pain. Mol Pain 7:57

Lesch KP et al (2008) Molecular genetics of adult ADHD: converging evidence from genome-wide association and extended pedigree linkage studies. J Neural Transm (Vienna) 115:1573–1585

Li G, Yin H, Kuret J (2004) Casein kinase 1 delta phosphorylates tau and disrupts its binding to microtubules. J Biol Chem 279:15938–15945

Li H, Liu L, Tang Y, Ji N, Yang L, Qian Q et al (2014) Sex-specific association of brain-derived neurotrophic factor (BDNF) Val66Met polymorphism and plasma BDNF with attention-deficit/ hyperactivity disorder in a drug-naïve Han Chinese sample. Psychiatry Res 217:191–197

Liu D, Zhen W, Yang Z, Carter JD, Si H, Reynolds KA (2006) Genistein acutely stimulates insulin secretion in pancreatic β-cells through a cAMP-dependent protein kinase pathway. Diabetes 55: 1043–1050

Liu JJ, Green P, JohnMann J, Rapoport SI, Sublette ME (2015a) Pathways of polyunsaturated fatty acid utilization: implications for brain function in neuropsychiatric health and disease. Brain Res 1597:220–246

Liu DY, Shen XM, Yuan FF, Guo OY, Zhong Y, Chen JG et al (2015b) The physiology of BDNF and its relationship with ADHD. Mol Neurobiol 52(3):1467–1476

Lopez-Vicente M et al (2019) Prenatal omega-6:omega-3 ratio and attention deficit and hyperactivity disorder symptoms. J Pediatr 209:204–211

Lowrey PL et al (2000) Positional syntenic cloning and functional characterization of the mammalian circadian mutation tau. Science 288:483–492

Luman M, Tripp G, Scheres A (2010) Identifying the neurobiology of altered reinforcement sensitivity in ADHD: a review and research agenda. Neurosci Biobehav Rev 34(5):744–754. https://doi.org/10.1016/j.neubiorev.2009.11.021

Luo SX, Timbang L, Kim J, Shang Y, Sandoval K, Tang AA, Whistler JL, Ding JB, Huang EJ (2016) TGF-β Signaling in dopaminergic neurons regulates dendritic growth, excitatory-inhibitory synaptic balance, and reversal learning. Cell Rep 17:3233–3245

Maitra S, Chatterjee M, Sinha S, Mukhopadhyay K (2017) Cyclin-dependent Kinase 5: novel role of gene variants identified in ADHD. Sci Rep 7:682–688

McLinden KA, Trunova S, Giniger E (2012) At the fulcrum in health and disease: Cdk5 and the balancing acts of neuronal structure and physiology. Brain Disord Ther 2012(Suppl 1):001

Missale C, Nash SR, Robinson SW, Jaber M, Caron MG (1998) Physiol Rev 78:189–225

Moors M et al (2012) Dickkopf 1 mediates glucocorticoid-induced changes in human neural progenitor cell proliferation and differentiation. Toxicol Sci 125(2):488–495

Murakami A, Kimura K, Nakano A (1999) The inactive form of a yeast casein kinase I suppresses the secretory defect of the sec12 mutant. Implication of negative regulation by the Hrr25 kinase in the vesicle budding from the endoplasmic reticulum. J Biol Chem 274:3804–3810

O'Brien LJ, Levac KD, Nagy LE (1998) Moderate dietary protein and energy restriction modulate cAMP-dependent protein kinase activity in rat liver. J Nutr 128:927–933

Oakes HV et al (2019) Neurogenesis within the hippocampus after chronic methylphenidate exposure. J Neural Transm 126:201–209

Odaka H, Adachi N, Numakawa T (2017) Impact of glucocorticoid on neurogenesis. Neural Regen Res 12:1028–1035

Ouchi Y, Banno Y, Shimizu Y, Ando S, Hasegawa H, Adachi K et al (2013) Reduced adult hippocampal neurogenesis and working memory deficits in the Dgcr8-deficient mouse model of 22q11.2 deletion-associated schizophrenia can be rescued by IGF2. J Neurosci 33:9408–9419

Panek HR et al (1997) Suppressors of YCK-encoded yeast casein kinase 1 deficiency define the four subunits of a novel clathrin AP-like complex. EMBO J 16:4194–4204

Papa M, Sergeant JA, Sadile AG (1998) Reduced transduction mechanisms in the anterior accumbal interface of an animal model of Attention-Deficit Hyperactivity Disorder. Behav Brain Res 94: 187–195

Pierce RC, Kalivas PW (1997) A circuitry model of the expression of behavioral sensitization to amphetamine-like psychostimulants. Brain Res Brain Res Rev 25:192–216

Plessen KJ, Bansal R, Zhu H, Whiteman R, Amat J, Quackenbush GA et al (2006) Hippocampus and amygdala morphology in attention-deficit/hyperactivity disorder. Arch Gen Psychiatry 63: 795–807

Qin S, Yang C, Huang W, Du S, Mai H, Xiao J et al (2018) Sulforaphane attenuates microglia-mediated neuronal necroptosis through down-regulation of MAPK/NF-kappa B signaling pathways in LPS-activated BV-2 microglia. Pharmacol Res 133:218–235

Ralph RJ, Paulus MP, Fumagalli F, Caron MG, Geyer MA (2001) J Neurosci 21:305–313

Rios M, Fan G, Fekete C, Kelly J, Bates B, Kuehn R et al (2001) Conditional deletion of brain-derived neurotrophic factor in the postnatal brain leads to obesity and hyperactivity. Mol Endocrinol 15:1748–1757

Robbins TW (2003) Dopamine and cognition. Curr Opin Neurol 16:S1–S2

Robinson-White A, Stratakis CA (2002) Protein kinase A signaling: "cross-talk" with other pathways in endocrine cells. Ann N Y Acad Sci 968:256–270

Sadasivan S et al (2012) Methylphenidate exposure induces dopamine neuron loss and activation of microglia in the basal ganglia of mice. PLoS One 7:e33693

Sánchez-Alegría K, Flores-León M, Avila-Muñoz E, Rodríguez-Corona N, Arias C (2018) PI3K signaling in neurons: a central node for the control of multiple functions. Int J Mol Sci 19:3725–3725

Schmitz F, Chao MV, Wyse ATS (2019) Methylphenidate alters Akt-mTOR signaling in rat pheochromocytoma cells. Int J Dev Neurosci 73:10–18

Schwartz PA, Murray BW (2011) Protein kinase biochemistry and drug discovery. Bioorg Chem 39:192–210

Schwarz R et al (2015) A preliminary study on methylphenidate-regulated gene expression in lymphoblastoid cells of ADHD patients. World J Biol Psychiatry 16:180–189

Shah K, Lahiri DK (2017) A tale of the good and bad: remodeling of the microtubule network in the brain by Cdk5. Mol Neurobiol 54:2255–2268

Shchemelinin I, Šefc L, Nečas E (2006) Protein kinases, their function and implication in cancer and other diseases. Folia Biol (Praha) 52:81–101

Shim S, Hwangbo Y, Kwon YJ, Lee H, Kim J, Yoon H, Hwang JA, Kim YK (2012) Association between glycogen synthase kinase-3β gene polymorphisms and attention deficit hyperactivity disorder in Korean children: a preliminary study. Prog Neuropsychopharmacol Biol Psychiatry 39:57–61

Simchon-Tenenbaum Y, Weizman A, Rehavi M (2015) Alterations in brain neurotrophic and glial factors following early age chronic methylphenidate and cocaine administration. Behav Brain Res 282:125–132

Su Y, Balice-Gordon RJ, Hess DM, Landsman DS, Minarcik J, Golden J, Hurwitz I, Liebhaber SA, Cooke NE (2004) Neurobeachin is essential for neuromuscular synaptic transmission. J Neurosci 24:3627–3636

Szentandrássy N, Pérez-Bido MR, Alonzo E, Negretti N, O'Neill SC (2007) Protein kinase A is activated by the *n*-3 polyunsaturated fatty acid eicosapentaenoic acid in rat ventricular muscle. J Physiol 582:349–358

Tripp G, Wickens JR (2008) Research review: dopamine transfer deficit: a neurobiological theory of altered reinforcement mechanisms in ADHD. J Child Psychol Psychiatry 49:691–704

Tsai SJ (2003) Attention-deficit hyperactivity disorder and brain-derived neurotrophic factor: a speculative hypothesis. Med Hypotheses 60:849–851

Wang YP, Mou DL, Song JF, Rao ZR, Li D, Ju G (2006) Aberrant activation of CDK5 is involved in the pathogenesis of OPIDN. J Neurochem 99:186–197

Wang LJ et al (2020) Gray matter volume and microRNA levels in patients with attention-deficit/hyperactivity disorder. Eur Arch Psychiatry Clin Neurosci 270(8):1037–1045

Weissflog L et al (2013) KCNIP4 as a candidate gene for personality disorders and adult ADHD. Eur Neuropsychopharmacol 23:436–447

Xu Y et al (2005) Functional consequences of a CKI delta mutation causing familial advanced sleep phase syndrome. Nature 434:640–644

Yabuki Y, Shioda N, Maeda T, Hiraide S, Togashi H, Fukunaga K (2014) Aberrant CaMKII activity in the medial prefrontal cortex is associated with cognitive dysfunction in ADHD model rats. Brain Res 1557:90–100

Yonezawa K et al (2018) Investigation into the plasma concentration of ω3 polyunsaturated fatty acids in Japanese attention-deficit hyperactivity disorder patients. J Neural Transm 125:1395–1400

Zarogoulidis P et al (2014) mTOR pathway: a current, up-to-date mini-review (review). Oncol Lett 8:2367–2370

Zhang-James Y, Fernàndez-Castillo N, Hess JL, Malki K, Glatt SJ, Cormand B, Faraone SV (2019) An integrated analysis of genes and functional pathways for aggression in human and rodent models. Mol Psychiatry 24:1655–1667

Zhong Y, Chen J, Li L, Qin Y, Wei Y, Pan S, Jiang Y, Chen J, Xie Y (2018) PKA-CREB-BDNF signaling pathway mediates propofol-induced long-term learning and memory impairment in the hippocampus of rats. Brain Res 1691:64–74

Zhou M, Rebholz H, Brocia C, Warner-Schmidt JL, Fienberg AA, Nairn AC et al (2010) Forebrain overexpression of CK1delta leads to down-regulation of dopamine receptors and altered locomotor activity reminiscent of ADHD. Proc Natl Acad Sci U S A 107:4401–4406

Zhou M, Gresack J, Cheng J, Uryu K, Brichta L, Greengard P, Flajolet M (2020) CK1δ over-expressing mice display ADHD-like behaviors, frontostriatal neuronal abnormalities and altered expressions of ADHD-candidate genes. Mol Psychiatry 25:3322–3336

Autism Spectrum Disorder (ASD) and Diet

8

Nahla Al Anqodi and Ruqaiya Moosa Al Balushi

Abstract

Autism spectrum disorder (ASD) is one of the diverse groups of neurodevelopmental disorders, and with a rising prevalence globally. ASD constitutes a significant public health concern. Challenges in ASD assessment pose barriers in terms of diagnosis, and it might be detected much later in life. It presents a significant health concern during the critical growing years of infancy for the child and the family, with direct psychological consequences. Individuals with ASD might lead an independent life or require lifelong support, accompanied by varying levels of impairment of individual functioning and quality of life. With improvement in the understanding of ASD over the last few decades, there was a revision of the diagnostic criteria in *the fifth edition of the Diagnostic and Statistical Manual of Mental Disorders* (DSM-5) and *the eleventh edition of the International Classification of Diseases* (ICD-11) manuals. Many developmental screening tools are available, and in the absence of specific medical investigations, physicians who make the first contact with patients, such as primary care physicians and pediatricians, must have a relatively high index of suspicion for this disorder. Early diagnosis must be coupled with prompt referral to specialists, developmental pediatricians, child psychiatrists, or psychologists to ensure the coordination of holistic care to the affected children and their families. Pharmacologic agents and non-pharmacologic therapies such as dietary interventions are available. The pharmacological agents include selective

N. Al Anqodi
Dietary Department, Directorate General of Medical Services, Royal Oman Police, Muscat, Oman

R. M. Al Balushi (✉)
Depratment of Food Science and Nutrition, College of Agricultural and Marine Sciences, Sultan Qaboos University, Muscat, Oman
e-mail: r.albalushi@squ.edu.om

M. W. Qoronfleh et al. (eds.), *Proteins Associated with Neurodevelopmental Disorders*, Nutritional Neurosciences, https://doi.org/10.1007/978-981-15-9781-7_8

serotonin reuptake inhibitors (SSRIs), antidepressants, and psychostimulants. The SSRIs have the most beneficial impact on challenging behaviors in children. Of the proposed dietary approaches, the elimination of diets such as the gluten- and casein-free diet (GF-CF diet) has been shown to address ASD symptoms in some studies. However, the evidence is not strong enough to support recommendations for this intervention in clinical practice. Novel evidence for nutritional interventions is emerging, and ongoing research might elucidate its exact role and the long-term effects on metabolism, metabolic dysfunction, and nutritional deficiencies to address the core symptoms of ASD.

Keywords

Autism spectrum disorder · Autism · ASD · Gluten-free casein-free diet · GF-CF diet · Selective serotonin reuptake inhibitors · SSRIs

8.1 Introduction

Autism spectrum disorder (ASD) is a neurodevelopmental disorder that occurs in the early years of life (Ali et al. 2011) with a high burden of healthcare cost (Buescher et al. 2014). Notably, in the last two decades, the number of diagnosed cases of ASD has grown remarkably, and this could be associated with the changes of concepts and diagnostic criteria and/or increased awareness (CDC 2020a, b, c). ASD is characterized by impaired social interactions, inability to communicate ideas and feelings, and restricted and repetitive behaviors or actions (*Autism Spectrum Disorders* n.d.). The symptoms and severity of ASD vary widely across the three core domains and there is a spectrum of comorbid medical and psychiatric conditions that commonly occur in these individuals including but not limited to sleep deficits, anxiety tantrums, epileptic seizures, and gastrointestinal (GI) problems, as shown in Fig. 8.1.

8.2 Epidemiology

The initial studies reported ASD to be relatively uncommon with a prevalence of only 4 cases per 10,000 children (Lotter 1966; Rutter 2005; Treffert 1970). Yet, in the last few years, the incidence and prevalence of ASD have increased considerably regardless of ethnicity, maternal age, or child gender. The increase might be a consequence of an improvement in public awareness, availability of services, and improvement in case finding.

The global prevalence in 2012 was reported to be 11.3 per 1000 children (Autism and Developmental Disabilities Monitoring Network Surveillance Year 2008 Principal Investigators and Centers for Disease Control and Prevention 2012). However, the current global prevalence is estimated to be 1 case per 270 people (GBD 2019 Diseases and Injuries Collaborators 2020). According to the CDC estimates, 1 in

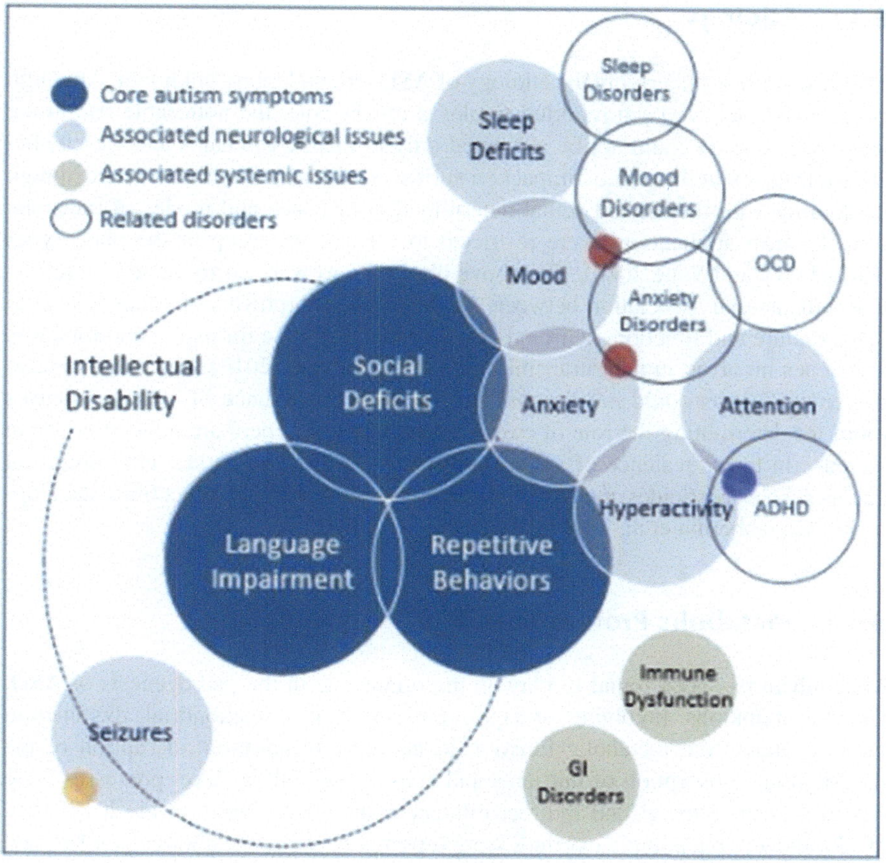

Fig. 8.1 ASD core symptoms and associated challenges often observed in ASD patients (*Helping Children with Autism* n.d.)

54 children is diagnosed with ASD in the United States (CDC 2020a, b, c); however, initial reports from differing number of sites had yielded prevalence figures of 1 in 150 in 2002 to 1 in 88 in 2008 (Autism and Developmental Disabilities Monitoring Network Surveillance Year 2008 Principal Investigators and Centers for Disease Control and Prevention 2012). Additionally, data from the United States has reported that ASD is about four times more common among boys than girls (CDC 2020a, b, c).

Studies from other parts of the world reported the prevalence of ASD to be 1.4, 29, and 59 per 10,000 children in Oman (Al-Farsi et al. 2011), the United Arab Emirates (Eapen et al. 2007), and Saudi Arabia (*Website* n.d.-a: www.kacst.edu.sa), respectively. The low prevalence figures could be attributed to the lack of diagnosis, misdiagnosis, or underreporting (Hassan 2019).

8.3 Etiology

The mechanisms underlying the etiology of ASD remain largely unknown. Although genetic factors have been reported to play a crucial role, the noticeable rise in the incidence of ASD could not be attributed only to them (Abrahams and Geschwind 2010). Data about the genetic impact on autism combines scientific evidence of high heritability with the limited actual determination of genes and modes of transmission. Clinical applications were restricted to a small subgroup of the phenotypes (El-Fishawy and State 2010). The above observations were corroborated by results that indicated an association between ASD and polymorphisms of genes related to cell structure and function, neuronal development, synaptic formation and function, and genes involved in neurotransmission (Parellada et al. 2014). Since the findings of genetic factors could not account for the entire prevalence of autism, it leaves room for determining the role of environmental factors. There are indications of an increase in the prevalence of autism following exposure to certain environmental agents such as pesticides and solvents, which could potentially affect brain development (Modabbernia et al. 2017; Parellada et al. 2014).

8.4 Metabolic Profiles in ASD

Metabolism has been found to play an important role in the pathogenesis of ASD. Genetic mutations involving synaptic proteins and mitochondrial dysfunction directly impact the metabolic function in neurons. In addition, disruption of the digestion and absorption of nutrients and subsequent cellular level processes were reported to have direct and indirect influences on ASD (Cheng et al. 2017; Frye 2015). Indeed, phenotypic overlaps were reported between metabolic disorders and ASD, including disorders of creatine, cholesterol, vitamins, purine and pyrimidine, and amino acid metabolism. However, caution should be exercised to determine specific syndromes as distinct from ASD appropriately.

8.5 Diagnosis

The diagnosis of ASD can pose particular challenges and at present there is no specific medical investigation that can diagnose ASD with certainty. The ASD diagnosis requires a multidisciplinary approach which involves a combination of standardized tools besides clinical expertise in neurocognitive, neurogenetic, speech, and motor assessment, and observational interviewing through administration of autism-specific behavior evaluation tools (Falkmer et al. 2013).

There are diverse tools that can facilitate the diagnosis of ASD, including the DISCO (Diagnostic Interview for Social and Communication Disorders), the ADI-R (Autism Diagnostic Interview-Revised), the ADOS (Autism Diagnostic Observation Schedule), and 3Di (Developmental, Dimensional and Diagnostic Interview) (Evers et al. 2021). The above tools assess against a set of criteria for autism, as per the diagnostic manual ICD-10, that take into account abnormalities in social interactions and communication, restricted, repetitive, stereotyped, repertoire of activities and interest. Further, the DSM-5 includes neurodevelopmental disorder and sensory issues. The DSM-5 employs a single diagnosis of ASD that replaced the subcategories of autistic disorder, Asperger's disorder, childhood disintegrative disorder, and pervasive developmental disorder (*DSM-5 and Autism: Frequently Asked Questions* n.d.).

8.6 DSM-5 Autism Spectrum Criteria

In 2013, *the fifth edition of the Diagnostic and Statistical Manual of Mental Disorders* (DSM-5) updated the previously published DSM-IV. The DSM-5 recognized ASD under the "Neurodevelopmental Disorders" reflecting brain development correlation to autism, and as a spectrum combining separate pervasive developmental disorder (PDD) diagnoses: autistic disorder, Asperger's syndrome, childhood disintegrative disorder, and pervasive developmental disorder not otherwise specified (PDD-NOS). A diagnosis of the autistic disorder requires the presence of at least 6 of 12 total symptoms from three domains (two social, at least one communication, and at least one behavioral), and onset before 36 months of age.

The features of the previously defined subcategories, a diagnosis of Asperger's disorder, are characterized by qualitative impairments in social interaction and presence of restricted interests and repetitive behaviors, but no cognitive, language, or nonsocial adaptive delays are noted in early development. Childhood disintegrative disorder is characterized by a loss of previously acquired language and social skills and persistent delay in these domains, and regression in social and emotional development with impaired ability to relate with others. Further, children with PDD-NOS have to meet at least two diagnostic criteria with one from the social domain, and the diagnosis requires impairment in reciprocal social skills, impairment in verbal and nonverbal communication, or presence of stereotyped behavior, interests, and activities.

According to the DSM-5, a formal ASD diagnosis is based on child observation as well as parental reports; it requires that individuals meet all three of the criteria in the category of social communication and interaction impairments, and at least two out of four criteria in the category of restricted and repetitive behaviors. Furthermore, it can be classified at one of the three levels of severity, mild, moderate, and severe, based on the degree of impairment.

8.7 Clinical Characteristics

The core features of ASD include the following (Doernberg and Hollander 2016):

1. Impairment in communication and intellect: Speech impairment ranges from speech delays, monotonous speech, echolalia, pronoun reversal, or poor

comprehension to a complete inability to speak. Impaired nonverbal communication involves poor eye contact, absence of index finger pointing, and difficulties in understanding facial expressions and descriptive gestures.

2. Impairment in social interactions and socio-emotional reciprocity: There are impaired ability to initiate conversation, lack of smiling, and lack of interest in peer interactions.
3. Restricted or repetitive patterns of behaviors, interest, or activities: The features range from simple stereotypical motor behaviors such as flapping of hands, repetitive use of objects, or repetitive speech (echolalia) to over- or under-sensitivity to various sensory stimuli, such as interest in spinning objects, and sensitivity to sound, light, smell, heat, or taste.

8.8 Associated Clinical Features

According to a surveillance study of children with ASD in the United States, there was a high prevalence of associated conditions that could mask ASD and lead to delayed diagnosis; 83% had one or more non-ASD developmental diagnosis, 10% had one or more psychiatric diagnosis, and 16% had one or more neurological diagnosis (Levy et al. 2010). In addition, epilepsy has been reported to have a higher prevalence rate in individuals with ASD than the general population (20%) (Besag 2018). Furthermore, variable rates of comorbid psychiatric diagnoses were reported; however, the comorbidities might be particularly elevated in individuals with ASD: anxiety (40%) (Zaboski and Storch 2018), depression (14.4%) (Hudson et al. 2019), bipolar disorder (6–21.4%) (Hossain et al. 2020), attention-deficit/hyperactivity disorder (ADHD) (81%) (Lecavalier et al. 2019), and obsessive-compulsive disorder (17%) (Leyfer et al. 2006; Özyurt and Beşiroğlu 2018).

Other commonly occurring comorbidities include GI disorders (9–91%) (Lefter et al. 2019), such as abdominal pain, constipation and diarrhea, and restrictive eating, sleep disturbances (60–86%) (Posar and Visconti 2020), and overweight (42.4%) and obesity (21.4%) (Criado et al. 2018). Finally, ASD has been reported to be associated with higher risks of fatal and nonfatal injuries and bullying (DiGuiseppi et al. 2018).

8.9 Evaluation

Early and timely detection and management form the core in the reduction of impairments and improvement of the outlook towards the disease and quality of life. A comprehensive behavioral assessment is vital, which can be conducted in various settings such as healthcare, community, and schools, and provides the opportunity for early detection as young as 18 months or less. However, the average age at diagnosis has been reported to be 3.1 years (Mandell 2005). Similarly, the CDC reported that the mean age at first diagnosis remains higher than 4 years (Developmental Disabilities Monitoring Network Surveillance Year 2010 Principal

Investigators and Centers for Disease Control and Prevention (CDC) 2014), and some children had not received a formal diagnosis until 8 years of age (Sheldrick et al. 2017). The pitfalls of delayed identification include functional loss, particularly during the adolescent period, with risks of lifelong impairments.

Evaluation in ASD begins with screening the general pediatric population to identify children at risk or demonstrating signs suggestive of ASD, following which a diagnostic evaluation is recommended. The American Academy of Pediatrics (AAP) guidelines recommend developmental surveillance at 9, 18, and 30 months of well-child visits and autism-specific screening at 18 months and again at 24 or 30 months (Lipkin et al. 2020). Early evaluations should include a comprehensive assessment of the general physical state, a neurological exam, assessment of child's behavior, cognition, speech and language, hearing and vision, and parental interviews. Clinical assessment should also seek out signs of associated comorbid conditions such as dysmorphic features and dermatological manifestations. Early red flags for ASD include poor eye contact, poor response to name, lack of showing and sharing, no gesturing by 12 months, loss of language or social skills, limited pretend play, odd or intensely focused interests, and rigidity (Hodges et al. 2020). Furthermore, school-age children might exhibit concrete or literal thinking, impairments in understanding emotions, and a lack of conversational skills or appropriate social approach (Hodges et al. 2020).

Besides a few ASD tools mentioned above, many different screening tools are available such as the Modified Checklist for Autism in Toddlers, Revised, with Follow-up (M-CHAT-R/F), Survey of Wellbeing of Young Children (SWYC), Ages and Stages Questionnaires (ASQ), Communication and Symbolic Behaviour Scales (CSBS), Parents Evaluation of Development Status (PEDS), Screening Tool for Autism in Toddlers and Young Children (STAT), Social Communication Questionnaire (SCQ), Social Responsiveness Scale (SRS), and Autism Spectrum Screening Questionnaire (ASSQ). In addition, screening should prompt timely referral to specialists such as pediatric neurologists, psychiatrists, or psychologists for a definitive diagnosis and management.

A systematic review on ASD screening reported that screening is an accurate means to identify children with ASD; however, results might differ across ages or settings (Levy et al. 2020). Of the various tools available, the M-CHAT is the most widely studied, and there is a need for further studies to compare different surveillance and screening tools and clinical outcomes. The M-CHAT includes 23 yes-or-no questions. Positive screening requires a "no" for any 3 of the 23 items or 2 of the 6 critical items (interest in other children, using the index finger to point, bringing objects to show parents, imitating, responding to one's name, and using one's eyes to follow an object across the room) (*Website* n.d.-b: *www.mchatscreen.com*).

There is a limited role of genetic testing at the moment aside from those for single-gene defects. Furthermore, there are no recommendations for other routine laboratory workups. Nevertheless, based on the individual's clinical evaluation, the following might be considered: complete blood count (CBC), ferritin, thyroid-stimulating hormone (TSH), liver and renal function tests, lactate, pyruvate, carnitine, amino acids, acylcarnitine profile, urine organic acids and/or urine glycosaminoglycans, lead levels, biochemical profile for the nutritional status, sleep study, and hearing tests. Few indications for neuroimaging include suspicion of tuberous sclerosis complex (TSC) or other neurocutaneous disorders, microcephaly, or abnormal neurologic examination findings (spasticity, severe hypotonia, or unilateral findings). Electroencephalography (EEG) should be done for patients with suspected seizures.

8.10 Treatment/Management

At present, there is no curative medical treatment for ASD. However, parental guidance and adaptive measures combined with cognitive behavioral therapy for accompanying conditions that take into consideration communication impairments appear to be a reasonable approach to tackle most challenges (DeFilippis and Wagner 2016).

In light of the variability of gut function and the role of the gut, microbiomes, and metabolism in many disease processes, the role of the GI system in the development of ASD has also been a topic of interest. Existing literature elucidated the high prevalence of GI problems and disorders in individuals with ASD (Lefter et al. 2019; Xu et al. 2018). Children with ASD have been found to exhibit higher levels of pro-inflammatory cytokines following exposure to food proteins from gluten, casein, and soy, compared with controls (Li et al. 2021). Microbiomes were reported to influence the systemic metabolic dysfunctions via bacterial metabolites, deficiencies of vitamins, and effects on the functioning of the immune system, which influence the body and brain functions (Buie 2015; Heintz-Buschart and Wilmes 2018). According to Rogers et al., diet is one of the factors that potentially influence brain function through shaping the gut microbiome (Rogers et al. 2016).

Moreover, challenges to maternal homoeostasis, such as infection, poor nutrition, or prenatal stress (PNS), were associated with neurodevelopmental disorders, including anxiety, autism, ADHD, depression, and schizophrenia (Finegold 2011). Disruption of the maternal microbiome, or "dysbiosis," appears to act as a link between external stressors and fetal development, either by altering normal developmental cues or through the presentation of inappropriate developmental stimuli. On the other hand, the microbiota also modulates a range of neurotrophins and proteins involved in brain development and plasticity. Indeed in animal models, early-life postnatal stress is linked with altered visceral pain sensitivity and impaired intestinal barrier function (Larauche et al. 2011).

Several theories have been put forward to explain the mechanism of action in the development of ASD, which involves the immune and the GI systems. These include the association of the blood-brain barriers in the gut where lipopolysaccharides and short-chain fatty acids (SCFA) as outcomes of the gut microbiota have been suggested to control the cytokine manufacture (Maigoro and Lee 2021). Similarly, the synthesis of neuropeptides such as serotonin, and peptides from gluten and casein, was assumed to foster the activity of the opioid system (Fig. 8.2). The neuropeptides were assumed to cause impairments in social and communication

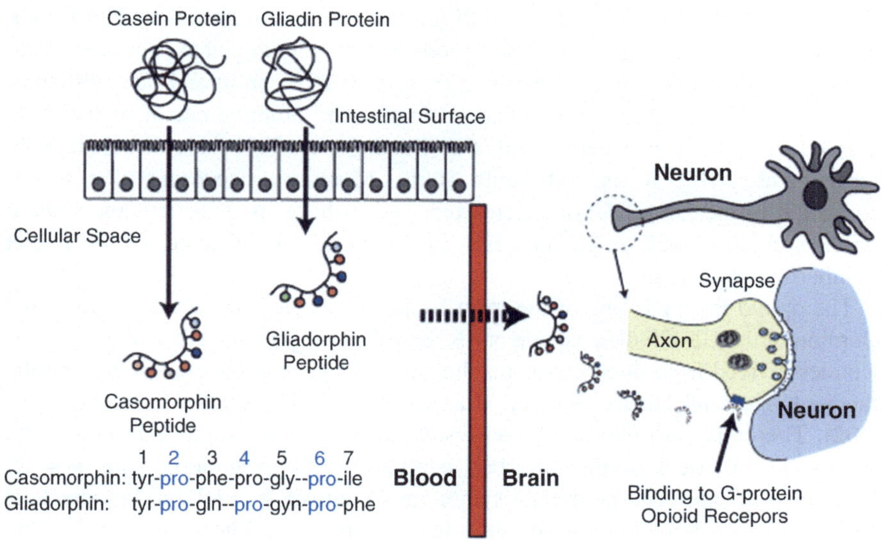

Fig. 8.2 Neuronal receptors for casein and gliadin peptides (*Autism, Casein and Gluten Food Sensitivity, Diet* n.d.)

behavior. Hence, the neuropeptides might possibly be implicated in the etiology of ASD (Keller et al. 2021).

The improved understanding of the mechanistic impairments in ASD shed light on the "fragile gut" and the role of food structure and protein digestion in ASD symptoms (Sanctuary et al. 2018). Symptoms are exacerbated through impairments in digestive enzymes, gut barrier integrity, and presence of antibodies against dietary proteins. Hence, the dysfunctions coupled with specific protein intake pave the way for potential dietary interventions to improve GI symptoms. It is hypothesized that children with ASD have reduced proteolytic enzyme activity that leads to colonic putrefaction, which in turn causes GI problems and exacerbates ASD. Taking into consideration the protein disturbances, diets with the elimination of several proteins provided insights into mechanisms of ASD and potential avenues of treatment. In particular, the gluten-free and casein-free diet (GF-CF diet) has shown a beneficial effect on the normalization of urinary peptide levels, reduced intestinal permeability, and improvement of behavioral symptoms (de Magistris et al. 2010; Knivsberg et al. 1995).

8.11 The Opioid Theory for ASD and Gluten- and Casein-Free Diet

ASD continues to pose challenges in determining an effective treatment in the management of ASD core and associated symptoms. The rise in awareness of ASD has led parents to seek alternative interventions for their children. One of the most popular interventions is the GF-CF (gluten- and casein-free) diet. The GF-CF diet consists of eliminating foods and beverages that contain gluten, a protein found in wheat, barley, and rye, and casein, a protein found in milk and dairy products (Hyman et al. 2016). According to various studies, the adoption rate of such diets in children with ASD is between 17 and 66% (Bandini et al. 2010; Herndon et al. 2009; Perrin et al. 2012; Wong and Smith 2006). Moreover, according to a survey conducted in the UK, 80% of parents reported to have used dietary intervention for their children with ASD, and 29% of them had administered a GF-CF diet (Winburn et al. 2014).

The opioid theory is one of the most frequently quoted theories to validate the adoption of the GF-CF diet. The theory is related to the possible entry of exogenous peptides derived from dietary proteins that enter the general circulation through the impaired intestinal barrier function (Lázaro et al. 2016; Shattock and Whiteley 2002). The entry into the circulation results in a systemic immune response and perhaps in an effect of the absorbed peptides on the central nervous system, influencing the core and peripheral symptoms of autism. In addition, the intestinal degradation of beta-casein releases bioactive peptides called beta-casomorphin that has an opioid activity similar to morphine due to its exogenous origin (Lázaro et al. 2016).

Globally, according to professionals and parents of children with ASD, elimination of such exogenous compounds through dietary exclusion generated some positive results in ASD and related behaviors. For example, some researchers who found peptiduria in the urine of children with ASD reported normal results of peptiduria after following an exclusion diet for a year (A. Knivsberg et al. 1995). In contrast, a study by Cass et al. on children with autism did not report any associations with opioid peptiduria (Cass et al. 2008).

Several systematic reviews have evaluated the efficacy of the GF-CF diet in managing ASD children (Elder et al. 2015; Keller et al. 2021; Marí-Bauset et al. 2016). Also, several studies have been conducted to highlight the role of GF-CF diets in ASD. However, the overall results till date lack solid support to promote this intervention (Alamri 2020; Baspinar and Yardimci 2020; Elder et al. 2015; Keller et al. 2021; Millward et al. 2004; Mulloy et al. 2010; Whiteley et al. 2012). For example, a randomized and double-blinded study with a sample of 15 children with ASD aged 2–16 years determined the efficacy of the GF-CF diet. They found no significant improvements in core symptoms (Elder et al. 2006). Likewise, another double-blinded randomized controlled trial with a sample of 14 children with autism, aged 3–5 years, over 12 weeks of the GF-CF diet failed to report significant changes in core symptoms of physiologic functioning, behavior problems, or autism (Hyman et al. 2016). Furthermore, another study that aimed to evaluate the effects of gluten and casein supplementation on 74 children with ASD for a short duration of 7 days did not report any increase of maladaptive behavior, GI symptom severity, or urinary intestinal fatty acid-binding protein (I-FABP) excretion, a marker of enterocyte damage (Pusponegoro et al. 2015). In contrast to studies reporting null findings, two studies in children aged 4–10 years with ASD, with longer durations of 2 years (Whiteley et al. 2010) and 1 year (Knivsberg et al. 2002) conducted on comparatively bigger sample sizes of 72 and 20 children, respectively, found significant improvements in the core ASD symptoms.

8.12 Potential Side Effects of GF-CF Diet in ASD

In a recent systematic review and meta-analysis which reported no improvement following a GF-CF diet, it was highlighted that there might be an increased risk of GI adverse events such as discomfort and secondary effects. The adverse events could be attributed to the selective eating patterns that could lead to eating disorders or malnutrition creating an extra burden for families (Keller et al. 2021). Moreover, there are tangible risks of decreased appetite, weight loss, and sleep disturbances. Low bone densities and lower serum folate and vitamin B12 levels with elimination diets have also been reported (Baspinar and Yardimci 2020). A case-control study on Spanish children with ASD on a GF-CF diet reported lower anthropometric measurements compared to those on a regular diet, which could have an effect on growth. Additionally, low levels of calcium, phosphorus, and sodium were reported, thus raising the risk of micronutrient deficiencies that required supplementation, especially vitamin D (Marí-Bauset et al. 2016). In this regard, certain studies that assessed GF-CF diets in children with ASD recognized concerns about the risk of deficiencies of vitamins and minerals. Hence, all participants on dietary interventions were supplemented with vitamins and minerals to reduce the risks (Elder et al. 2006; Whiteley et al. 2010).

Significant vitamin D deficiencies have been reported in studies. In a study by Mostafa et al., 48% of the ASD children were vitamin D insufficient, and 40% were vitamin D deficient. On the other hand, none of the healthy children was deficient, and only 20% were insufficient (Mostafa and Al-Ayadhi 2012). Furthermore, in another study, 56% of children with ASD were reported to be on supplements compared to 78% of ASD children on GF-CF diet. The most important deficiencies reported were vitamin D, calcium, potassium, pantothenic acid, and choline (Stewart et al. 2015). Hence, care should be exercised when recommending the diet to individuals and their families, and an informed decision considering the thorough pros and cons should be the best course of action.

8.13 Nutritional Supplements

Presently, nutritional supplements such as omega-3 polyunsaturated fatty acids and vitamins including vitamin B6, vitamin B12, vitamin C, vitamin D, folic acid, and folinic acid have gained interest with studies reporting beneficial effects. A meta-analysis reported that the dietary supplements were more effective than placebo in treating anxiety-affect, behavioral problems, impulsiveness, hyperactivity, irritability, language, and social-autistic symptoms and stereotypic, restricted, and repetitive behaviors (Fraguas et al. 2019). In particular, omega-3 supplementation improved language and social-autistic symptoms, and vitamin supplementation improved global severity, language, stereotypic, restricted, repetitive behaviors, hyperactivity, and irritability. Potential mechanisms of the beneficial effects include increase in the omega-3/omega-6 ratio in the erythrocyte membrane. Further, the bioactive ingredient had an impact on neural development via anti-inflammatory and detoxification properties (Campisi et al. 2018; Parletta et al. 2016).

8.14 Pharmacological Therapy

Although most children with ASD receive pharmacological agents, there is limited evidence to demonstrate the benefits that outweigh the adverse effects (Oswald and Sonenklar 2007). Medications might be helpful to treat behavioral problems and comorbid symptoms of irritability, aggression, and hyperactivity in individuals of ASD. To date, no medicine holistically addresses the core characteristics of ASD. But, at present, there are two FDA-approved atypical antipsychotic medications, risperidone and aripiprazole, as first-line treatment for hyperactivity, impulsivity, and agitation (LeClerc and Easley 2015). Further, some studies have shown that selective serotonin reuptake inhibitors (SSRIs), such as fluoxetine and sertraline, might be helpful to address repetitive behaviors, anxiety, and OCD, and mirtazapine to address sleep disturbances (Persico et al. 2021). However, long-term evidence is uncertain and the usefulness of medications might be limited by their side effects. SSRIs are also often prescribed to treat comorbid symptoms in ASD, but clinical trials are yet to demonstrate their effectiveness. A review of nine randomized controlled trials assessed various SSRIs, including fluoxetine and citalopram, but failed to show a positive result in symptom reduction (Williams et al. 2010).

8.15 Novel Dietary Approaches

Several potential treatment options are in the pipeline. A novel metabolic intervention includes sulforaphane (SFN), an isothiocyanate cruciferous vegetable derivative, which interacts with heat-shock proteins and has been found to improve the behavioral symptoms of ASD (Houghton 2019; Yagishita et al. 2019). The use of SFN therapy imitates the metabolic effects of fever and influences negative oxidative stress, inflammation, as well as DNA damage which play a significant etiologic role in ASD. A recent systematic review showed an association between SFN use and ASD symptoms and a significant improvement in behavior and social and cognitive functions (McGuinness and Kim 2020).

The use of bacteria as therapeutic agents, such as probiotics, has been suggested as potential treatment alternatives in children with ASD with GI symptoms (Cerdó et al. 2017). Probiotics constitute lactic acid-producing bacteria, such as lactobacilli, lactococci, and bifidobacteria, or yeasts such as *Saccharomyces boulardii*. According to a systematic review, probiotics alone only improved certain GI symptoms; however, in combination with the exclusion diet, there was a reduction in anti-sociability scores (Ng et al. 2019). Nonetheless, there is room for further research with well-designed trials, and at present, there is limited evidence to recommend probiotics to address the GI and/or behavioral symptoms in children with ASD.

Another therapeutic avenue involves digestible proteins and enzymes. Elimination diets are usually accompanied by the substitution of dietary proteins by alternative sources of protein such as maize, legumes, or other plant-based proteins, which are also indigestible. Hence, it has been suggested that protein sources should be replaced by highly digestible protein. Combination treatments that utilize an elimination diet and high-quality protein diets could influence beneficial bacterial colonization and enzymatic fermentation and help normalize gut inflammation and permeability with subsequent improvement in ASD symptoms (Ng et al. 2019; Sanctuary et al. 2018).

Considering the digestive enzymes, supplementation with proteases is hypothesized to positively influence the GI and behavioral symptoms in ASD children. In contrast to a previous trial (Munasinghe et al. 2010), a more recent double-blind randomized control trial demonstrated that digestive enzyme intervention for 3 months showed improvement in emotional response, general impression autistic score, general behavior, and GI symptoms (Saad et al. 2015).

8.16 Conclusions

ASD is a neurodevelopmental disorder with a marked rise in prevalence globally. It is characterized by deficits in social communication and presence of restricted interests and repetitive behaviors. Recent changes to the diagnostic criteria occurred with the transition to the new diagnostic manual (DSM-5). Clinical evaluation begins with the developmental screening of the general pediatric population to identify at-risk children, followed by referral to a specialist for a definitive diagnosis and comprehensive neuropsychological assessment. Children with ASD should also be screened for common comorbid diagnoses. As determined from the evidence available on the efficacy, elimination of gluten and casein from the diet cannot be based entirely on ASD as an indication. There should be explicit and thorough informed consent. Hence, recommendations on dietary restrictions are currently limited to individuals with intolerance or an allergen, regardless of a diagnosis of ASD. The domain of dietary interventions in ASD is an evolving field and requires further exploration for the generalizability of findings and inclusion in the practice recommendations. Other research recommendations include determining the precise associations between gut function and ASD symptoms for characterizing personalized dietary interventions and protein sources.

References

Abrahams BS, Geschwind DH (2010) Connecting genes to brain in the autism spectrum disorders. Arch Neurol 67(4):395–399. https://doi.org/10.1001/archneurol.2010.47

Alamri ES (2020) Efficacy of gluten- and casein-free diets on autism spectrum disorders in children. Saudi Med J 41(10):1041–1046

Al-Farsi YM, Al-Sharbati MM, Al-Farsi OA, Al-Shafaee MS, Brooks DR, Waly MI (2011) Brief report: prevalence of autistic spectrum disorders in the Sultanate of Oman. J Autism Dev Disord 41(6):821–825

Ali A, Waly MI, Al-Farsi YM, Essa MM, Al-Sharbati MM, Deth RC (2011) Hyperhomocysteinemia among Omani autistic children: a case-control study. Acta Biochim Pol 58(4):547–551

Autism and Developmental Disabilities Monitoring Network Surveillance Year 2008 Principal Investigators, & Centers for Disease Control and Prevention (2012) Prevalence of autism spectrum disorders—Autism and Developmental Disabilities Monitoring Network, 14 sites, United States, 2008. Morb Mortal Wkly Rep Surveill Summ 61(3):1–19

Autism, Casein and Gluten Food Sensitivity, Diet (n.d.). https://peds.ufl.edu/divisions/genetics/programs/autism_card/casein.htm. Accessed 24 May 2021

Autism Spectrum Disorders (n.d.). https://www.who.int/news-room/fact-sheets/detail/autism-spectrum-disorders. Accessed 24 May 2021

Bandini LG, Anderson SE, Curtin C, Cermak S, Evans EW, Scampini R, Maslin M, Must A (2010) Food selectivity in children with autism spectrum disorders and typically developing children. J Pediatr 157(2):259–264

Baspinar B, Yardimci H (2020) Gluten-free casein-free diet for autism spectrum disorders: can it be effective in solving behavioural and gastrointestinal problems? Eurasian J Med 52(3):292–297

Besag FM (2018) Epilepsy in patients with autism: links, risks and treatment challenges. Neuropsychiatr Dis Treat 14:1–10

Buescher AVS, Cidav Z, Knapp M, Mandell DS (2014) Costs of autism spectrum disorders in the United Kingdom and the United States. JAMA Pediatr 168(8):721–728

Buie T (2015) Potential etiologic factors of microbiome disruption in autism. Clin Ther 37(5): 976–983

Campisi L, Imran N, Nazeer A, Skokauskas N, Azeem MW (2018) Autism spectrum disorder. Br Med Bull 127(1):91–100

Cass H, Gringras P, March J, McKendrick I, O'Hare AE, Owen L, Pollin C (2008) Absence of urinary opioid peptides in children with autism. Arch Dis Child 93(9):745–750

CDC (2020a, August 11) Autism and developmental disabilities monitoring (ADDM) network. https://www.cdc.gov/ncbddd/autism/addm.html

CDC (2020b, August 11) What is autism spectrum disorder? https://www.cdc.gov/ncbddd/autism/facts.html

CDC (2020c, September 25) Data & statistics on autism spectrum disorder. https://www.cdc.gov/ncbddd/autism/data.html

Cerdó T, Ruíz A, Suárez A, Campoy C (2017) Probiotic, prebiotic, and brain development. Nutrients 9(11):1247. https://doi.org/10.3390/nu9111247

Cheng N, Rho JM, Masino SA (2017) Metabolic dysfunction underlying autism spectrum disorder and potential treatment approaches. Front Mol Neurosci 10:34

Criado KK, Sharp WG, McCracken CE, De Vinck-Baroody O, Dong L, Aman MG, McDougle CJ, McCracken JT, Eugene Arnold L, Weitzman C, Leventhal JM, Vitiello B, Scahill L (2018) Overweight and obese status in children with autism spectrum disorder and disruptive behavior. Autism 22(4):450–459

de Magistris L, Familiari V, Pascotto A, Sapone A, Frolli A, Iardino P, Carteni M, De Rosa M, Francavilla R, Riegler G, Militerni R, Bravaccio C (2010) Alterations of the intestinal barrier in patients with autism spectrum disorders and in their first-degree relatives. J Pediatr Gastroenterol Nutr 51(4):418–424

DeFilippis M, Wagner KD (2016) Treatment of autism spectrum disorder in children and adolescents. Psychopharmacol Bull 46(2):18–41

Developmental Disabilities Monitoring Network Surveillance Year 2010 Principal Investigators, & Centers for Disease Control and Prevention (CDC) (2014) Prevalence of autism spectrum disorder among children aged 8 years—autism and developmental disabilities monitoring network, 11 sites, United States, 2010. Morb Mortal Wkly Rep Surveill Summ 63(2):1–21

DiGuiseppi C, Levy SE, Sabourin KR, Soke GN, Rosenberg S, Lee L-C, Moody E, Schieve LA (2018) Injuries in children with autism spectrum disorder: study to explore early development (SEED). J Autism Dev Disord 48(2):461–472

Doernberg E, Hollander E (2016) Neurodevelopmental disorders (ASD and ADHD): DSM-5, ICD-10, and ICD-11. CNS Spectr 21(4):295–299

DSM-5 and Autism: Frequently Asked Questions (n.d.). https://www.autismspeaks.org/dsm-5-and-autism-frequently-asked-questions. Accessed 24 May 2021

Eapen V, Mabrouk AA, Zoubeidi T, Yunis F (2007) Prevalence of pervasive developmental disorders in preschool children in the UAE. J Trop Pediatr 53(3):202–205

Elder JH, Shankar M, Shuster J, Theriaque D, Burns S, Sherrill L (2006) The gluten-free, casein-free diet in autism: results of a preliminary double blind clinical trial. J Autism Dev Disord 36(3):413–420

Elder JH, Kreider CM, Schaefer NM, de Laosa MB (2015) A review of gluten- and casein-free diets for treatment of autism: 2005-2015. Nutr Diet Suppl 7:87–101

El-Fishawy P, State MW (2010) The genetics of autism: key issues, recent findings, and clinical implications. Psychiatr Clin North Am 33(1):83–105

Evers K, Maljaars J, Carrington SJ, Carter AS, Happé F, Steyaert J, Leekam SR, Noens I (2021) How well are DSM-5 diagnostic criteria for ASD represented in standardized diagnostic instruments? Eur Child Adolesc Psychiatry 30(1):75–87

Falkmer T, Anderson K, Falkmer M, Horlin C (2013) Diagnostic procedures in autism spectrum disorders: a systematic literature review. Eur Child Adolesc Psychiatry 22(6):329–340. https://doi.org/10.1007/s00787-013-0375-0

Finegold SM (2011) State of the art; microbiology in health and disease. Intestinal bacterial flora in autism. Anaerobe 17(6):367–368. https://doi.org/10.1016/j.anaerobe.2011.03.007

Fraguas D, Díaz-Caneja CM, Pina-Camacho L, Moreno C, Durán-Cutilla M, Ayora M, González-Vioque E, de Matteis M, Hendren RL, Arango C, Parellada M (2019) Dietary interventions for autism spectrum disorder: a meta-analysis. Pediatrics 144(5). https://doi.org/10.1542/peds.2018-3218

Frye RE (2015) Metabolic and mitochondrial disorders associated with epilepsy in children with autism spectrum disorder. Epilepsy Behav 47:147–157

GBD 2019 Diseases and Injuries Collaborators (2020) Global burden of 369 diseases and injuries in 204 countries and territories, 1990-2019: a systematic analysis for the Global Burden of Disease Study 2019. Lancet 396(10258):1204–1222

Hassan A (2019) Arab views on autism. In: Encyclopedia of autism spectrum disorders, pp 1–5. https://doi.org/10.1007/978-1-4614-6435-8_102309-1

Heintz-Buschart A, Wilmes P (2018) Human gut microbiome: function matters. Trends Microbiol 26(7):563–574

Helping Children with Autism (n.d.). https://www.makingtherapyfun.com/conditions-we-treat/autism-spectrum-disorder/. Accessed 24 May 2021

Herndon AC, DiGuiseppi C, Johnson SL, Leiferman J, Reynolds A (2009) Does nutritional intake differ between children with autism spectrum disorders and children with typical development? J Autism Dev Disord 39(2):212–222

Hodges H, Fealko C, Soares N (2020) Autism spectrum disorder: definition, epidemiology, causes, and clinical evaluation. Transl Pediatr 9(Suppl 1):S55–S65

Hossain MM, Khan N, Sultana A, Ma P, McKyer ELJ, Ahmed HU, Purohit N (2020) Prevalence of comorbid psychiatric disorders among people with autism spectrum disorder: an umbrella review of systematic reviews and meta-analyses. Psychiatry Res 287:112922

Houghton CA (2019) Sulforaphane: its "coming of age" as a clinically relevant nutraceutical in the prevention and treatment of chronic disease. Oxid Med Cell Longev 2019:1–27. https://doi.org/10.1155/2019/2716870

Hudson CC, Hall L, Harkness KL (2019) Prevalence of depressive disorders in individuals with autism spectrum disorder: a meta-analysis. J Abnorm Child Psychol 47(1):165–175

Hyman SL, Stewart PA, Foley J, Cain U, Peck R, Morris DD, Wang H, Smith T (2016) The gluten-free/casein-free diet: a double-blind challenge trial in children with autism. J Autism Dev Disord 46(1):205–220. https://doi.org/10.1007/s10803-015-2564-9

Keller A, Rimestad ML, Friis Rohde J, Holm Petersen B, Bruun Korfitsen C, Tarp S, Briciet Lauritsen M, Händel MN (2021) The effect of a combined gluten- and casein-free diet on children and adolescents with autism spectrum disorders: a systematic review and meta-analysis. Nutrients 13(2). https://doi.org/10.3390/nu13020470

Knivsberg A, Reichelt KL, Nodland M, Hoien T (1995) Autistic syndromes and diet: a follow-up study. Scand J Educ Res 39(3):223–236. https://doi.org/10.1080/0031383950390304

Knivsberg AM, Reichelt KL, Høien T, Nødland M (2002) A randomised, controlled study of dietary intervention in autistic syndromes. Nutr Neurosci 5(4):251–261

Larauche M, Mulak A, Taché Y (2011) Stress-related alterations of visceral sensation: animal models for irritable bowel syndrome study. J Neurogastroenterol Motil 17(3):213–234

Lázaro CP, Pondé MP, Rodrigues LEA (2016) Opioid peptides and gastrointestinal symptoms in autism spectrum disorders. Rev Bras Psiquiatr (Sao Paulo, Brazil: 1999) 38(3):243–246

Lecavalier L, McCracken CE, Aman MG, McDougle CJ, McCracken JT, Tierney E, Smith T, Johnson C, King B, Handen B, Swiezy NB, Eugene Arnold L, Bearss K, Vitiello B, Scahill L (2019) An exploration of concomitant psychiatric disorders in children with autism spectrum disorder. Compr Psychiatry 88:57–64

LeClerc S, Easley D (2015) Pharmacological therapies for autism spectrum disorder: a review. P T 40(6):389–397

Lefter R, Ciobica A, Timofte D, Stanciu C, Trifan A (2019) A descriptive review on the prevalence of gastrointestinal disturbances and their multiple associations in autism spectrum disorder. Medicina 56(1). https://doi.org/10.3390/medicina56010011

Levy SE, Giarelli E, Lee L-C, Schieve LA, Kirby RS, Cunniff C, Nicholas J, Reaven J, Rice CE (2010) Autism spectrum disorder and co-occurring developmental, psychiatric, and medical conditions among children in multiple populations of the United States. J Dev Behav Pediatr 31(4):267–275

Levy SE, Wolfe A, Coury D, Duby J, Farmer J, Schor E, Van Cleave J, Warren Z (2020) Screening tools for autism spectrum disorder in primary care: a systematic evidence review. Pediatrics 145 (Suppl 1):S47–S59

Leyfer OT, Folstein SE, Bacalman S, Davis NO, Dinh E, Morgan J, Tager-Flusberg H, Lainhart JE (2006) Comorbid psychiatric disorders in children with autism: interview development and rates of disorders. J Autism Dev Disord 36(7):849–861

Li H, Liu H, Chen X, Zhang J, Tong G, Sun Y (2021) Association of food hypersensitivity in children with the risk of autism spectrum disorder: a meta-analysis. Eur J Pediatr 180(4): 999–1008

Lipkin PH, Macias MM, Chen BB, Coury D, Gottschlich EA, Hyman SL, Sisk B, Wolfe A, Levy SE (2020) Trends in pediatricians' developmental screening: 2002–2016. Pediatrics 145(4): e20190851. https://doi.org/10.1542/peds.2019-0851

Lotter V (1966) Epidemiology of autistic conditions in young children. Soc Psychiatry 1(3): 124–135. https://doi.org/10.1007/bf00584048

Maigoro AY, Lee S (2021) Gut microbiome-based analysis of lipid A biosynthesis in individuals with autism spectrum disorder: an in silico evaluation. Nutrients 13(2):688. https://doi.org/10.3390/nu13020688

Mandell DS (2005) Factors associated with age of diagnosis among children with autism spectrum disorders. Pediatrics 116(6):1480–1486. https://doi.org/10.1542/peds.2005-0185

Marí-Bauset S, Llopis-González A, Zazpe I, Marí-Sanchis A, Suárez-Varela MM (2016) Nutritional impact of a gluten-free casein-free diet in children with autism spectrum disorder. J Autism Dev Disord 46(2):673–684

McGuinness G, Kim Y (2020) Sulforaphane treatment for autism spectrum disorder: a systematic review. EXCLI J 19:892–903

Millward C, Ferriter M, Calver S, Connell-Jones G (2004) Gluten- and casein-free diets for autistic spectrum disorder. Cochrane Database Syst Rev. https://doi.org/10.1002/14651858.cd003498.pub2

Modabbernia A, Velthorst E, Reichenberg A (2017) Environmental risk factors for autism: an evidence-based review of systematic reviews and meta-analyses. Mol Autism 8:13

Mostafa GA, Al-Ayadhi LY (2012) Reduced serum concentrations of 25-hydroxy vitamin D in children with autism: relation to autoimmunity. J Neuroinflammation 9:201

Mulloy A, Lang R, O'Reilly M, Sigafoos J, Lancioni G, Rispoli M (2010) Gluten-free and casein-free diets in the treatment of autism spectrum disorders: a systematic review. Res Autism Spectr Disord 4(3):328–339. https://doi.org/10.1016/j.rasd.2009.10.008

Munasinghe SA, Oliff C, Finn J, Wray JA (2010) Digestive enzyme supplementation for autism spectrum disorders: a double-blind randomized controlled trial. J Autism Dev Disord 40(9): 1131–1138

Ng QX, Loke W, Venkatanarayanan N, Lim DY, Soh AYS, Yeo WS (2019) A systematic review of the role of prebiotics and probiotics in autism spectrum disorders. Medicina 55(5). https://doi.org/10.3390/medicina55050129

Oswald DP, Sonenklar NA (2007) Medication use among children with autism spectrum disorders. J Child Adolesc Psychopharmacol 17(3):348–355. https://doi.org/10.1089/cap.2006.17303

Özyurt G, Beşiroğlu L (2018) Autism spectrum symptoms in children and adolescents with obsessive compulsive disorder and their mothers. Noro Psikiyatr Ars 55(1):40–48

Parellada M, Penzol MJ, Pina L, Moreno C, González-Vioque E, Zalsman G, Arango C (2014) The neurobiology of autism spectrum disorders. Eur Psychiatry 29(1):11–19. https://doi.org/10.1016/j.eurpsy.2013.02.005

Parletta N, Niyonsenga T, Duff J (2016) Omega-3 and omega-6 polyunsaturated fatty acid levels and correlations with symptoms in children with attention deficit hyperactivity disorder, autistic spectrum disorder and typically developing controls. PLoS One 11(5):e0156432

Perrin JM, Coury DL, Hyman SL, Cole L, Reynolds AM, Clemons T (2012) Complementary and alternative medicine use in a large pediatric autism sample. Pediatrics 130(Suppl 2):S77–S82. https://doi.org/10.1542/peds.2012-0900e

Persico AM, Ricciardello A, Lamberti M, Turriziani L, Cucinotta F, Brogna C, Vitiello B, Arango C (2021) The pediatric psychopharmacology of autism spectrum disorder: a systematic review—part I: the past and the present. Prog Neuropsychopharmacol Biol Psychiatry 110: 110326

Posar A, Visconti P (2020) Sleep problems in children with autism spectrum disorder. Pediatr Ann 49(6):e278–e282

Pusponegoro HD, Ismael S, Firmansyah A, Sastroasmoro S, Vandenplas Y (2015) Gluten and casein supplementation does not increase symptoms in children with autism spectrum disorder. Acta Paediatr 104(11):e500–e505. https://doi.org/10.1111/apa.13108

Rogers GB, Keating DJ, Young RL, Wong M-L, Licinio J, Wesselingh S (2016) From gut dysbiosis to altered brain function and mental illness: mechanisms and pathways. Mol Psychiatry 21(6): 738–748

Rutter M (2005) Aetiology of autism: findings and questions*. J Intellect Disabil Res 49(4): 231–238. https://doi.org/10.1111/j.1365-2788.2005.00676.x

Saad K, Eltayeb AA, Mohamad IL, Al-Atram AA, Elserogy Y, Bjørklund G, El-Houfey AA, Nicholson B (2015) A randomized, placebo-controlled trial of digestive enzymes in children with autism spectrum disorders. Clin Psychopharmacol Neurosci 13(2):188–193

Sanctuary MR, Kain JN, Angkustsiri K, German JB (2018) Dietary considerations in autism spectrum disorders: the potential role of protein digestion and microbial putrefaction in the gut-brain axis. Front Nutr 5:40

Shattock P, Whiteley P (2002) Biochemical aspects in autism spectrum disorders: updating the opioid-excess theory and presenting new opportunities for biomedical intervention. Expert Opin Ther Targets 6(2):175–183

Sheldrick RC, Maye MP, Carter AS (2017) Age at first identification of autism spectrum disorder: an analysis of two US surveys. J Am Acad Child Adolesc Psychiatry 56(4):313–320

Stewart PA, Hyman SL, Schmidt BL, Macklin EA, Reynolds A, Johnson CR, James SJ, Manning-Courtney P (2015) Dietary supplementation in children with autism spectrum disorders: common, insufficient, and excessive. J Acad Nutr Diet 115(8):1237–1248

Treffert DA (1970) Epidemiology of infantile autism. In. Archives of General Psychiatry 22(5):431. https://doi.org/10.1001/archpsyc.1970.01740290047006

Website (n.d.-a) Retrieved 24 May 2021, from Aljarallah A, Alwaznah T, Alnasari S, Alhazmi M (2006) A study of autism and developmental disorders in Saudi children. Riyadh: Report, King Abdulaziz City for Science and Technology. http://www.kacst.edu.sa. Accessed 16 May 2021

Website (n.d.-b). www.mchatscreen.com. Accessed 23 May 2021

Whiteley P, Haracopos D, Knivsberg A-M, Reichelt KL, Parlar S, Jacobsen J, Seim A, Pedersen L, Schondel M, Shattock P (2010) The ScanBrit randomised, controlled, single-blind study of a gluten- and casein-free dietary intervention for children with autism spectrum disorders. Nutr Neurosc 13(2):87–100. https://doi.org/10.1179/147683010x12611460763922

Whiteley P, Shattock P, Knivsberg A-M, Seim A, Reichelt KL, Todd L, Carr K, Hooper M (2012) Gluten- and casein-free dietary intervention for autism spectrum conditions. Front Hum Neurosci 6:344

Williams K, Wheeler DM, Silove N, Hazell P (2010) Selective serotonin reuptake inhibitors (SSRIs) for autism spectrum disorders (ASD). Cochrane Database Syst Rev. https://doi.org/10.1002/14651858.cd004677.pub2

Winburn E, Charlton J, McConachie H, McColl E, Parr J, O'Hare A, Baird G, Gringras P, Wilson DC, Adamson A, Adams S, Le Couteur A (2014) Parents' and child health professionals' attitudes towards dietary interventions for children with autism spectrum disorders. J Autism Dev Disord 44(4):747–757

Wong HHL, Smith RG (2006) Patterns of complementary and alternative medical therapy use in children diagnosed with autism spectrum disorders. J Autism Dev Disord 36(7):901–909

Xu G, Snetselaar LG, Jing J, Liu B, Strathearn L, Bao W (2018) Association of food allergy and other allergic conditions with autism spectrum disorder in children. JAMA Netw Open 1(2): e180279

Yagishita Y, Fahey JW, Dinkova-Kostova AT, Kensler TW (2019) Broccoli or sulforaphane: is it the source or dose that matters? In. Molecules 24(19):3593. https://doi.org/10.3390/molecules24193593

Zaboski BA, Storch EA (2018) Comorbid autism spectrum disorder and anxiety disorders: a brief review. Future Neurol 13(1):31–37

Therapeutic Approaches for Attention Deficit-Hyperactivity Disorder

9

Arokiasamy Justin-Thenmozhi, Thamilarasan Manivasagam, and Anupom Borah

Abstract

Attention deficit-hyperactivity disorder (ADHD) is characterized by inattention, impulsivity, and hyperactivity symptoms that affect about 5% of children worldwide. ADHD pathology has been associated with the disruption in dopaminergic, serotoninergic, cholinergic, and adrenergic pathways. Currently used pharmaceutical agents effectively reduce the symptoms in the short term with various side effects and are not effective over long periods. Therefore, more interest has developed towards the development of drugs from natural products, including herbal medicines/phytochemicals. In Ayurveda, the progression of ADHD is curable when treated by using both the internal and external medications with a variety of therapeutic procedures, including Swedana (sudation), Snehan (oleation), Shodhana procedures like Virechana (purgation), Nasya, Lepa (external application), Dhumapana (medicated smoke), Vamana (emesis), Basti (medicated enema), Parisheka (oil bath), Anjana (Collyrium application), Abhyanga (massage), and Shamana Chikitsa (internal medication). In the Siddha medicinal text, ADHD symptoms have been associated to Sanni noi called

A. Justin-Thenmozhi
Department of Biochemistry, School of Biological Sciences, Madurai Kamaraj University, Madurai, Tamil Nadu, India

Department of Biochemistry and Biotechnology, Faculty of Science, Annamalai University, Chidambaram, Tamil Nadu, India

T. Manivasagam (✉)
Department of Biochemistry and Biotechnology, Faculty of Science, Annamalai University, Chidambaram, Tamil Nadu, India

A. Borah
Cellular and Molecular Neurobiology Laboratory, Department of Life Sciences and Bioinformatics, Assam University, Silchar, Assam, India

M. W. Qoronfleh et al. (eds.), *Proteins Associated with Neurodevelopmental Disorders*, Nutritional Neurosciences, https://doi.org/10.1007/978-981-15-9781-7_9

Alathidu Sanni. A study indicated that the exposure of the ADHD child to various varmams nullified the symptoms and behavior. According to the traditional Chinese medicinal theories, ADHD can be treated using Chinese herbal medicine (CHM), with adjuvant acupuncture, tuina, tai chi chuan, and diet. A wide range of nutritional supplements such as vitamins, amino acids, essential fatty acids, and minerals were projected as possible adjuncts and alternative ADHD treatment strategies. Due to the multifactorial pathology of ADHD, the multi-therapeutic approach is needed to manage the disorder. Moreover, combined therapies involving both pharmaceutical agents and natural products/nutrients might also help improve the overall functioning by targeting the symptoms of ADHD.

Keywords

Attention deficit-hyperactivity disorder · ADHD · Natural products · Phytochemicals · Herbal medicine · Nutritional supplements · Pharmacological agents · Therapy

9.1 Introduction

Attention deficit-hyperactivity disorder (ADHD) is a behavioral and psychiatric disease affecting 5% of children worldwide. About 60% of cases persist into adolescence and adulthood stages. It is mainly characterized by invasive symptoms of inattention, impulsivity, and hyperactivity. It has been associated with conduct rebellious disorder, obsessive-compulsive disorder, personality disorder, depression, enhanced threat of obesity, autism, and epilepsy (Cabarkapa et al. 2019). In addition, children, adolescents, and adults with ADHD exhibit worse connection with family members; enhanced threat of accidental injuries; lower class performance; enhanced school denial and grade retention; earlier sexual commitment resulting in teenage pregnancy; addiction to marijuana, tobacco, and other drugs; reduced job performance; and increased psychological problems resulting in poor quality of life (Peasgood et al. 2016; Dalsgaard et al. 2015).

9.2 Anatomical Changes in ADHD

Neuroimaging studies indicated that the structural abnormalities in the cortex might lead to ADHD. Abnormal cortical development might be due to mitochondrial dysfunction, impaired energy, oxidative stress, inflammation, and apoptosis. Langleben et al. (2001) reported reduced flow of blood to the prefrontal cortex (PFC). In addition, structural abnormalities and alterations in their functions were found in the affected children. The main morphological aspect linked with ADHD was the structural thinning of the PFC and precentral area of the cortex. About 50% of the cortical thickness was attained in 7.5 years of normal-developing children, while the same was developed in 10-year-old ADHD children, indicating a significant lack of cortical development. Children with ADHD showed a reduction in cortical volume due to diminished cortical folding and surface area, which worsened due to the aging process and drug intake.

Batty et al. (2015) indicated a significant decline in the volume of thalamus and dorsolateral PFC in ADHD children. They suggested that the pathological changes found in cortex development were linked with the maturation of thalamus. Apart from the cortex and thalamus, decreased striatum size (responsible for hyperactivity and memory) and a decline in the volume of the cerebellum (related to cognitive and affective function) were found in ADHD children (Hill et al. 2003).

9.3 Neurobiology of ADHD

In 1971, Wender was the first scientist who indicated that the abnormalities in the neurotransmission system could lead to the symptoms of ADHD. There are several neurotransmitters such as dopamine, noradrenaline, and serotonin that are reported to have a wide range of functions. Dopamine exerts key roles in the arousal, control of motor and executive functions, reinforcement, motivation, and reward by activating various signaling pathways by binding with the dopaminergic receptors found in the regions of the dopaminergic pathways of the human brain. Serotonin (5-hydroxytryptamine, 5-HT), an important neurotransmitter, is reported to be involved in mood, movement, social behavior, pain appreciation, endocrine secretions, sleep-wake cycle, cardiac outputs, and memory. Serotonergic neurons are mainly found in the raphe nuclei present in the medulla, pons, and a few regions of the brain stem, which extends primarily to the cerebral cortex and limbic areas. Noradrenaline is another neurotransmitter responsible for the arousal behavior (learning, pleasure, and anxiety) and activity of human beings. Noradrenergic neurons are present abundantly in the locus coeruleus nucleus of medulla and connected to the cerebral cortex and limbic areas by several pathways. ADHD has been associated with the disruption of dopaminergic, serotoninergic, cholinergic, and adrenergic pathways (Yuan et al. 2018).

9.4 Dopaminergic Hypothesis

Levy (1991) showed the relationship between the dopamine (DA) depletion in cortex and striatum and the ADHD symptoms, and proposed the dopamine theory of ADHD.

The dynamic developmental theory The malfunction of dopamine transmission in frontal-limbic circuits leads to lowered levels of tonic dopamine that results in a sharp and delayed reinforcement gradient and reduced extinction (Sagvolden et al. 2005). ADHD children tend to react superior to immediate rewards than delayed rewards.

Dopamine transfer deficit theory Although there is a presence of stimulant level of dopamine, the alterations of phasic dopamine response to reinforcement were found. Schultz (1998) indicated that there is a reduced transfer of dopaminergic signals from actual rewards to earlier actions that consistently envisage the future reward. The lack of reinforcement of attending by anticipation of dopamine release leads to inattention symptoms; that is, the children cannot pay more attention to information, commit careless errors, and neglect on-task behavior. Inadequate phasic dopamine response to rewards results in impulsivity (delay between the target behavior and actual reinforcement) and hyperactivity (children skip the seats in the classroom).

Four diverse proofs support the role of DA in the cause of ADHD as follows:

1. Data from experimental studies indicated that the symptoms of ADHD were linked with the alterations of the dopaminergic system in animals.
2. Support from pharmacological studies pointed out that the drugs used for the treatment of ADHD act in the DA system.
3. Evidence from brain imaging studies revealed that the patients showed problems in the brain regions having dopaminergic nerves.
4. Proof from genetic experiments indicated that alterations in the genes associated with the synthesis, transport, and degradation of DA were mainly found in the patients.

The facts that are against the strength of the dopamine hypothesis are as follows:

1. Psychostimulants and inhibitors of the noradrenaline transporter nullified the symptoms of ADHD.
2. The key discovery in ADHD is its linkage with polymorphism of the D4 receptor gene. However, few studies indicated that ADHD is a disorder raised due to defects in few regions of the brain.
3. There is also a possibility of dull dopamine responses in few ADHD patients even after the administration of methylphenidate due to higher baseline dopamine tone.

9.5 Serotoninergic Theory

Clinical and preclinical studies have indicated the role of serotonin in the inflection of impulsivity, attention, and hyperactivity symptoms. There is a connection available between the serotoninergic neurons of raphe nuclei and the dopaminergic neurons of midbrain regions (substantia nigra and central tegmental area). Further, the serotoninergic projections are also connected to the dopaminergic terminals of striatum, PFC, and nucleus accumbens. In dopamine transporter (DAT) knockout mice, without inducing the levels of dopamine, hyperactivity was inhibited after the administration of inhibitors of serotonin reuptake or precursors of this neurotransmitter, indicating the importance of 5-HT. Moreover, elevated brain serotonergic activity in ADHD children was measured by indirect methods, i.e., enhancement in the levels of 5-hydroxyindoleacetic acid (metabolite of 5-HT) in the cerebrospinal fluid, diminished release of circulatory prolactin (hormonal secretion is regulated by 5-HT), and lowered levels of circulatory 5-HT and platelet imipramine H3 B_{max} (substance used for labeling serotonin binding). All the above experiments highlighted the key role of serotoninergic system in the pathophysiology of ADHD (Oades 2010).

9.6 Cholinergic Theory

As per the cholinergic hypothesis, the changes in the function of the nicotinic acetylcholine receptor (nAChR) may result in ADHD symptoms, which is relieved by the activation of the receptors. But till now, nAChR medications are not approved for the treatment of ADHD. However, both the clinical and animal studies indicated that the activation of nAChRs could improve the symptoms and recover the executive functions (Potter et al. 2014).

Levin et al. (2001) studied the role of a placebo, nicotine, methylphenidate (MPH), and their combined treatment in adult patients for 21 days. Variability in response time was found to be enhanced in nicotine alone, and nicotine- and methylphenidate-treated groups. A decline in depressive symptoms was found initially (15th day) in the nicotine and combination groups and finally (21st day) in MPH-treated individuals. Other studies by Conners et al. (1998) and Gehricke et al. (2006) indicated that the administration of nicotine nullified the problems in learning, difficulty in concentrating, and daydreaming/zoning out in most of the subjects. ABT-089, -418, and -894 and AZD-1446 and -3480 (agonists of $\alpha4\beta2$ nicotinic receptors) were developed and tested for ADHD. Administration of ABT-089 improved the symptoms such as Clinical Global Impression-Severity (CGI scale), working memory, and impulsive responding in ADHD children (Wilens et al. 2006).

9.7 Pharmaceutical Agents

ADHD is connected with disturbances in the function of brain catecholamines. The therapeutic agents enhanced the brain catecholamine levels, thereby attenuating the symptoms of ADHD. The drugs are categorized into stimulants and non-stimulants. Methylphenidate and dextroamphetamine are the predominantly used stimulant drugs that mimic the structure of dopamine and norepinephrine, thereby restoring the catecholamine levels and correcting the underlying abnormalities during ADHD. Atomoxetine, antidepressants (bupropion, phenelzine, and imipramine), and norepinephrine-specific reuptake inhibitors are non-stimulant drugs (inferior to stimulants) that have also been demonstrated to enhance the brain catecholamine levels resulting in the improvement of behavior (Table 9.1).

Table 9.1 Major pharmacological agents employed in ADHD management

Medication	Mode of action	Symptoms managed	Disadvantages	Common side effects
Amphetamines (AMPH)	Due to their structural similarities, (1) amphetamines are transported by the dopamine (DA), serotonin (SE), and norepinephrine (NE) transporters, thus diminishing the reuptake of synaptic dopamine, serotonin, and norepinephrine; (2) they bind trace amine-associated receptor 1 (TAAR1), phosphorylate DAT, and internalize into neurons to stop internal transport of dopamine; (3) they also induce the release of these neurotransmitters from the vesicles of presynaptic neurons into the cytoplasm	Short attention span, impulsive behavior, and hyperactivity	Only cure the symptoms in 70% of adults and 70–80% of children	Headache, stomach discomfort, and higher blood pressure Other symptoms—Loss of appetite, weight loss, nervousness, and insomnia
Methylphenidate (MPH)	By binding with dopamine and norepinephrine transporters, the drug reduces the reuptake of these neurotransmitters	Excessive hyperactivity, impulsivity, and inattention	Reduced growth It can create bipolar illness or Tourette's syndrome	Loss of appetite, weight loss, jitteriness, irritability, sleep disturbance, stomach upset, constipation, and heartburn
Atomoxetine	By binding with norepinephrine transporter (NET), it inhibits NE reuptake in all regions of the brain except cortex As there is a very low expression of DAT found in	Slows down the hyperactivity and impulsive behavior, enhances the attention and concentration span	Data obtained on short-term use, whereas disadvantages of long-term use have not been assessed so far	Nausea, vomiting, insomnia, abdominal pain, reduced appetite, weight loss, headache, and sedation Dysmenorrhea, urinary retention, erectile dysfunction, and diminished libido in adults

	PFC, the reuptake of dopamine NET is stopped			
Extended-release clonidine	Induces adrenergic receptors (α2)	Has the calming effect and reduces the hyperactivity, impulsivity, aggression, overarousal, and sleep disorder Increases the day-to-day activity and performance	Unhelpful for inattentive symptoms Not suitable for treating adult ADHD patients	Dizziness, headache, vomiting, fatigue, dry mouth, and irritability Pain in upper abdomen, constipation Lowered blood pressure and erectile dysfunction
Extended-release guanfacine	Induces adrenergic receptors (α2)	Reduces the inattention and hyperactivity-impulsivity in ADHD youngsters	Not suitable for children with heart, hepatic, or kidney problems It can interact with other medicines such as phenytoin clonidine, rifampin, carbamazepine, ketoconazole, and valproic acid	Dizziness, headache, vomiting, fatigue, dry mouth, and irritability Pain in upper abdomen, constipation Lowered blood pressure and erectile dysfunction
Bupropion	Inhibits DAT and NET weakly	Minor improvement of ADHD symptoms	Insufficient evidence is present. More studies are required for confirming its therapeutic effect in adults and children	Headache, insomnia, gastrointestinal tract problems, body pains, chest pain, and dry mouth
Imipramine	Inhibits NET and serotonin transporter (SET)	Hyperactivity and impulsivity	As it is an antidepressant maximum efficacy is reached after 2–4 weeks	Nausea, dry mouth, drowsiness, and weakness Anxiety Changes in appetite
Modafinil	Inhibits DAT weakly than other psychostimulants	Treat symptoms of ADHD in adults 18 years of age and older	According to FDA, its use can be abused. The safety and effectiveness of children have not been established	Insomnia and decreased appetite Serious skin reactions

Disadvantages of pharmaceutical agents The pharmaceutical agents are effective in reducing the symptoms of ADHD for a short period, but not for more extended periods (Leucht et al. 2012). Few adverse effects, including drug abuse or dependency, might also occur. Safety is the major problem for a few patients, particularly those having cardiovascular disorders. The causes for discontinuation of drugs in the middle of the treatment include the apparent need for effectiveness, aversion during drug intake, and stigmatization. The choice of the drug for a specific patient is made on a trial-and-error basis, as the cause of ADHD is not fully known. Although these drugs improve ADHD symptoms (Ahn et al. 2016), 20–30% of affected individuals are nonresponders or unable to bear the side effects.

9.8 Traditional Therapeutic Approaches

Due to the adverse effects and lowered efficacy of current pharmacological drugs, research interests have developed towards the development of drugs from natural products including herbal medicines/phytochemicals (Bader and Adesman 2012; Searight et al. 2012). Since it is natural, complementary and alternative medicines are gaining attention by the caregivers. About 50% of caregivers choose these types of natural supplementations alone or in synergy with other drugs (Sinha and Efron 2005).

9.8.1 Ayurveda

Ayurveda has mainly accepted the undividable and inter-reliant relationship of sarira (body) and manas (psyche) and their individuality in humans. In Ayurveda, it is defined that the combination of body, soul, mind, and senses forms life. Acharya Charaka, the Ayurvedic text, narrates the "manas" as the entity that accounts for thinking. Ayurveda is the only ancient system of medicine that clarifies approximately all the behavioral and psychiatric diseases as Apasmara and Unmada. Unmada is described as the brain and psychiatric illness featuring uneven remembrance, mind, mental power, consciousness, and knowledge with bad manners (Shukla 2005).

Etiology The formation and development of the fetus (Garbha) are mainly dependent on the normalcy of Shonita (ovum) and Shukra (sperm). The lifestyle and diet of the pregnant lady are responsible for the normal growth and well-being of a child. Emotional instability such as excessive grief, fear, anger, and irritability; excess physical workload by brittle persons; control of natural urges; and unhygienic, improperly processed diet with irregular dietary habits are a few factors that affect the physical, mental, or both status of the baby and should be evaded by the pregnant women (Shukla 2005). The disturbance in the psychological factors and dietary patterns during the antenatal period would eventually affect the fetus and lead to psychological and neurobehavioral disorders. Depending upon the psychological state of the parents, mana of fetus attains satwa, raja, and tama characters (Shukla 2005). Satwa Guna (positivity quality of mind) is likely to get diminished and this leads to Tridosha (three humors of body) that accumulates in Hridaya (heart/mental faculty), reducing the mental function and finally leading to Unmada (Gaud 2014).

The psychological problems arise due to an imbalance in the three humors that regulate the functions of body. Fear, insomnia, anxiety, and mental instability arise due to Vata imbalance; anger and irritability form due to Pitta imbalance, and depression and lethargy occur from Kapha imbalance (Sharma 2008). As Vata is the primary form of Tridosha that regulates the brain functions, their dysfunction results in hyperactivity. Children will be unable to control their stimuli and thoughts and will not be able to obey their parents or have fair activities. In most cases, Pitta dosha prevails over Vata dosha resulting in irritability, anger, not liking hot things or wearing clothes, and loving of cold air and water.

Treatment The progression of Unmada is curable by treating it with both the internal and external medications with a variety of therapeutic procedures including Swedana (sudation), Snehan (oleation), Shodhana procedures like Virechana (purgation), Nasya, Lepa (external application), Dhumapana (medicated smoke), Vamana (emesis), Basti (medicated enema), parisheka (oil bath), Anjana (Collyrium application), Abhyanga (massage), and Shamana Chikitsa (internal medication) (Kleigmn 2017). The preservation of Agni (digestive fire/metabolism) is done by using formulations such as Agnitundi Vati, Mustarishta, and Abhayarishta. It is of utmost importance, as in the absence of appropriate metabolism, actions of drugs can be unexpected. To manage the Vata and Pitta dosha, medicated ghee containing cognitive improvers (internal oleation therapy) and oil processed with soothing and cooling drugs (external oleation therapy) are used. As lipophilic agents can cross the blood-brain barrier, medicated ghee might act as a carrier for cognitive modifiers (Sharma 1990). The external oil massage stimulated the touch receptors and assisted in soothing the hyperactive children.

For treating Pitta-dominant disease, formulations such as Avipattikara Churna are the primary choice of drug. Virechana (purgation) is used to eliminate the accumulated metabolic wastes such as nitrogen compounds that may induce hyperactivity and aggressiveness in children. Basti is primarily used for Vata type of diseases by removing the toxins and preserving dosha balance. Acharya Charaka indicated that Rasayanas are reported to have compounds that maintain the brilliant quality of rasadidhatus (body tissues) that improve life span, inhibit aging processes, enhance intelligence, heal the disease, and build a strong and healthy body (Gaud 2014).

The drugs used to treat neurological diseases possess vatapitagna and kaphavatagna properties. Teekshna drugs such as Pippali (*Piper longum*), Twak (*Cinnamomum zeylanicum* Breyn.), Nidigdhika (*Solanum surattense* Burm.f.) and Vidanga (*Embelia ribes*), Bala (*Sida cordifolia*), Draksha (*Vitis vinifera*), Neelkamal (*Nymphaea stellate*), Manjishtha (*Rubia cordifolia*), Anshumati (*Desmodium gengeticum*), Swadanshatra (*Tribulus terrestris*), and Prapaundarika, (*Nelumbo nucifera* gaeris) have vatapitagna properties. Saindhava lavana (rock salt-potassium chloride) (Sharma 2006), Madhuka (*Madhuca indica*), Rasna (*Pluchea lanceolata*), Brihati (*Solanum indicum* Linn), Twak (*Cinnamomum zeylanicum*), Nidigdhika (*Solanum surattense*), and Til oil (*Sesamum indicum*) containing drugs (Pandeya 2005) have Pittahara action and are used to diminish the Pitta-dominant symptoms like aggressive and agitated behavior.

9.8.2 Herbs

Bacopa monnieri (L.) Wettst. (B. monnieri) or Brahmi is an excellent cognitive enhancing Ayurvedic herb. More than 20 clinical trials studied the efficacy of Brahmi in children and adolescents with ADHD. Treatment with Bacopa improved the attention, cognition, intelligence, and behavior by altering sentence repetition, logical memory, delayed response learning, cognitive control, word recall (non-meaningful), digit span test, picture recall, and hyperactive and impulsive subscale in children and adolescents diagnosed with ADHD (Kean et al. 2016). Combination of Bacopa with *Centella asiatica* (primary ingredients), among few herbs, significantly improved the behavior compared to placebo in children and adolescents with ADHD (Dean et al. 2017).

Previous studies indicated that the daily administration of *Ginkgo biloba* extract (240 mg) led to modest improvements in the behavior (Uebel-von Sandersleben et al. 2014). In contrast, another study reported that the administration of the same extract (80–120 mg) showed an improvement in the teacher and maternal ratings of attention compared to placebo in ADHD children (Shakibaei et al. 2015). Another study compared the daily administration of methylphenidate (20–30 mg) with that of ginkgo extract (80–120 mg). Although the administration of methylphenidate is better than ginkgo, the side effects such as insomnia, diminished appetite, and headache were found in the former group as compared to ginkgo (Salehi et al. 2010). One study led by Niederhofer (2010) showed the beneficiary effect of St. John's wort, whereas another study by Weber et al. (2008) failed to nullify the symptoms in children with ADHD. Treatment of leaf extract of *Passiflora incarnata* (0.4 mg/kg/day) showed equal efficacy as methylphenidate (1 mg/kg/day) in improving symptoms without the adverse effects of methylphenidate (Akhondzadeh et al. 2005). Another clinical trial involving *Valeriana officinalis* root failed to show improvement in the children with ADHD (Razlog et al. 2012). Ayurvedic polyherbal medications such as Saraswat Churna, Saraswataghritha, Saraswatha Arishta, Manasamitra Vati, Brento syrup, Brahmivati, Memovit granules, Maha Kalyanaka Ghrita, Panchagavya Ghrita, Braintone syrup, and Shankhapushpi syrup provided better relief for ADHD.

9.8.3 Siddha

Siddha system is mainly practiced in South India and is an ancient traditional system of medicine. In this system, various modes of therapies such as varmam, podithimiral, kombukattal, and thokkanam are available. Varmam is mainly used to treat numerous diseases that are linked with musculoskeletal and neurological problems. In the Siddha medicinal text, ADHD symptoms are connected to Sanni noi particularly called Alathidu Sanni, which is formed due to change in Vatham, Pitham, and Kabam. Varmam therapy is entirely anatomical and is regulated by Varmam treatment. A clinical study by Sasikumar et al. (2020) indicated that the exposure of ADHD children to Thilarthavarmamnatchathira varmam and Pidari varmam nullified the symptoms and behavior.

9.9 Traditional Chinese Medicines (TCM)

TCM is a broad medical system, followed for more than 2000 years. Although no specific name or diagnostic symptoms for ADHD are available in TCM literatures, the disease and their symptoms were depicted as injudicious behavior, forgetfulness, dysphoria, etc. According to TCM, it is a disease that affects emotion, thought, and mind. Yin controls calmness, while Yang controls movement; coordination is produced if the equilibrium between yin and yang occurs. According to yin-yang theory, both are conflicting and restricting, interdependent, and equally promoting each other. Both of them in the body are in balance, and their imbalance leads to diseases. The primary pathology lies in the disturbance of yin-yang, which results in the dysfunction of the Zang-fu (organs). The functions of theses 5 Zang-fu (liver, lung, kidney, heart, and spleen) are to synthesize and accumulate blood, qi, and other body fluids. The main functions of these organs are to maintain the mind and spirit. Ni et al. (2014) indicated that the TCM doctors recommend individual-based therapies for every patient, consisting Chinese herbal medicine, with adjuvant therapies like acupuncture, tuina (massage), tai chi chuan (breathing and exercise), and diet.

9.9.1 Chinese Herbal Medicine (CHM)

CHM formulae mainly consist of various herbs, minerals, and animal drugs that are utilized by TCM clinicians as therapeutic agents for ADHD. Treatment with Duodongning granule, a CHM containing Shudihuang (Rehmanniae radix preparata), Gancao (Glycyrrhizae radix et rhizoma), Renshen (Ginseng radix et rhizome), Wuweizi (*Schisandra chinensis*), Gouqizi (*Fructus lycii*), and Fuling (*Poria cocos*), offered a similar effect in hyperactivity, social performance, and academic success. The outcomes were similar to MPH (10 mg/day) treatment but with fewer side effects (Li and Chen 1999). Combined administration of Yizhi mixture (Guiban (Testudinis carapacis et plastri), Gouteng (Uncariae ramulus cum uncis), Shudihuang (Rehmanniae radix preparata), and Lujiaoshuang (Cervicornude gelatinatum) and Jingling oral liquid (Shanyao (Rhizome Dioscoreae)), Shichangpu (Rhizoma Acori Tatarinowii), Longgu (Os Draconis), Yuanzhi (Polygalae radix), and Shudihuang (Rehmanniae radix preparata) with MPH offered more neuroprotective effect than the individual treatment because of their synergistic action with fewer side effects in ADHD children (Ding et al. 2002; Wang et al. 2011). Ningdong granule (Dangshen (Codonopsis radix)), Baishao (Paeoniae alba radix), Maidong (Ophiopogonis radix), and Tianma (Gastrodiae rhizome) treatment was more effective and safe than MPH treatment by increasing circulatory homovanillic acid concentration (increased dopamine metabolism) in ADHD children (Li et al. 2011). Ni et al. (2014) indicated that about 94 herbs are used as CHM formulae in 39 different studies for nullifying the symptoms of ADHD.

9.9.2 Acupuncture

Acupuncture (De-qi) is done by inserting fine needles (sterilized) at acupoints (particular body surface) followed by lifting, twisting, and rotating them to obtain the desired psychological effect responses. Acupuncture searches the meridian and controls the yin-yang and Zang-fu. Experiments by Chai (1999) and Shi (2002) indicated that the children with ADHD who received acupuncture showed a comparable effect as MPH in the short-term exposure. Further studies reported higher efficacy in long term with no side effects.

9.9.3 Tuina (Chinese Medical Massage)

It is a naturopathy process involving nonpharmaceutical and noninvasive methods such as rubbing, pushing, pressing, kneading, transporting, nipping, rotating, and foulage gently and delicately on specific meridians and acupoints. Tuina searches meridians, enhances qi and blood flow, fortifies the body resistance, and normalizes yin-yang. Wang and Shi (2005) applied tuina on the head, neck, chest, back, and abdomen of 33 children with ADHD. After 18 treatments, 10 children were cured, and 18 children showed significant improvement, while 5 had no effect. Another study by Zhuo (2006) compared the impact of tuina with MPH and indicated that both of them showed a similar effect, while tuina group demonstrated a reduced reappearance rate even 6 months after the treatment period.

9.9.4 Tai chi chuan (Breathing and Exercise)

Tai chi chuan (Taiji or tai chi) treatment is also an ancient form of TCM, which emerged 300 years ago. The underlying mechanism is balancing yin-yang and changing deficiency excess, which is also the basis for TCM. It involves the sluggish movements of active and stagnant forms. By moving, breathing, and exercising, Tai chi chuan induced and regulated qi and blood. By uniting deep, full, and shallow breathing with slow, gentle, and graceful movements, one can acquire inner rest and calm mind by transfer of the disturbing thoughts into attention.

Wen (2009) demonstrated that the ADHD children undergoing Tai chi chuan training program for 12 weeks showed reduction in symptoms of ADHD and improvement in vestibular function, learning abilities, and proprioception, as compared to the controls. Hernandez-Reif et al. (2001) observed that ADHD adolescents practising Tai chi chuan (two times a week for 5 weeks) showed improvement in hyperactivity, conduct disorder, anxiety, improper emotions, and daydreaming as compared to the control ADHD adolescent subjects.

9.10 Nutritional Supplements

Numerous nutritional supplements such as amino acids, vitamins, essential fatty acids, and minerals were projected as potential adjunct and alternative ADHD treatments.

9.10.1 Vitamins

Vitamins were used as probable adjuncts or substitutes for ADHD treatments because supplementation to normal children improved attention and concentration. Combination therapy of vitamin B6 with magnesium for 8 weeks enhanced the symptoms in ADHD children (Mousain-Bosc et al. 2006), which might be due to the synthesis of serotonin by vitamin B6. Another combination therapy involving vitamin C (antioxidant) and alpha-linolenic acid-rich flax oil (ALA, a precursor fatty acid needed for the synthesis of docosahexaenoic acid, required for the development of the brain) alleviated hyperactivity scores in ADHD patients (Joshi et al. 2006).

9.10.2 Minerals

Mineral deficiencies were reported to be involved in the cause of this disease. Therefore, supplementation might alleviate the ADHD symptoms. One clinical study involving the supplementation of zinc sulfate reported reduction in the symptoms of ADHD, while another study showed better efficacy and fewer side effects when compared to MPH. Few studies showed beneficial effect (Bilici et al. 2004; Akhondzadeh et al. 2004). Iron acts as a cofactor for the norepinephrine and dopamine biosynthesis, and anemic children are more prone to attention deficits. Few trials with iron supplementation offered beneficial effects, while others showed variable results in ADHD patients (Rucklidge et al. 2009). Another strategy involving the supplementation of all the minerals and vitamins improved the ADHD symptoms and mood due to their synergistic effect that affected the interrelated and defective biochemical pathways in ADHD patients (Rucklidge and Kaplan 2014).

9.10.3 Amino Acids

Numerous amino acids such as glycine, L-tyrosine, taurine, L-theanine, GABA, 5-hydroxytryptophan, acetyl-L-carnitine, and *S*-adenosyl-L-methionine acted as precursors for the synthesis of neurotransmitters. A study by Torrioli et al. (2008) reported that the supplementation of acetyl-L-carnitine significantly diminished the hyperactivity and poor social behavior in ADHD children by modulating neural transmission through elevated acetylcholine synthesis and stimulated the release of dopamine. Few experiments involving the supplementation of theanine, an amino acid available in green and black tea, revealed the positive effect on ADHD symptoms (Lardner 2014).

9.10.4 Essential Fatty Acids

Supplementation of essential fatty acid (EFA-omega-3 and 6) mixture of docosahexaenoic acid (DHA), γ-linolenic acid, eicosapentaenoic acid (EPA), vitamin E, *cis*-linoleic acid, arachidonic acid (AA), and thyme oil (Bloch and Qawasmi 2011; Sonuga-Barke et al. 2013) and a mixture of phosphatidylserine containing omega-3, EPA, and DHA (Manor et al. 2012) abolished the symptoms of ADHD. Few studies (Milte et al. 2012; Manor et al. 2012) indicated the beneficial role of EFA supplementation in ADHD, while others (Raz et al. 2009) showed controversial results.

9.11 Conclusions

Although there are numerous therapeutic options available for ADHD, few of them have adverse effects. Multi-therapeutic strategies are extremely prominent as they are more "suitable" and specific to each patient. Combination therapies involving both pharmaceutical agents and natural products might assist in improving the overall performance by nullifying the ADHD symptoms. Presently, only a handful of experiments were performed using this approach. More studies are needed in future to explore proof for long-term efficacy and safety.

References

Ahn J, Ahn HS, Cheong JH, Peña ID (2016) Natural product-derived treatments for Attention-Deficit/Hyperactivity Disorder: safety, efficacy, and therapeutic potential of combination therapy. Neural Plast 2016:1320423

Akhondzadeh S, Mohammadi MR, Khademi M (2004) Zinc sulfate as an adjunct to methylphenidate for the treatment of attention deficit hyperactivity disorder in children: a double blind and randomized trial. BMC Psychiatry 4:9

Akhondzadeh S, Mohammadi MR, Momeni F (2005) *Passiflora incarnata* in the treatment of attention-deficit hyperactivity disorder in children and adolescents. Therapy 2:609–614

Bader A, Adesman A (2012) Complementary and alternative therapies for children and adolescents with ADHD. Curr Opin Pediatr 24:760–769

Batty MJ, Palaniyappan L, Scerif G, Groom MJ, Liddle EB, Liddle PF, Hollis C (2015) Morphological abnormalities in prefrontal surface area and thalamic volume in attention deficit/hyperactivity disorder. Psychiatry Res Neuroimaging 233:225–232

Bilici M, Yıldırım F, Kandil S, Bekaroğlu M, Yildirmiş S, Değer O, Ulgen M, Yildiran A, Aksu H (2004) Double-blind, placebo-controlled study of zinc sulfate in the treatment of attention deficit hyperactivity disorder. Prog Neuropsychopharmacol Biol Psychiatry 28:181–190

Bloch MH, Qawasmi A (2011) Omega-3 fatty acid supplementation for the treatment of children with attention-deficit/hyperactivity disorder symptomatology: systematic review and meta-analysis. J Am Acad Child Adolesc Psychiatry 50:991–1000

Cabarkapa S, King JA, Dowling N, Ng CH (2019) Co-morbid obsessive-compulsive disorder and attention deficit hyperactivity disorder: neurobiological commonalities and treatment implications. Front Psychiatry 10:557

Chai TQ (1999) Clinical observation of acupuncture in the treatment of 155 cases of attention deficit hyperactivity disorder. J Chin Acupunct 1:5–8

Conners CK, Sitarenios G, Parker JD, Epstein JN (1998) The revised Conners' Parent Rating Scale (CPRS-R): factor structure, reliability, and criterion validity. J Abnorm Child Psychol 26:257–268

Dalsgaard S, Leckman JF, Mortensen PB, Nielsen HS, Simonsen M (2015) Effect of drugs on the risk of injuries in children with attention deficit hyperactivity disorder: a prospective cohort study. Lancet Psychiatry 2:702–709

Dean JD, Downey LA, Stough C (2017) Systematic overview of *Bacopa monnieri* (L) Wettst dominant poly-herbal formulas in children and adolescents. Medicines (Basel) 4:86

Ding GA, Yu GH, Chen SF (2002) Assessment on effect of treatment for childhood hyperkinetic syndrome by combined therapy of yizhi mixture and Ritalin. Chin J Integr Tradit West Med 22: 255–227

Gaud B (2014) Agnivesha, Charaka, Dridhabala, Charaka Samhita, Chikitsa Sthana, Unmadachikitsitvyakhynam, 1st edn. Rashtriya Ayurveda Vidyapeeth, New Delhi, p 5

Gehricke JG, Whalen CK, Jamner LD, Wigal TL, Steinhoff K (2006) The reinforcing effects of nicotine and stimulant medication in the everyday lives of adult smokers with ADHD: a preliminary examination. Nicotine Tob Res 8:37–47

Hernandez-Reif M, Field TM, Thimas E (2001) Attention deficit hyperactivity disorder: benefits from tai chi. J Bodyw Mov Ther 5:120–123

Hill DE, Campbell RA, Hart B, Yeo RA, Vigil J, Brooks W (2003) Magnetic resonance imaging correlates of attention deficit/hyperactivity disorder in children. Neuropsychology 17:496–506

Joshi K, Lad S, Kale M et al (2006) Supplementation with flax oil and vitamin C improves the outcome of Attention Deficit Hyperactivity Disorder (ADHD). Prostaglandins Leukot Essent Fatty Acids 74(1):17–21

Kean JD, Downey LA, Stough C (2016) A systematic review of the Ayurvedic medicinal herb *Bacopa monnieri* in child and adolescent populations. Complement Ther Med 29:56–62

Kleigmn B (2017) Learning disorder. Nelson text book of pediatrics, Chapter no 33. Elsevier, Relixindia Private Limited, p 203

Langleben DD, Austin G, Krikorian G, Ridlehuber HW, Goris ML, Strauss HW (2001) Interhemispheric asymmetry of regional cerebral blood flow in prepubescent boys with attention deficit hyperactivity disorder. Nucl Med Commun 22:1333–1340

Lardner AL (2014) Neurobiological effects of the green tea constituent theanine and its potential role in the treatment of psychiatric and neurodegenerative disorders. Nutr Neurosci 17:145–155

Leucht S, Hierl S, Kissling W, Dold M, Davis JM (2012) Putting the efficacy of psychiatric and general medicine medication into perspective: review of meta-analyses. Br J Psychiatry 2:97–106

Levin ED, Conners CK, Silva D, Canu W, March J (2001) Effects of chronic nicotine and methylphenidate in adults with attention deficit/hyperactivity disorder. Exp Clin Psychopharmacol 9:83–90

Levy F (1991) The dopamine theory of attention deficit hyperactivity disorder (ADHD). Aust N Z J Psychiatry 25:277–283

Li XR, Chen ZJ (1999) Clinical comparative observation on duodongning and Ritalin in treating child hyperkinetic syndrome. Chin J Integr Tradit West Med 19:410–411

Li JJ, Li ZW, Wang SZ, Qi FH, Zhao L, Lv H, Li AY (2011) Ningdong granule: a complementary and alternative therapy in the treatment of attention deficit/hyperactivity disorder. Psychopharmacology 216:501–509

Manor I, Magen A, Keidar D, Rosen S, Tasker H, Cohen T, Richter Y, Zaaroor-Regev D, Manor Y, Weizman A (2012) The effect of phosphatidylserine containing Omega3 fatty-acids on attention-deficit hyperactivity disorder symptoms in children: a double-blind placebo-controlled trial, followed by an open-label extension. Eur Psychiatry 27:335–342

Milte CM, Parletta N, Buckley JD, Coates AM, Young RM, Howe PRC (2012) Eicosapentaenoic and docosahexaenoic acids, cognition, and behavior in children with attention-deficit/hyperactivity disorder: a randomized controlled trial. Nutrition 28:670–677

Mousain-Bosc M, Roche M, Polge A, Pradal-Prat D, Rapin J, Bali JP (2006) Improvement of neurobehavioral disorders in children supplemented with magnesium-vitamin B6. I. Attention deficit hyperactivity disorders. Magnes Res 19:46–52

Ni X, James YZ, Han X, Lei S, Sun J, Zhou R (2014) Traditional Chinese medicine in the treatment of ADHD: a review. Child Adolesc Psychiatr Clin N Am 23:853–881

Niederhofer H (2010) St. John's wort may improve some symptoms of attention-deficit hyperactivity disorder. Nat Prod Res 24:203–205

Oades RD (2010) The role of serotonin in Attention-Deficit Hyperactivity Disorder (ADHD). In: Müller C, Jacobs B (eds) Handbook of behavioral neurobiology of serotonin. Elsevier V.B. Publications, pp 565–584

Pandeya K (2005) Charaka Samhita, Hindi Commentary, Chaukhambha Bharati Academy, p 97

Peasgood T, Bhardwaj A, Biggs K, Brazier JE, Coghill D, Cooper CL, Daley D, De Silva C, Harpin V, Hodgkins P, Nadkarni A, Setyawan J, Sonuga-Barke EJ (2016) The impact of ADHD on the health and wellbeing of ADHD children and their siblings. Eur Child Adolesc Psychiatry 25:1217–1231

Potter AS, Schaubhut G, Shipman M (2014) Targeting the nicotinic cholinergic system to treat attention-deficit/hyperactivity disorder: rationale and progress to date. CNS Drugs 28:1103–1113

Raz R, Carasso RL, Yehuda S (2009) The influence of short-chain essential fatty acids on children with attention-deficit/hyperactivity disorder: a double-blind placebo-controlled study. J Child Adolesc Psychopharmacol 19:167–177

Razlog R, Pellow J, White SJ (2012) A pilot study on the efficacy of *Valeriana officinalis* mother tincture and *Valeriana officinalis* 3 x in the treatment of attention deficit hyperactivity disorder. Health SA Gesondheid 17:1–7

Rucklidge JJ, Kaplan BJ (2014) Broad-spectrum micronutrient treatment for attention-deficit/hyperactivity disorder: rationale and evidence to date. CNS Drugs 28:775–785

Rucklidge JJ, Johnstone J, Kaplan BJ (2009) Nutrient supplementation approaches in the treatment of ADHD. Expert Rev Neurother 9:461–476

Sagvolden T, Johansen EB, Aase H, Russell VA (2005) A dynamic developmental theory of attention-deficit/hyperactivity disorder (ADHD) predominantly hyperactive/impulsive and combined subtypes. Behav Brain Sci 28:397–419

Salehi B, Imani R, Mohammadi MR, Fallah J, Mohammadi M, Ghanizadeh A, Tasviechi AA, Vossoughi A, Rezazadeh S, Akhondzadeh S (2010) Ginkgo biloba for attention-deficit/hyperactivity disorder in children and adolescents: a double blind, randomized controlled trial. Prog Neuropsychopharmacol Biol Psychiatry 34:76–80

Sasikumar P, Arul Mozhi P, Meenakshi Sundaram M, Meenakumari RM (2020) Therapeutic management of varmam therapy for Attention Deficit Hyperactivity Disorder (ADHD) in children—a review. IOSR J Dent Med Sci 19:19–21

Schultz W (1998) Predictive reward signal of dopamine neurons. J Neurophysiol 80:1–27

Searight HR, Robertson K, Smith T, Perkins S, Searight BK (2012) Complementary and alternative therapies for pediatric attention deficit hyperactivity disorder: a descriptive review. ISRN Psychiatry 2012:8

Shakibaei F, Radmanesh M, Salari E, Mahaki B (2015) Ginkgo biloba in the treatment of attention-deficit/hyperactivity disorder in children and adolescents. A randomized, placebo-controlled, trial. Complement Ther Clin Pract 21:61–67

Sharma HM (1990) Butter oil (ghee) – myths and facts. Indian J Clin Pract 1:31–33

Sharma PV (2006) Dravyaguna Vijnana, vol II. Chaukhambha Bharati Academy, p 253

Sharma PV (2008) Charaksamhita edited by Chaukhambha oriented publication, Varanasi, reprint edition, Sutra sthana Chap 8, p 54

Shi XM (2002) Acupuncture. Chinese Press of Traditional Chinese Medicine, Beijing, pp 154–155

Shukla K (2005) Agnivesha, Charaka, Dridhabala, Charaka Samhita, Nidaan Sthana, unmada nidaan adhyaya. R T Vaidya Vidhyadhar

Sinha D, Efron D (2005) Complementary and alternative medicine use in children with attention deficit hyperactivity disorder. J Paediatr Child Health 41:23–26

Sonuga-Barke EJ, Brandeis D, Cortese S, Daley D, Ferrin M, Holtmann M, Stevenson J, Danckaerts M, Oord SV, Döpfner M, Dittmann RW, Simonoff E, Zuddas A, Banaschewski A, Buitelaar J, Coghill D, Hollis C, Konofal E, Lecendreux M, Wong ICK, Sergeant J, European ADHD Guidelines Group (2013) Nonpharmacological interventions for ADHD: systematic review and meta-analyses of randomized controlled trials of dietary and psychological treatments. Am J Psychiatry 170:275–289

Torrioli MG, Vernacotola S, Peruzzi L, Tabolacci E, Mila M, Militerni R, Musumeci S, Ramos FJ, Frontera M, Sorge G, Marzullo E, Romeo G, Vallee L, Veneselli E, Cocchi E, Garbarino E, Moscato U, Chiurazzi P, D'Iddio S, Calvani M, Neri G (2008) A double-blind, parallel, multicenter comparison of L-acetylcarnitine with placebo on the attention deficit hyperactivity disorder in fragile X syndrome boys. Am J Med Genet A 146:803–812

Uebel-von Sandersleben H, Rothenberger A, Albrecht B, Rothenberger LG, Klement S, Bock N (2014) Ginkgo biloba extract EGb 761 in children with ADHD. Z Kinder Jugendpsychiatr Psychother 42:337–347

Wang YL, Shi XP (2005) Observation on the effects of massage on treatment of attention deficit hyperactivity disorder in children. Chin J Tissue Eng Res 9:216–217

Wang MJ, Wei H, Zhang Y (2011) Clinical observation of Jingling oral liquid combined with methylphenidate in the treatment of ADHD with transient Tic disorder. Chin Tradit Patent Med 33:1638–1639

Weber W et al (2008) Hypericum perforatum (St John's wort) for attention-deficit/hyperactivity disorder in children and adolescents: a randomized controlled trial. JAMA 299:2633–2641

Wen HX (2009) Study of the effect of Taijiquan on children who have a tendency to ADHD. Fujian Normal University, Fuzhou

Wender PH, Epstein RS, Kopin IJ, Gordon EK (1971) Urinary monoamine metabolites in children with minimal brain dysfunction. Am J Psychiatry 127:1411–1415

Wilens TE, Verlinden MH, Adler LA, Wozniak PJ, West SA (2006) ABT-089, a neuronal nicotinic receptor partial agonist, for the treatment of attention-deficit/hyperactivity disorder in adults: results of a pilot study. Biol Psychiatry 59:1065–1070

Yuan H, Yang M, Han X, Ni X (2018) The therapeutic effect of the Chinese herbal medicine, Rehmanniae Radix Preparata, in Attention Deficit Hyperactivity Disorder via reversal of structural abnormalities in the cortex. Evid Based Complement Alternat Med 2018:1–9

Zhuo Y (2006) Clinical study of tui-na and ear acupoint-sticking in the treatment of ADHD. Jiliin J Tradit Chin Med 26:41–42

Influence of Amino Acids on Autism and Attention-Deficit Hyperactive Disorder

10

Pathan Shajahan Begum, Meerza Abdul Razak, and Senthilkumar Rajagopal

Abstract

Recent reports illustrated the increasing universal occurrence of autism and attention-deficit hyperactivity disorder (ADHD). Much of the pathogenesis and etiology of autism and ADHD remain unclear. Autism has a prevalence of 1:59 children in the United States, while ADHD affects almost 3% of adults worldwide. Presently, there are no reliable diagnostic biomarkers for autism and ADHD. Also, the most acceptable documented treatment is stimulant medication for autism and ADHD. The side effects and adverse reactions connected with stimulant medications are significant and grave, hampering growth and possibly being life-threatening. The pharmacological treatment choices are available; however, the relative benefits and adverse effects of individual treatments remain largely unknown. Inadequate availability of behavioral therapies and worries over adverse effects of pharmacological treatments provoked research for alternative strategies for management of autism and ADHD involving amino acid supplementation.

Amino acids have a significant function in brain functioning and development. In addition, many of the amino acids have been demonstrated to exert direct or indirect effects on the levels of specific neurotransmitters. Thus, amino acids are

P. S. Begum
Department of Zoology, Silver Jubilee Government College, Kurnool, India

M. A. Razak
Department of Life Sciences, Dr. Buddollas Institute of Life Sciences, Tirupathi, India

S. Rajagopal (✉)
Department of Biotechnology, School of Applied Sciences, REVA University, Bengalore, India
e-mail: senthilkumar.r@reva.edu.in

M. W. Qoronfleh et al. (eds.), *Proteins Associated with Neurodevelopmental Disorders*, Nutritional Neurosciences, https://doi.org/10.1007/978-981-15-9781-7_10

anticipated to be effective in the management of autism and ADHD. There are several scientific reports about the significant functions of some amino acids in neurobiology and treatment of autism and ADHD. In this chapter, the authors exemplify the abnormalities in the profile of few amino acids in autism and ADHD patients, the pathways affected by amino acid imbalance in the brain, changes in the neuroactive amino acids which might have a role in pharmacotherapy and pathogenesis, amino acid supplementation therapy, and dysregulated amino acid metabolism which might considerably interfere with autism and ADHD.

Keywords

Autism · ASD · Attention-deficit hyperactive disorder · ADHD · Amino acids · Amino acid supplementation · Therapy

10.1 Autism: An Introduction

Paul Eugen Bleuler, one of the Swiss psychiatrists, used the word "autism" for the first time to define the symptoms of schizophrenia. He derived the word "autism" from the Greek word αὐτός which means autos (self). Hans Asperger applied Bleuler's term autism in the modern sense to define child psychology (Park et al. 2016; Bleuler 1912). Leo Kanner described three girls and eight boys who had the inborn incapability to form the usual biologically provided affective interaction with individuals and introduced the tag premature infantile autism. Leo Kanner and Hans Asperger were considered as persons who planned the origin of the modern study of autism (Park et al. 2016; Asperger 1944; Kanner 1943). Autism refers to the neurodevelopmental disease that affects 1–2% of children, as per the literature available. Autism is categorized by the diverse levels of severity and occurs in all ethnic groups. Noto et al. described that 1 among 88 children aged 8 years is affected by autism with males more at risk than females; male-to-female ratio is 3–4:1 (Autism and Developmental Disabilities Monitoring Network Surveillance Year 2008 Principal Investigators, Centers for Disease Control and Prevention 2012; Noto et al. 2014). Blumberg et al. showed that the prevalence of autism increased to 75% from 2007 to 2012 in the United States (Blumberg et al. 2013).

Autism is characterized by compromised social interaction abilities along with constricting repetitive behavior (Zheng et al. 2017; American Psychiatric Association 2013). The environmental factors and genetic predisposition unquestionably have effects on the pathophysiology of autism. However, the exact mechanisms connected to the pathophysiology of autism remain unidentified, and more absolute approaches for early diagnosis or treatment are missing (Zheng et al. 2017; Brian et al. 2016). Diagnostic and Statistical Manual of Mental Disorders (DSM-5) suggests that autism symptoms must appear in the early childhood (American Psychiatric Association 2013). Getting diagnosis at the initial stage of development could be very helpful for early medication and therapy benefiting both patients and their families (Zwaigenbaum et al. 2015; Sacrey et al. 2015). On the other hand, behavioral irregularities are often unnoticed in the initial stage of autism even though experienced professionals are involved in pediatric health care (Sacrey et al. 2015). Hence, many scientists have been researching to find quantitative diagnostic strategies that could help for early and more precise autism diagnosis. Several interacting aspects are probably contributing to the etiology of autism and these effective factors are illustrated in several outstanding articles (Park et al. 2016; Muller et al. 2016).

Several children with autism are selective eaters; they do not prefer a variation in their food routines. Consumption problems are threat factors for nutritional deficiencies. Food selectivity and food rejection were reported in 90% of these children (Ghanizadeh 2013; Kral et al. 2013). Children with autism show a shortage of essential plasma amino acids demonstrating their deprived protein consumption. However, the level of at least one essential amino acid such as lysine, leucine, phenylalanine, and valine deficiency in children with autism was 58%. The percentage in the control group was 4% (Arnold et al. 2003). Numerous lines of confirmation revealed that alterations in amino acids associated with central brain functions may perform a crucial role in the pharmacotherapy and pathogenesis of numerous psychiatric disorders that show symptoms like cognitive impairment and complications with social interactions in common with autism (Durrant and Heresco-Levy 2014; Yuksel and Ongur 2010). Many clinical and preclinical studies reported amino acids in the etiology of autism but several of these investigations have concentrated on GABA glutamate and glutamine (Robertson et al. 2016).

Other amino acids could also be involved in the etiology of autism and it might be key to carry out detailed studies in which several of these amino acids are investigated simultaneously. Due to the active role of amino acids in the pathogenesis and treatment of autism, observing variations in their concentrations in body fluids is also vital provided that they are related to the initial diagnosis and intervention in patients with autism. This chapter reviews the literature on measurement of several amino acids and their impact on body fluids in autism.

10.2 Impact of Amino Acids on Autism

10.2.1 Glutamate

Glutamate, which is vastly concentrated in the brain, is the key excitatory neurotransmitter (Naaijen et al. 2015). Glutamate generally has a protecting effect on cognitive function and neural plasticity; however, excess of glutamate might be neurotoxic and causes death of glia and neurons. Also, it might play a crucial role in the pathogenesis of psychiatric disorders such as autism (Manev et al. 1989). It was reported that glutamate is responsible for neuroinflammation in autism. Homocysteine and glutamate are targets for therapy of aggression and irritability in patients with autism (Ghanizadeh and Namazi 2010). A hyperglutamatergic theory of autism has also been proposed (Blaylock and Strunecka 2009). Fatemi et al. demonstrated that the levels of GAD 67 kDa and GAD 65 kDa (glutamic acid decarboxylases), both of which are involved in converting glutamate to gamma aminobutyric acid, are reduced in the brains of patients with autism leading to increased levels of glutamate in the brain (Fatemi et al. 2002).

Few investigations employed magnetic resonance spectroscopy to investigate amino acid levels in patients with autism and have described that patients with autism have elevated glutamate levels in the brain (Hassan et al. 2013). Cochran et al. proved that compared with healthy controls, autism patients had decreased GABA levels, more glutamine levels, and no change in glutamate levels in brain (Cochran et al. 2015). However, van Elst et al. described that glutamate and glutamine levels were diminished in autism brain (van Elst et al. 2014). Research reports on glutamate levels in the plasma of proteins with autism compared to healthy controls are inconsistent with some reporting decreased levels and some increased levels. In addition, increased levels have been reported in serum and decreased levels in urine samples and platelets (Nadal-Desbarats et al. 2014; El-Ansary 2016).

10.2.2 D-Serine

Currently, D-serine in the brain has been the theme of wide-ranging research (Sacchi et al. 2016). D-serine is a significant amino acid in glutamatergic transmission and is an effective agonist at NMDA receptors in few mammalian brain parts and perhaps plays a role in the pathogenesis of numerous neurological and psychiatric disorders such as Alzheimer's disease, bipolar disorder, schizophrenia, and addiction (Fuchs et al. 2005; Ozeki et al. 2016; Paula-Lima et al. 2013; Liu et al. 2016). D-serine has very high affinity for synaptic NMDA receptors while glycine has more affinity for extrasynaptic NMDA receptors (Vizi et al. 2013). Through scientific investigation on urine samples, Kałuzna-Czaplinska et al. stated that L-serine levels were diminished in autism (Kałuzna-Czaplinska et al. 2014).

Noto et al. demonstrated that L-serine levels were elevated (Noto et al. 2014). Moreover, Ming et al. and Evans et al. described that the pooled serine levels were decreased (Ming et al. 2012; Evans et al. 2008). In one of the studies in which pooled serine levels were stated, the D-serine and L-serine were not assessed independently. There is a rareness of significant research on body fluid levels of D-serine in autism. Shinohe et al. presented that L-serine levels and D-serine in serum were no dissimilar between healthy controls and adult patients with autism (Shinohe et al. 2006). When compared with healthy controls Tirouvanziam et al. presented that pooled serine levels in plasma were declined in autism (Tirouvanziam et al. 2011).

10.2.3 Glycine

Glycine and GABA are key inhibitory neurotransmitters in the central nervous system (Ito 2016). They show their effect on receptors linked to chloride channels which play a crucial role in general function of the central nervous system. During early development, glycine and GABA depolarize membrane potentials, which act as excitatory neurotransmitters (Kaila et al. 2014). They swing from excitatory to inhibitory neurotransmitters during birth and in maturation and if that does not occur it may lead to neurological disorders including autism (Tyzio et al. 2014). However, in few regions of the brain, glycine functions as a co-agonist at NMDA glutamate receptors and it has been recommended that the glycine/D-serine site on the NMDA receptor could be a main objective for autism (Basu et al. 2009). By comparing with healthy controls, glycine levels in serum and plasma of autism individuals have been described to be unaffected, and levels in urine samples reported to be improved (Shinohe et al. 2006; Tirouvanziam et al. 2011).

10.2.4 Gamma-Aminobutyric Acid (GABA)

The equilibrium between GABA and glutamate, inhibitory and excitatory neurotransmitters, correspondingly, is very significant for the function of brain, and numerous neurological and psychiatric disorders may be due to the disproportion between glutamate and GABA (Rojas et al. 2014). Decreased GABAergic action in animal and human models of autism has been projected to be one of the causes for an inequity between inhibition and excitation (Gogolla et al. 2009). When compared with healthy controls, GABA levels in plasma have been described to be elevated in autism patients.

Dhossche et al. described that plasma GABA levels be likely to decline with age in autism (Dhossche et al. 2002). When compared with healthy controls, GABA levels in platelets have been described to be reduced in autism whereas those in urine samples increased (Cohen 2002). Neuroimaging techniques reported diminished GABA in brains of autism patients (Gaetz et al. 2014). Rojas et al. described that the left perisylvian GABA levels were reduced in patients with autism and their unaffected siblings. In recent times, investigation using oxytocin to cure animal models of autism reported that oxytocin may increase excitatory GABA and develop hyperglutamatergic activity (Rojas et al. 2014; Young and Barrett 2015).

10.2.5 Glutamine

Glutamate is deposited in the form of glutamine in astrocytes until it is relocated to presynaptic terminals and transformed back to glutamate (Magistretti and Pellerin 1999). Ghanizadeh and Namazi (2010) stated that a glutamine (GLN) synthetase inhibitor can enhance inflammation in autism (Ghanizadeh and Namazi 2010). Shimmura et al. (2011) proposed that the level of glutamine in plasma could be a diagnostic test for identifying autism in children, particularly those with a usual intelligence quotient (IQ) (Shimmura et al. 2011). In research of glutamine levels in autism patients compared to healthy controls, platelet and plasma levels have been stated to be decreased, urine levels to be either decreased or increased, and serum levels to be same (Ghanizadeh 2013).

10.2.6 Tryptophan

Serotonin is a significant neurotransmitter and tryptophan is the precursor of neurotransmitter serotonin (Zhang et al. 2015). Dysfunction of serotonin systems is associated with few forms of autism and could lead to social communication impairments (Yang et al. 2014). Whole-blood serotonin has been described to be increased in minimum 25% of autism children (Muller et al. 2016). On the other hand, decreasing tryptophan in the diet can weaken social behavior in mice and autism patients and increasing tryptophan in the diet has been reported to improve social behavior in mice (McDougle et al. 1996). When compared with healthy controls, tryptophan levels in plasma have been stated to be declined in autism patients; however Noto et al. (2014) stated levels to be elevated in urine samples and Kałuzna-Czaplinska et al. (2014) described them to be declined (Noto et al. 2014; Kałuzna-Czaplinska et al. 2014). Joanna Kałuzna-Czaplińska's study exposed both decreased and increased levels of tryptophan in autism children without magnesium and vitamin B supplementation. This may lead to more severity of autism symptoms. As per the recent literature, irregular levels of tryptophan cause more abnormalities, including serotonin pathway. Amino acids participating in it play a crucial role in the appropriate functioning of the body responsible for basic physiological and mental activities (Kałużna-Czaplińska et al. 2017).

10.2.7 Taurine

Taurine is a neuromodulator and osmoregulator that inhibits vasopressin and has been described to be exhausted in the urine of autistic children. On the other hand, in other investigations on taurine levels in autism patients compared to healthy controls, plasma levels have been stated to be decreased or increased, and taurine levels in urine samples to be increased or decreased (Good 2011). Even though the results described on taurine levels in urine and plasma samples are inconsistent, there is a reliable view that taurine plays a defending role in patients with autism. Kuwabara et al. (2013) revealed higher plasma taurine levels in adults with autism and projected that taurine is compensatory against pathogenesis of autism, such as that triggered by oxidative stress (Kuwabara et al. 2013).

10.2.8 Other Amino Acids

Arginine is a needed precursor for the production of nitric oxide and proteins and it can spare glutamine, increase brain blood flow, and detoxify ammonia (Good 2011). When compared with healthy controls, arginine levels in the plasma of autism patients have been stated to be no different or increased (Kuwabara et al. 2013). The metabolism of homocysteine is linked closely with vitamin B12 and folic acid. Desai et al. (2016) presented that a deficiency of folic acid may lead to pathogenesis of autism (Desai et al. 2016). Bala et al. (2016) and James et al. (2004) stated that the concentration of homocysteine in plasma with autism patients is small (Bala et al. 2016; James et al. 2004). Bala et al. (2016) described low plasma levels of vitamin B12 in autism patients when compared to values in healthy controls (Bala et al. 2016). However, other research reports stated that homocysteine levels were elevated in autism patients when compared with healthy controls. Ali et al. (2011) and Pascal et al. (2006) reported levels in serum, Tu et al. (2012) reported levels in plasma, and Noto et al. (2014) and Puig-Alcaraz et al. (2015) reported levels in urine samples, and all described to be increased (Zheng et al. 2017). Puig-Alcaraz et al. (2015) reported that increased urinary levels of homocysteine interrelated directly with the severity of insufficiency in communication skills in autism patients (Puig-Alcaraz et al. 2015).

Table 10.1 shows the reported levels of neuroactive amino acids in patients with autism in comparison with healthy controls and includes some other amino acids such as tyrosine, phenylalanine, citrulline, valine, methionine, proline, threonine, aspartate, asparagine, and histidine not mentioned previously. Valine, leucine, and isoleucine are all termed branched-chain amino acids (BCAAs) and share a common transport system with large, neutral amino acids (LNAAs), such as tyrosine, tryptophan, and phenylalanine which are the precursors of the catecholamine and the neurotransmitter amine 5-hydroxytryptamine (5-HT, serotonin) (Fernstrom 2005). Arnold et al. (2003) described that the level of the essential amino acids phenylalanine, leucine, valine, and lysine in autism was 58% when compared to healthy controls (Arnold et al. 2003). However, there is a scarcity of research on the levels of branched-chain amino acids in autism; several of the investigations stated a reduction of branched-chain amino acid levels in autistic individuals (Table 10.1), signifying that upcoming research in this area is necessary.

Generally, the results on amino acid levels in autism described in the literature are, with the possible exception of the branched-chain amino acids, unsatisfying and inconsistent. Table 10.1 is a summary of reported differences between healthy controls and autism patients in the levels of amino acids. Branched-chain amino acids are essential amino acids that make up about 1/3 of muscle protein, and these insufficiencies may disturb connective tissue and muscle integrity in autism subjects (Evans et al. 2008). It has been recommended that BCAA scarcities may be connected to deprived nutrition due to uncommon food preferences in autism children (Arnold et al. 2003; Woods et al. 2008).

Regrettably, only some investigations on amino acids in saliva with autism patients have been done. It may be advantageous to utilize saliva sampling collectively with standardized conditions to identify amino acids in autism patients usually in the future. This is a noninvasive testing that can be swiftly done more regularly than other sampling, thus facilitating more effective monitoring. Currently, a few studies have concentrated on saliva samples to identify cortisol which is a good indicator of stress pressure and behavior recovery in patients with autism. Due to the

Table 10.1 Reported assessment of the levels of amino acids in healthy controls and autism patients (Zheng et al. 2017)

Amino acid	Sample source	Level increased/decreased
Alanine	Plasma	Increased
		No difference
	Urine	Increased
		Decreased
Aspartate	Plasma	Increased
		No difference
	Platelets	Decreased
	Urine	Decreased
Asparagine	Plasma	Increased
		Decreased
	Urine	Decreased
Arginine	Plasma	Increased
		No difference
Citrulline	Plasma	Decreased
Glycine	Plasma	No difference
	Serum	No difference
	Urine	Increased
		Decreased
GABA	Plasma	Increased
	Platelets	Decreased
	Urine	Increased
	Neuroimaging (brain)	Decreased
Histidine	Plasma	Increased
	Urine	Decreased
Tyrosine	Plasma	Increased
		Decreased
	Urine	Increased
		Decreased
Phenylalanine	Plasma	Increased
		Decreased
	Urine	Increased
		Decreased
Methionine	Plasma	Increased
		Decreased
	Cerebrospinal fluid	Decreased
	Urine	No difference
Proline	Plasma	No difference
	Urine	Decreased
Threonine	Plasma	Decreased
	Urine	Decreased
Isoleucine	Plasma	Decreased
	Cerebrospinal fluid	Decreased

(continued)

Table 10.1 (continued)

Amino acid	Sample source	Level increased/decreased
	Urine	Decreased
Valine	Plasma	Decreased
		No difference
Lysine	Plasma	Increased
		No difference
Leucine	Plasma	Decreased
	Cerebrospinal fluid	Decreased
	Urine	Decreased
Homocysteine	Plasma	Increased
		Decreased
	Serum	Increased
	Urine	Increased
Serine (D- and L-)	Plasma	Decreased
	Urine	Decreased
L-Serine	Serum	No difference
	Urine	Increased
		Decreased
D-Serine	Serum	No difference
Tryptophan	Plasma	Decreased
	Urine	Increased
		Decreased
Taurine	Plasma	Increased
		Decreased
	Urine	Increased
		Decreased
Glutamate	Plasma	Increased
		Decreased
	Serum	Increased
	Platelets	Decreased
	Urine	Decreased
	Neuroimaging (brain)	Increased
		Decreased
		No difference
Glutamine	Plasma	Decreased
	Serum	No difference
	Platelets	Decreased
	Urine	Increased
		Decreased
	Neuroimaging (brain)	Increased
		Decreased

easiness of collecting saliva, it is suitable for caregivers to help infants, toddlers, and patients to collect samples at any place. There should be reduced emotional fluctuations compared to collecting blood samples and thus perhaps increased exactness of results. In research data on amino acid levels in autism subjects reported in the literature, there has been substantial difference in terms of factors such as gender, age, intelligence quotient, number of subjects, and psychoactive medication being taken. Upcoming studies could be improved by standardizing these issues and analyzing levels of numerous amino acids simultaneously.

10.3 The Many Pathways Affected by Amino Acid Imbalance in the Brain

The recent investigation by Tărlungeanu et al. recognizes a form of autism resulting from a failure of the brain to appropriately take in amino acids, because of ever-expanding number of genes that are mutated in autism is exemplifying us how imbalances in fundamental cellular processes pathway can lead to disease (Maynard and Manzini 2017). SLC7A5 gene is an amino acid (AA) transporter located on the membrane of endothelial cells creating the blood-brain barrier. Tărlungeanu et al. (2016) recognized that this gene is mutated in a syndromic form of autism connected with modifications in branched-chain amino acids in the brain (Maynard and Manzini 2017). Branched-chain amino acids are metabolized through the branched-chain amino acid aminotransferase (BCAT) and the branched-chain alphaketo acid transferase (BCKD) to transport metabolites into multiple pathways, including the Krebs cycle or tricarboxylic acid (TCA) cycle. BCKD kinase (BCKDK) mutations straight away disturb this pathway and cause a decline in branched-chain amino acid levels connected with autism disease. Branched-chain amino acid decrease can also be detected through the free-transfer tRNA sensing pathway or through the leucine sensor Sestrin 2 to modulate the mTor pathway leading to dysregulation of protein synthesis.

It was identified that malfunctioning in amino acid metabolism and homeostasis is interconnected to autism identified an easily manipulation through dietary supplementation, which should be investigated in other procedures of non-syndromic and syndromic autism. The molecular mechanisms by which decreased branched-chain amino acid levels interrupt motor function and social behavior remain poorly understood. The work from Tărlungeanu et al. (2016) unlocks several new paths for research on how amino acids contribute to normal and diseased brain development (Maynard and Manzini 2017).

10.4 The Interaction Between Amino Acids and the Gut Microbiome in Relation to Autism

Over the past few years, wide-ranging studies have exposed that the gut microbiota composition plays a significant role in the recycling and metabolism of nitrogen-containing compounds (Macfarlane and Macfarlane 2012). One of the key nitrogenous compounds consumed by gut bacteria are amino acids. Amino acids obtained from either the host or the food move in the gut and are metabolized by gut bacteria into a large range of products. Therefore, it can be speculated that the gut microorganisms control the amino acid composition and pool that is accessible to the host (Neis et al. 2015). This is witnessed by a report representing that gut bacteria modify the amino acid distribution in the gastrointestinal tract (Macfarlane et al. 1988). Variations in amino acid availability may lead to fluctuations in signaling pathways that are sensitive to amino acids, including the inflammatory pathways and mTOR signaling. For example, studies reported that people with autism have decreased branched-chain amino acid levels in their cerebrospinal fluid, plasma, and urine (Tirouvanziam et al. 2012). A shortage of any of the branched-chain amino acids is revealed to source immune cell impairments in numerous animal models, causing an additional pro-inflammatory immune status (Zhang et al. 2017), which is also seen in people with autism. A possible elucidation for the uncharacteristic levels of branched-chain amino acids might be a higher transformation of these amino acids by gut proteobacteria as these belong to the best extensive branched-chain amino acid-fermenting bacteria and are reported to be more abundant in individuals with autism (Williams et al. 2012).

Gut microbiota do not only alter amino acid accessibility, but also they play a vital role in the accessibility of amino acid-derived metabolites. For example, levels of threonine are reported to be decreased in patients with autism (Bala et al. 2016). Gut bacteria use threonine for the synthesis of short-chain fatty acids, mostly propionate. Research investigations have discovered an elevated large quantity of propionate in individuals with autism (Wang et al. 2012). In fact, increased propionate synthesis is proposed to be involved in the pathogenesis of autism, as this short-chain fatty acid is reported to cross the blood-brain barrier and is capable to trigger autism-like behavior in adult rodents, may be through activation of microglia (Ossenkopp et al. 2012). The gut microbiome also has a significant role in the regulation of tryptophan metabolism (Gao et al. 2018). Abnormalities in either endogenous or bacterial tryptophan metabolism have been connected through a diverse range of immune-related disorders, including autism (Nikolaus et al. 2017). For example, a current scientific investigation illustrated a noteworthy decrease of serotonin bioavailability in the gastrointestinal tract of a mouse model for autism; this decrease was accompanied by a downregulation of the gene Tph1, representing a lesser synthesis of serotonin from dietary tryptophan (Golubeva et al. 2017). Fascinatingly, serotonin production from tryptophan was certainly linked with the abundance of the Blautia bacteria present in the intestine. In both humans with autism and mouse models of autism, the abundance of Blautia bacteria is described to be lower when compared to healthy controls. This variation in the microbiota might cause the abnormal tryptophan metabolism and serotonin bioavailability that is regularly detected in autism patients as well as in in vivo models for autism and thus may initiate autism pathogenesis (Inoue et al. 2016).

10.5 Attention-Deficit Hyperactivity Disorder

Attention-deficit hyperactivity disorder (ADHD), a neurodevelopmental disorder, is characterized by the basic indications of inattentiveness, impulsivity, and hyperactivity. ADHD is presently considered as the most common neuropsychiatric illness among children (Harpin 2005). In addition, ADHD is most often detected during childhood; it may also affect an ADHD patient throughout life. It is essential to improve effective treatments for ADHD, given its serious familial, social, and academic consequences, along with the risk of inviting comorbid disorders and later substance abuse (Hinshaw et al. 2015). ADHD has been linked with abnormalities in catecholaminergic function in the central nervous system. The usage of drugs that rise levels of catecholamine in the brain obtained extensive support as these medications have been proved to lessen ADHD symptoms. Medication directed in the management of ADHD is classified as non-stimulant and stimulant medications (ADD/ADHD treatment in children: finding treatments that work for kids and teens 2015).

The extensively used or recommended pharmacological treatment for ADHD is stimulant medications. Dextroamphetamine and methylphenidate are good examples of these stimulant drugs, which are anatomically comparable to endogenous catecholamines and whose function elevates extracellular norepinephrine and dopamine levels, thus improving the underlying irregularities in catecholaminergic functions and re-establishing neurotransmitter imbalance (Storebo et al. 2011). Non-stimulant medication of norepinephrine specific reuptake inhibitors, atomoxetine, and the antidepressants phenelzine, bupropion, and imipramine. Non-stimulant medication has also been demonstrated to raise catecholamine levels in the brain causing behavioral improvement (Pellow et al. 2011). On the other hand, non-stimulant drugs have been reported to be of low grade to stimulant treatments on efficiency endpoints. Although pharmacological medication usually improves ADHD conditions for most children, 20–30% of affected patients are incapable to withstand adverse side effects of these drugs or nonresponders (Sinha and Efron 2005). Few of the side effects of stimulant medications include abdominal pain, decreased appetite, headache, motor tics, insomnia, and nausea. Concerns related to long-term usage hazards also discourage parents to treat their children with stimulant drugs. Moreover, the high possibility for diversion, dependence, and abuse, specifically to drugs categorized as stimulants, restricts the usage of stimulant drugs, as ADHD has also been connected with more risk of substance-use disorder (Ahn et al. 2016).

It is usually recognized evidence that a key element in the development of ADHD is the status of the monoamine system to include epinephrine, norepinephrine serotonin, and dopamine. In reply, the pharmaceutical industry has proved to the satisfaction of the US Food and Drug Administration (FDA) that particular drugs that affect the monoamine systems meet FDA efficiency standards. Some of these drugs include methylphenidate, amphetamine, lisdexamfetamine dimesylate, neutral sulfate salts of dextroamphetamine, and atomoxetine (Ahn et al. 2016). Severe reactions and side effects related with ADHD prescription drugs are significant, grave, and possibly lethal. The following is a partial list of these side effects connected with the ADHD group of drugs as a whole, which include hallucinations, delusional thinking, myocardial infarction, risk of stroke, increased risk of suicidal ideation, higher incidence of infection, speech disorder, acute increase of symptoms of behavior disturbance, severe liver injury, risk of drug dependence, decrease of seizure threshold, serious skin rashes, and development of anemia and/or leukopenia. Agitation, tension, anxiety, increased hostility and aggression, reduced libido, and photosensitivity reaction are also some of the effects (Ahn et al. 2016).

This book chapter reviews the effects of new techniques of treatment involving the usage of amino acids that do what pharmacological drugs are incapable to do. This original method has the capability to increase the overall number of neurotransmitter molecules in the central nervous system leading to efficiency observations that look better than those of ADHD prescription drugs without the possibility for neurotransmitter neurotoxicity issues, depletion, and severe effectively life-threatening drug negative effects connected with ADHD prescription drugs.

10.6 Impact of Amino Acids on ADHD

Many amino acids were proven to exert direct or indirect influence on the levels of specific neurotransmitters. Thus, they have been scientifically reported to be effective in curing ADHD. Amino acids such as *S*-adenosyl-L-methionine (SAMe), GABA, 5-hydroxytryptophan (5-HTP), taurine, acetyl-L-carnitine (ALC), L-theanine, glycine, and L-tyrosine are all regarded as potential complementary ADHD interventions. A major part of research studies on amino acid supplementation concentrated on ALC, an amino acid derivative. One such research investigation using ALC described that ALC supplementation significantly reduced the symptoms of ADHD. The action was exerted primarily on poor social behavior and hyperactivity in 51 trial children aged between 6 and 13 years (Torrioli et al. 2008). The influence of ALC resulted in the modulation of neural transmission by elevating acetylcholine synthesis, initiating its release and release of dopamine in the striatum of several brain parts, other than carnitine metabolism. Moreover, a randomized, double-blind placebo-controlled research carried out in 112 ADHD children between ages 5 and 12 years reported contradictory results that no noteworthy influence of ALC was observed in ADHD patients (Arnold et al. 2007).

Theanine is an amino acid analogue of L-glutamate and L-glutamine that is found plenty in both green and black teas. The *n*-ethylglutamic acid (non-proteinaceous constituent) has reaped increasing focus currently due to its claimed central nervous system effects. Due to its capability to pass through the blood-brain barrier, theanine has a diversity of potential pharmacological effects, the most important effect being

Table 10.2 Alteration of essential amino acid levels in the serum of ADHD cases and neurotypical controls (Ahn et al. 2016)

Amino acid	Control, μM[a]	ADHD, μM[b]	*F*-value	*P*-value
Histidine	85.0 (50–120.6)	60.7 (45.2–94.6)	3.140	0.081
Valine	167.2 (136.9–212.3)	167.2 (136.9–212.3)	0.060	0.907
Tryptophan	56.5 (43.3–70.6)	58.1 (44.6–71.5)	0.011	0.794
Threonine	108.9 (90.5–129)	111.4 (86.8–143.1)	0.768	0.605
Phenylalanine	55.8 (47.5–68.2)	57.8 (48.4–66.4)	0.009	0.923
Methionine	54.2 (46.2–69.1)	51.6 (44.1–60.2)	0.416	0.396
Lysine	170.6 (147.5–198.4)	161.5 (120.8–212.1)	1.242	0.360
Leucine	118.6 (103.6–133.7)	117.5 (96.2–138.5)	0.105	0.862
Isoleucine	56.1 (47.6–63.9)	54.1 (43.5–70.8)	0.029	0.923

[a]Data are presented as the median (interquartile range)
[b]*ADHD* attention-deficit/hyperactivity disorder

the anxiolytic effect. The pharmacological effects of theanine have been attributed to the control of serotonin and dopamine and enhanced synthesis of inhibitory neurotransmitters (Lardner 2014). In addition, it has been described that theanine showed enhancement in selective focus during the execution of mental responsibilities through modulation of alpha brain wave activity. At present, many studies have been scrutinizing the therapeutic possibilities of theanine in ADHD. Theanine has also been recommended for obsessive-compulsive disorder, bipolar disorder, and panic disorder besides anxiety disorders and ADHD (Lardner 2014).

In a study, the fibroblast cells were cultured from skin biopsies acquired from 14 boys diagnosed with ADHD and from 13 matched controls without a diagnosis of a developmental disorder. Transportation of the amino acids tryptophan, alanine, and tyrosine, across the cell membrane, was evaluated by the cluster tray method (Johansson et al. 2011). The maximal transport capacity (V_{max}), kinetic parameters, and affinity constant (K_m) were also measured. Difference between the two groups was assessed by Mann-Whitney U-test and Student's unpaired t-test. The ADHD group had considerably diminished V_{max} ($p = 0.039$) and K_m (more affinity) ($p = 0.010$) of tryptophan transportation when compared to healthy controls. They also had a noteworthy more V_{max} of alanine transport ($p = 0.031$), but the K_m of alanine transport did not vary considerably (Johansson et al. 2011). There were no major alterations in any of the kinetic parameters in relation to tyrosine transportation in fibroblasts for the ADHD group. Tryptophan utilizes similar transport systems in both blood-brain barrier (BBB) and fibroblasts. Therefore, a reduced transport capacity of tryptophan indicates that very less tryptophan is being transported across the blood-brain barrier in the ADHD group and this could lead to incomplete serotonin availability in the brain which in turn might result in an imbalance in both the catecholaminergic and serotonergic neurotransmitter systems, which are highly interrelated. The physiological significance of an increased transport capacity of alanine to the brain is not clear yet (Johansson et al. 2011).

Amino acids have an essential role in brain functioning and development. Glutamate, in particular among the amino acids, or their precursors, including γ aminobutyric acid, glutamate, and glutamine, are well recognized to be crucial in neuronal signaling as neurotransmitters. Respectively, disturbance of amino acid metabolism leads to major neurological disorders, particularly in initial ontogenesis (Skalny et al. 2020, 2021). Table 10.2 illustrates altered essential amino acid levels in the serum of ADHD cases and neurotypical controls (Ahn et al. 2016).

By considering the role of imbalanced neurochemistry in ADHD pathogenesis and the role of amino acids in central nervous development, it is postulated that dysregulated amino acid metabolism might be considerably interrelated to ADHD. Nevertheless, there is no consensus in the currently available data. A previous investigation by Bornstein et al. reported considerably lower plasma levels of tryptophan (Trp), histidine (His), phenylalanine (Phe), tyrosine (Tyr), and isoleucine (Ile) in patients with ADHD when compared with the healthy controls (Skalny et al. 2021; Tinkov et al. 2021).

Improvement in ADHD symptoms was attributed to tryptophan, phenylalanine, and tyrosine levels. Simultaneously, no considerable alterations in urinary and blood levels of tryptophan, phenylalanine, and tyrosine levels were seen in children with ADHD. Based on the abovementioned variations and the usage of amino acid supplementation therapy in ADHD management, detailed analyses of amino acid profiles in ADHD are mandatory. The results might assist in interpreting the conflicting results (Skalny et al. 2021; Holecek 2020; International Statistical Classification of Diseases and Related Health Problems 10th Revision (ICD-10): Chapter V Mental and behavioral disorders (F00-F99) 2020; Razak et al. 2017).

There are several existing treatment choices for ADHD. A few of them might cause hazards to the patients. Amino acid supplementation therapy explained in this chapter has been proven to be effective in ADHD treatment with favorable therapeutic outcomes and fewer harmful side effects. However, it is well known that ADHD is a serious disorder caused by several reasons and therefore the application of amino acid-based treatments alone might not show reliable change in ADHD symptoms. As discussed earlier, more definite medical benefit might be obtained by using a multimodal treatment method such as combination therapy of diverse amino acids and conventional pharmacological treatments, amino acids, and/or micronutrients and also behavioral therapy.

10.7 Conclusions

In the recent years, the outcomes of several scientific papers pointed out the efficacy of employing amino acid profiling and amino acid supplementation for successful diagnosis and treatment of numerous neurological diseases, including autism and ADHD. Neurotransmission among neurons, which can happen in the duration of a few milliseconds, depends on the organized release of small-molecule neurotransmitters, several of which are amino acids. Moreover, due to the shared nature of the amino acid transport systems, lack of evenness in the levels of few essential amino acids might disturb others (Razak et al. 2017). Despite the important advances made in understanding neurotransmission in current decades, much remains unknown, particularly in view of the role of amino acids in autism and ADHD. Certainly, several amino acids, including some essential and nonessential amino acids, are known to exert neurotransmitter-like effects, and their deficiency could lead to autism and ADHD (Holecek 2020; International Statistical Classification of Diseases and Related Health Problems 10th Revision (ICD-10): Chapter V Mental and behavioral disorders (F00-F99) 2020). Yet, significant mechanistic queries about the release and neurological significance remain unanswered. This chapter discussed the therapeutic properties of amino acids and the application of these chemicals in the management of diseases like autism and ADHD. The future prospective of amino acid-based therapeutics in the treatment of neurological diseases and the diverse effects of amino acids in the treatment of autism and ADHD are yet to be researched extensively. Research of plasma amino acid profiling was reviewed in this chapter with the objective of evaluating the role of amino acids in the diagnosis of autism and ADHD. Review of available literature revealed that the levels of many plasma amino acids have altered considerably between healthy subjects and patients with autism and ADHD. This chapter also highlighted the possible applications of amino acid detection methods as diagnostic tools for autism, ADHD, and other neurological disorders.

References

ADD/ADHD treatment in children: finding treatments that work for kids and teens (2015). http://www.helpguide.org/articles/add-adhd/attention-deficit-disorder-adhd-treatment-inchildren.htm

Ahn J, Ahn HS, Cheong JH, Dela Peña I (2016) Natural product-derived treatments for attention-deficit/hyperactivity disorder: safety, efficacy, and therapeutic potential of combination therapy. Neural Plast 2016:1320423

Ali A, Waly MI, Al-Farsi YM et al (2011) Hyperhomocysteinemia among Omani autistic children: a case-control study. Acta Biochim Pol 58:547–551

American Psychiatric Association (2013) Diagnostic and statistical manual of mental disorders, 5th edn. American Psychiatric, Washington, DC

Arnold GL, Hyman SL, Mooney RA, Kirby RS (2003) Plasma amino acids profiles in children with autism: potential risk of nutritional deficiencies. J Autism Dev Disord 33:449–454

Arnold LE, Amato A, Bozzolo H (2007) Acetyl-L-carnitine (ALC) in attention-deficit/hyperactivity disorder: a multi-site, placebo-controlled pilot trial. J Child Adolesc Psychopharmacol 17:791–801

Asperger H (1944) Die "Autistischen Psychopathen" im Kindesalter. Arch Psychiatr Nervenkr 117: 76–136

Autism and Developmental Disabilities Monitoring Network Surveillance Year 2008 Principal Investigators, Centers for Disease Control and Prevention (2012) Prevalence of autism spectrum disorders—autism and developmental disabilities monitoring network, 14 sites, United States, 2008. Morb Mortal Wkly Rep Surveill Summ 61:1–19

Bala KA, Dogan M, Mutluer T et al (2016) Plasma amino acid profile in autism spectrum disorder (ASD). Eur Rev Med Pharmacol Sci 20:923–929

Basu AC, Tsai GE, Ma C-L et al (2009) Targeted disruption of serine racemase affects glutamatergic neurotransmission and behavior. Mol Psychiatry 14:719–727

Blaylock RL, Strunecka A (2009) Immune-glutamatergic dysfunction as a central mechanism of the autism spectrum disorders. Curr Med Chem 16:157–170

Bleuler E (1912) The theory of schizophrenic negativism. The Journal of Nervous and Mental Disease Publishing Company, New York

Blumberg SJ, Bramlett MD, Kogan MD et al (2013) Changes in prevalence of parent-reported autism spectrum disorder in school aged US children: 2007 to 2011–2012. Natl Health Stat Rep 65:1–11

Brian J, Bryson SE, Smith IM et al (2016) Stability and change in autism spectrum disorder diagnosis from age 3 to middle childhood in a high-risk sibling cohort. Autism 20:888–892

Cochran DM, Sikoglu EM, Hodge SM et al (2015) Relationship among glutamine, γ-aminobutyric acid, and social cognition in autism spectrum disorders. J Child Adolesc Psychopharmacol 25: 314–322

Cohen BI (2002) The significance of ammonia/gamma-aminobutyric acid (GABA) ratio for normality and liver disorders. Med Hypotheses 59:757–758

Desai A, Sequeira JM, Quadros EV (2016) The metabolic basis for developmental disorders due to defective folate transport. Biochimie 126:31–42

Dhossche D, Applegate H, Abraham A et al (2002) Elevated plasma gamma-aminobutyric acid (GABA) levels in autistic youngsters: stimulus for a GABA hypothesis of autism. Med Sci Monit 8:1–6

Durrant AR, Heresco-Levy U (2014) d-Serine in neuropsychiatric disorders: new advances. Adv Psychiatry 2014:1–16

El-Ansary A (2016) Data of multiple regressions analysis between selected biomarkers related to glutamate excitotoxicity and oxidative stress in Saudi autistic patients. Data Brief 7:111–116

Evans C, Dunstan HR, Rothkirch T et al (2008) Altered amino acid excretion in children with autism. Nutr Neurosci 11:9–17

Fatemi SH, Halt AR, Stary JM et al (2002) Glutamic acid decarboxylase 65 and 67 kDa proteins are reduced in autistic parietal and cerebellar cortices. Biol Psychiatry 52:805–810

Fernstrom JD (2005) Branched-chain amino acids and brain function. J Nutr 135:1539–1546

Fuchs SA, Berger R, Klomp LWJ, de Koning TJ (2005) d-Amino acids in the central nervous system in health and disease. Mol Genet Metab 85:168–180

Gaetz W, Bloy L, Wang DJ et al (2014) GABA estimation in the brains of children on the autism spectrum: measurement precision and regional cortical variation. Neuroimage 86:1–9

Gao J, Xu K, Liu H, Liu G, Bai M, Peng C et al (2018) Impact of the gut microbiota on intestinal immunity mediated by tryptophan metabolism. Front Cell Infect Microbiol 8:13

Ghanizadeh A (2013) Increased glutamate and homocysteine and decreased glutamine levels in autism: a review and strategies for future studies of amino acids in autism. Dis Markers 35(5): 281–286

Ghanizadeh A, Namazi MR (2010) A possible research and clinical implication of association of homocysteine and glutamate levels in brain for management of aggression. Int J Dev Neurosci 28:125

Gogolla N, Leblanc JJ, Quast KB et al (2009) Common circuit defect of excitatory-inhibitory balance in mouse models of autism. J Neurodev Disord 1:172–181

Golubeva AV, Joyce SA, Moloney G, Burokas A, Sherwin E, Arboleya S (2017) Microbiota-related changes in bile acid & tryptophan metabolism are associated with gastrointestinal dysfunction in a mouse model of autism. EBioMedicine 24:166–178

Good P (2011) Do salt cravings in children with autistic disorders reveal low blood sodium depleting brain taurine and glutamine? Med Hypotheses 77:1015–1021

Harpin VA (2005) The effect of ADHD on the life of an individual, their family, and community from preschool to adult life. Arch Dis Child 90:i2–i7

Hassan TH, Abdelrahman HM, Abdel Fattah NR et al (2013) Blood and brain glutamate levels in children with autistic disorder. Res Autism Spectr Disord 7:541–548

Hinshaw SP, Arnold LE, The MTA Cooperative Group (2015) Attention-deficit hyperactivity disorder, multimodal treatment, and longitudinal outcome: evidence, paradox, and challenge. Wiley Interdiscip Rev Cogn Sci 6:39–52

Holecek M (2020) Histidine in health and disease: metabolism, physiological importance, and use as a supplement. Nutrients 12:848

Inoue R, Sakaue Y, Sawai C, Sawai T, Ozeki M, Romero-Perez GA (2016) A preliminary investigation on the relationship between gut microbiota and gene expressions in peripheral mononuclear cells of infants with autism spectrum disorders. Biosci Biotechnol Biochem 80: 2450–2458

International Statistical Classification of Diseases and Related Health Problems 10th Revision (ICD-10): Chapter V Mental and behavioral disorders (F00-F99) (2020) Behavioral and emotional disorders with onset usually occurring in childhood and adolescence (F90-F98). https://icd.who.int/browse10/2019/en#/F90-F98. Accessed 8 Dec 2020

Ito S (2016) GABA and glycine in the developing brain. J Physiol Sci 66:375–379

James SJ, Cutler P, Melnyk S et al (2004) Metabolic biomarkers of increased oxidative stress and impaired methylation capacity in children with autism. Am J Clin Nutr 80:1611–1617

Johansson J, Landgren M, Fernell E et al (2011) Altered tryptophan and alanine transport in fibroblasts from boys with attention-deficit/hyperactivity disorder (ADHD): an in vitro study. Behav Brain Funct 7:40

Kaila K, Price TJ, Payne JA et al (2014) Cation-chloride cotransporters in neuronal development, plasticity and disease. Nat Rev Neurosci 15:637–654

Kałuzna-Czaplinska J, Zurawicz E, Struck W, Markuszewski M (2014) Identification of organic acids as potential biomarkers in the urine of autistic children using gas chromatography/mass spectrometry. J Chromatogr B 966:70–76

Kałużna-Czaplińska J, Jóźwik-Pruska J, Chirumbolo S et al (2017) Tryptophan status in autism spectrum disorder and the influence of supplementation on its level. Metab Brain Dis 32:1585–1593

Kanner L (1943) Autistic disturbances of affective contact. Nerv Child 2:217–250

Kral TV, Eriksen WT, Souders MC, Pinto-Martin JA (2013) Eating behaviours, diet quality, and gastrointestinal symptoms in children with autism spectrum disorders: a brief review. J Pediatr Nurs 28(6):548–556. https://doi.org/10.1016/j.pedn.2013.01.008

Kuwabara H, Yamasue H, Koike S et al (2013) Altered metabolites in the plasma of autism spectrum disorder: a capillary electrophoresis time-of-flight mass spectroscopy study. PLoS One 8:73814

Lardner AL (2014) Neurobiological effects of the green tea constituent theanine and its potential role in the treatment of psychiatric and neurodegenerative disorders. Nutr Neurosci 17:145–155

Liu Z-Q, Gu X-H, Yang Y-J et al (2016) d-Serine in the nucleus accumbens region modulates behavioral sensitization and extinction of conditioned place preference. Pharmacol Biochem Behav 143:44–56

Macfarlane GT, Macfarlane S (2012) Bacteria, colonic fermentation, and gastrointestinal health. J AOAC Int 95:50–60

Macfarlane GT, Allison C, Gibson SA, Cummings JH (1988) Contribution of the microflora to proteolysis in the human large intestine. J Appl Bacteriol 64:37–46

Magistretti PJ, Pellerin L (1999) Cellular mechanisms of brain energy metabolism and their relevance to functional brain imaging. Philos Trans R Soc Lond Ser B Biol Sci 354:1155–1163

Manev H, Favaron M, Guidotti A, Costa E (1989) Delayed increase of Ca^{2+} influx elicited by glutamate: role in neuronal death. Mol Pharmacol 36:106–112

Maynard T, Manzini M (2017) Balancing act: maintaining amino acid levels in the autistic brain. Neuron 93(3):476–479

McDougle CJ, Naylor ST, Cohen DJ et al (1996) Effects of tryptophan depletion in drug-free adults with autistic disorder. Arch Gen Psychiatry 53:993–1000

Ming X, Stein TP, Barnes V et al (2012) Metabolic perturbance in autism spectrum disorders: a metabolomics study. J Proteome Res 11:5856–5862

Muller CL, Anacker AMJ, Veenstra-Vanderweele J (2016) The serotonin system in autism spectrum disorder: from biomarker to animal models. Neuroscience 321:24–41

Naaijen J, Lythgoe DJ, Amiri H et al (2015) Fronto-striatal glutamatergic compounds in compulsive and impulsive syndromes: a review of magnetic resonance spectroscopy studies. Neurosci Biobehav Rev 52:74–88

Nadal-Desbarats L, Aïdoud N, Emond P et al (2014) Combined 1HNMR and 1H–13C HSQC-NMR to improve urinary screening in autism spectrum disorders. Analyst 139:3460–3468

Neis EP, Dejong CH, Rensen SS (2015) The role of microbial amino acid metabolism in host metabolism. Nutrients 7:2930–2946

Nikolaus S, Schulte B, Al-Massad N, Thieme F, Schulte DM, Bethge J et al (2017) Increased tryptophan metabolism is associated with activity of inflammatory bowel diseases. Gastroenterology 153:1504–1516.e1502

Noto A, Fanos V, Barberini L et al (2014) The urinary metabolomics profile of an Italian autistic children population and their unaffected siblings. J Matern Fetal Neonatal Med 27:46–52

Ossenkopp KP, Foley KA, Gibson J, Fudge MA, Kavaliers M, Cain DP et al (2012) Systemic treatment with the enteric bacterial fermentation product, propionic acid, produces both conditioned taste avoidance and conditioned place avoidance in rats. Behav Brain Res 227: 134–141

Ozeki Y, Sekine M, Fujii K et al (2016) Phosphoserine phosphatase activity is elevated and correlates negatively with plasma d-serine concentration in patients with schizophrenia. Psychiatry Res 237:344–350

Park HR, Lee JM, Moon HE, Lee DS, Kim BN, Kim J, Kim DG, Paek SH (2016) A short review on the current understanding of autism spectrum disorders. Exp Neurobiol 25:1–13

Pascal SP, Nemeş B, Vlase L et al (2006) High levels of homocysteine and low serum paraoxonase 1 arylesterase activity in children with autism. Life Sci 78:2244–2248

Paula-Lima AC, Brito-Moreira J, Ferreira ST (2013) Deregulation of excitatory neurotransmission underlying synapse failure in Alzheimer's disease. J Neurochem 126:191–202

Pellow J, Solomon EM, Barnard CN (2011) Complementary and alternative medical therapies for children with attention deficit/hyperactivity disorder (ADHD). Altern Med Rev 16:323–337

Puig-Alcaraz C, Fuentes-Albero M, Calderon J et al (2015) Increased homocysteine levels correlate with the communication deficit in children with autism spectrum disorder. Psychiatry Res 229: 1031–1037

Razak MA, Begum PS, Viswanath B, Rajagopal S (2017) Multifarious beneficial effect of nonessential amino acid, glycine: a review. Oxidative Med Cell Longev 2017:1716701

Robertson CE, Ratai E-M, Kanwisher N (2016) Reduced GABAergic action in the autistic brain. Curr Biol 26:80–85

Rojas DC, Singel D, Steinmetz S et al (2014) Decreased left perisylvian GABA concentration in children with autism and unaffected siblings. Neuroimage 86:28–34

Sacchi S, Binelli G, Pollegioni L (2016) G72 primate-specific gene: a still enigmatic element in psychiatric disorders. Cell Mol Life Sci 73:2029–2039

Sacrey L-AR, Bennett JA, Zwaigenbaum L (2015) Early infant development and intervention for autism spectrum disorder. J Child Neurol 30:1921–1929

Shimmura C, Suda S, Tsuchiya KJ et al (2011) Alteration of plasma glutamate and glutamine levels in children with high-functioning autism. PLoS One 6:25340

Shinohe A, Hashimoto K, Nakamura K et al (2006) Increased serum levels of glutamate in adult patients with autism. Prog Neuropsychopharmacol Biol Psychiatry 30:1472–1477

Sinha D, Efron D (2005) Complementary and alternative medicine use in children with attention deficit hyperactivity disorder. J Paediatr Child Health 41:23–26

Skalny AV, Skalny AA, Lobanova YN, Korobeinikova TV, Ajsuvakova OP, Notova SV, Burtseva TI, Skalnaya MG, Tinkov AA (2020) Serum amino acid spectrum in children with autism spectrum disorder (ASD). Res Autism Spectr Disord 77:101605

Skalny AV, Mazaletskaya AL, Zaitseva IP, Skalny AA, Spandidos DA, Tsatsakis A, Lobanova YN, Skalnaya MG, Aschner M, Tinkov AA (2021) Alterations in serum amino acid profiles in children with attention deficit/hyperactivity disorder. Biomed Rep 14:47

Storebo OJ, Skoog M, Damm D, Thomsen PH, Simonsen E, Gluud C (2011) Social skills training for attention deficit hyperactivity disorder (ADHD) in children aged 5 to 18 years. Cochrane Database Syst Rev 12:CD008223

Tărlungeanu DC, Elena D, Dotter CP, Majdi K et al (2016) Impaired amino acid transport at the blood brain barrier is a cause of autism spectrum disorder. Cell 167(6):1481–1494

Tinkov AA, Skalnaya MG, Skalny AV (2021) Serum trace element and amino acid profile in children with cerebral palsy. J Trace Elem Med Biol 64:126685

Tirouvanziam R, Obukhanych TV, Laval J et al (2011) Distinct plasma profile of polar neutral amino acids, leucine, and glutamate in children with autism spectrum disorders. J Autism Dev Disord 42:827–836

Tirouvanziam R, Obukhanych TV, Laval J, Aronov PA, Libove R, Banerjee AG et al (2012) Distinct plasma profile of polar neutral amino acids, leucine, and glutamate in children with autism spectrum disorders. J Autism Dev Disord 42:827–836

Torrioli MG, Vernacotola S, Peruzzi L (2008) A double-blind, parallel, multicentre comparison of L-acetylcarnitine with placebo on the attention deficit hyperactivity disorder in fragile X syndrome boys. Am J Med Genet 146:803–812

Tu W-J, Chen H, He J (2012) Application of LC-MS/MS analysis of plasma amino acids profiles in children with autism. J Clin Biochem Nutr 51:248–249

Tyzio R, Nardou R, Ferrari DC et al (2014) Oxytocin-mediated GABA inhibition during delivery attenuates autism pathogenesis in rodent offspring. Science 343:675–679

van Elst LT, Maier S, Fangmeier T et al (2014) Disturbed cingulate glutamate metabolism in adults with high-functioning autism spectrum disorder: evidence in support of the excitatory/inhibitory imbalance hypothesis. Mol Psychiatry 19:1314–1325

Vizi ES, Kisfali M, Lorincz T (2013) Role of nonsynaptic GluN2Bcontaining NMDA receptors in excitotoxicity: evidence that fluoxetine selectively inhibits these receptors and may have neuroprotective effects. Brain Res Bull 93:32–38

Wang L, Christophersen CT, Sorich MJ, Gerber JP, Angley MT, Conlon MA (2012) Elevated fecal short chain fatty acid and ammonia concentrations in children with autism spectrum disorder. Dig Dis Sci 57:2096–2102

Williams BL, Hornig M, Parekh T, Lipkin WI (2012) Application of novel PCR based methods for detection, quantitation, and phylogenetic characterization of Sutterella species in intestinal biopsy samples from children with autism and gastrointestinal disturbances. mBio 3:e00261–e00211

Woods DL, Kovach CR, Raff H et al (2008) Using saliva to measure endogenous cortisol in nursing home residents with advanced dementia. Res Nurs Health 31:283–294

Yang C-J, Tan H-P, Du Y-J (2014) The developmental disruptions of serotonin signaling may involved in autism during early brain development. Neuroscience 267:1–10

Young LJ, Barrett CE (2015) Can oxytocin treat autism? Science 347:825–826

Yuksel C, Ongur D (2010) Magnetic resonance spectroscopy studies of glutamate-related abnormalities in mood disorders. Biol Psychiatry 68:785–794

Zhang WQ, Smolik CM, Barba-Escobedo PA et al (2015) Acute dietary tryptophan manipulation differentially alters social behavior, brain serotonin and plasma corticosterone in three inbred mouse strains. Neuropharmacology 90:1–8

Zhang S, Zeng X, Ren M, Mao X, Qiao S (2017) Novel metabolic and physiological functions of branched chain amino acids: a review. J Anim Sci Biotechnol 8:10

Zheng HF, Wang WQ, Li XM, Rauw G, Baker GB (2017) Body fluid levels of neuroactive amino acids in autism spectrum disorders: a review of the literature. Amino Acids 49(1):57–65

Zwaigenbaum L, Bryson SE, Brian J et al (2015) Stability of diagnostic assessment for autism spectrum disorder between 18 and 36 months in a high-risk cohort. Autism Res 9:790–800

Autism and the Scaffolding Protein Neurobeachin

11

Sawsan Mohammed and M. Walid Qoronfleh

Abstract

Autism is a group of early-onset, diverse, lifelong, neurodevelopmental conditions that affect social interaction, verbal and nonverbal communications, and behavior. It is a multifactorial condition, which results from the interaction of environmental and genetic factors. A number of proteins are identified as autism-related proteins such as neurobeachin (NBEA), a large multi-domain scaffolding protein belonging to BEACH domain-containing proteins (BDCPs), which is a family of proteins found in all eukaryotes. The human's BDCP family includes nine proteins that share the characteristic of containing BEACH and WD repeat domains. Mutation of each one of the genes encoding these proteins leads to a distinct disease condition. NBEA mutations includes deletion, translocation, inversion, or duplication and is associated with autism spectrum disorders (ASD).

Keywords

Autism · ASD · Neurobeachin · Scaffolding proteins · Membrane protein trafficking · Neuronal junction · Synapses · Neural development

S. Mohammed (✉)
Clinical Academic Education Department, Qatar University, College of Medicine, Doha, Qatar

M. W. Qoronfleh (✉)
Q3CG Research Institute (QRI), Research and Policy Division, Ypsilanti, MI, USA

M. W. Qoronfleh et al. (eds.), *Proteins Associated with Neurodevelopmental Disorders*, Nutritional Neurosciences, https://doi.org/10.1007/978-981-15-9781-7_11

11.1 Introduction

The term autism was used for the first time in 1912 by Paul Eugen Bleuler, a Swiss psychiatrist, to describe the symptoms of schizophrenia. In 1938, Hans Asperger used "autistic psychopathy" to define a group of children with abnormal behavior and unusual movements, later known as Asperger's syndrome. The term autism in the modern sense was used for the first time in 1943 by the child psychiatrist Dr. Leo Kanner. Autism can be defined as a group of early-onset (mostly at the age of 2–3 years), diverse, lifelong neurodevelopmental conditions that can substantially affect social interaction, communication (both verbal and nonverbal), as well as behavior. Autism, Asperger's syndrome, Rett syndrome, and pervasive developmental disorder-not otherwise specified (PDD-NOS) are generally known as autism spectrum disorders (ASDs) (Park et al. 2016; Hodges et al. 2020; Styles et al. 2020).

Autism is a multifactorial condition, which results from the interaction of genetic and environmental factors such as exposure to teratogenic medications like thalidomide and valproic acid during pregnancy. ASD is reported globally with no overall differences between racial, ethnic, and socioeconomic groups. However, it affects more males than females with a ratio of 4.3:1 and a higher risk of severe disease condition is seen in male patients. When autism was first recognized, patients mainly were diagnosed with mental retardation of different severities or learning disabilities. The outcome for such individuals has improved tremendously, especially with early and intensive individual therapy. Today, some autistic children can attend regular schools (Styles et al. 2020; Levy et al. 2009; Chaste and Leboyer 2012; Baio et al. 2018; Begovac et al. 2009; Howlin et al. 2004; Freitag 2007).

While there is no exact known etiology for ASD, a number of risk factors associated with its development have been identified. The risk factors include genetic, epigenetic, and environmental factors, in addition to neuroanatomical abnormalities. Multiple studies have linked the development of ASD to brain protein mutations such as Shank3 protein, which usually acts as an information receiver protein. Researchers also showed that mutation of nSR100 protein leads to abnormal social and environmental interaction in mutated mice models. Major histocompatibility complex class I (MHC-I) is an essential cell surface protein for the adaptive immune system. This molecule was found to play a critical role in restricting the connection between brain cells and is linked to ASD in many studies. An interesting quantitative analysis study showed that the combination of a group of five proteins (complement C3, complement C5, integrin alpha-IIb (ITGA2B), talin-1 (TLN1), and vitamin D-binding protein GC) that were earlier found to be involved in different pathways linked to the pathophysiology of ASD was also validated by enzyme-linked immunosorbent assay (ELISA) and could distinguish children with ASD from normal controls. Mutations in the fragile X mental retardation 1 (FMR1) gene have been linked to the most common inherited type of ASD in humans. It has been found that mutated oocytes deficient in FMR1 typically result in embryos with severe neural defect conditions (Wieczorek et al. 2017; Alexiou et al. 2018; Shen et al. 2018; Greenblatt and Spradling 2018; Wang et al. 2017).

11.2 Scaffolding Proteins

Scaffold proteins are key regulators of several signaling cascades. They are also defined as adaptor or linker proteins. Several scaffolding protein groups are found naturally. Their function is not fully identified, but these proteins are known to bind and/or interact with multiple other proteins forming new complexes, known as signalsome or transducisome, in the signaling pathways. Scaffold proteins help to convey the messages between the nucleus and cell membrane in a faster manner (Fig. 11.1) (Hata and Iida 2009; Garbett and Bretscher 2014).

Scaffold proteins can act in several ways. Its key function is to tether the signaling components increasing the signaling pathway efficiency and specificity by enhancing the physical assembly and effective concentration of the relevant components in the scaffold complex, which avoids the unwanted interaction between other proteins in the pathway. Another function is to localize the signaling components and target them to precise cell compartments. They also regulate the signal transduction by managing the positive and the negative feedback signals. Finally, they sequester the correct signaling proteins from other competing proteins and prevent the deactivation and/or degradation of the activated signaling components (Fig. 11.2). In short, scaffold proteins increase the efficacy and specificity of the signaling pathways and act as catalysts (Engström et al. 2010; Shaw and Filbert 2009; Locasale et al. 2007).

11.3 BEACH Domain-Containing Proteins

BEACH domain-containing proteins (BDCPs) are a group of related scaffolding proteins that belong to a new family, the Beige and Chediak-Higashi (BEACH) family. BDCPs were initially identified during a study to characterize the human type of Chediak-Higashi syndrome (CHS) using the *Beige* murine model. BEACH is conserved in all eukaryotes, including mammals, plant cells, and yeast. In humans,

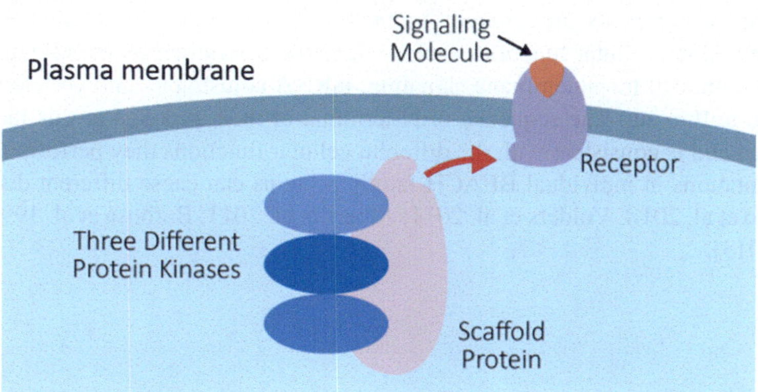

Fig. 11.1 Scaffold proteins are proteins that bind multiple other proteins simultaneously, forming new complexes that enhance signaling efficiency and fidelity

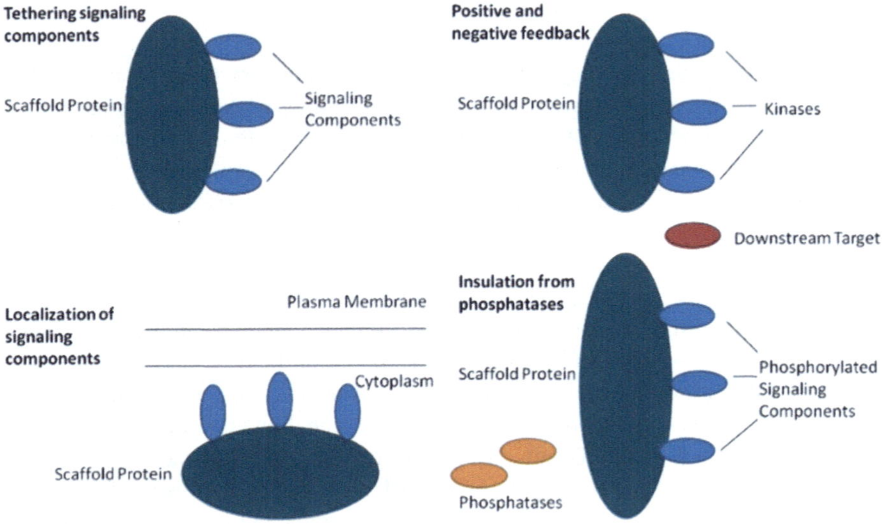

Fig. 11.2 Scaffolding protein main functions (Shaw and Filbert 2009)

BDCP family includes nine proteins. Lysosomal trafficking regulator protein (LYST) is the first member with the BEACH domain to be discovered, followed by the identification of the other family members, lipopolysaccharide-responsive, beige-like anchor protein (LRBA), neurobeachin (NBEA), neurobeachin-like 1 (NBEAL1), neurobeachin-like 2 (NBEAL2), WD repeat domain 81 (WDR81), neutral sphingomyelinase activation-associated factor (NSMAF), WD and FYVE zinc finger domain-containing protein 3 (WDFY3), and WD and FYVE zinc finger domain-containing protein 4 (WDFY4). The proteins belonging to this family are mostly large in size and share the presence of PH-like domain in their C-terminal for membrane association, followed by the BEACH domain, which is vital for their functions in vesicle trafficking, membrane dynamics, and receptor signaling. The WD sequence repeats are thought to facilitate protein-protein interactions and regulate various cellular functions such as division, determination of cell fate, gene transcription, cell transmembrane signaling, mRNA adjustment, and vesicle formation and trafficking. The sequence of N-terminal of most proteins in this family is unrelated and is consistent with the different cellular functions they perform. Therefore, mutations in individual BEACH family proteins can cause different disorders (Repetto et al. 2018; Volders et al. 2011; Albers et al. 2011; Barbosa et al. 1996; Teh et al. 2015).

11.4 Neurobeachin

Neurobeachin (NBEA) is a cytosolic multi-domain scaffold protein with no less than seven different protein motifs. The mammalian neurobeachin is a neuron-specific polypeptide of 327 kDa with 38% of hydrophobic amino acids. It is characterized by the presence of the PH-BEACH sequence in its C-terminal followed by multiple WD40 repeats. In addition to BEACH, PH-like (pleckstrin homology) domain, and four C-terminal tryptophan-aspartic acid WD repeats, the NBEA protein domains uniquely include an A-kinase anchoring protein (AKAP) motif that binds to protein kinase A (PKA), which is essential in establishing cell microdomains. It contains in its N-terminal a Concanavalin A-like lectin (ConA-like) CALL domain, which is thought to play a role in intracellular sorting due to the similarity to the clostridial neurotoxin N-terminal heavy chain and also contains an armadillo repeat domain which was presented as a domain of unknown function 4704 (DUF4704). Another DUF domain (DUF1088) with assumed nuclear localization signals is also present (Fig. 11.3).

Neurobeachin is concentrated at the trans-Golgi complex, post-Golgi vesicles, and synaptic contacts. NBEA is expressed throughout the cell of many tissues but at different levels. It is largely expressed in human brain tissues, mostly in the plasma membrane of the postsynaptic area. It can be expressed at medium levels in the tissues of the spleen, thymus and prostate glands, testis, and ovaries. Lower levels of

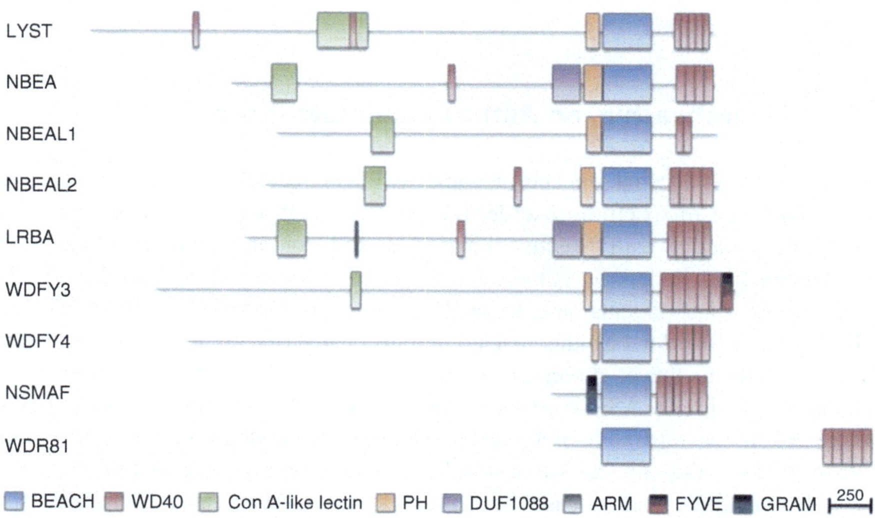

Fig. 11.3 Schematic diagram displaying standard protein domains in human BDCPs. The BEACH domain is aligned for all nine proteins, and the drawing is to scale, where the scale bar represents 250 amino acids. Note the similarity in number and positions of the WD40 repeat domains following and the PH domains preceding the BEACH domain in 7 out of the 9 proteins. Other recognized domains include ConA-like lectin, DUF1088, ARM, FYVE, and GRAM domains (Cullinane et al. 2013)

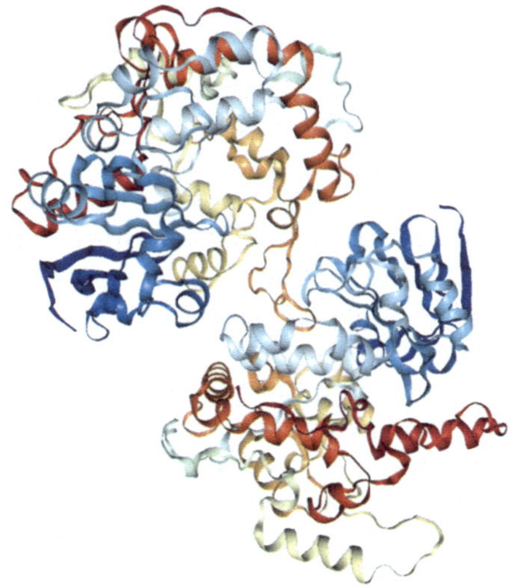

Fig. 11.4 Crystal structure of the PH-BEACH domain of human neurobeachin. Blue is the N-terminus, and red is the C-terminus. Source: Protein Data Bank (PDB, https://www.rcsb.org)

expression are found in the heart, kidneys, pancreas, skeletal muscle tissues, and intestine. Neurobeachin and more than ten other mammalian proteins belonging to the same family were also found in invertebrates, plants, yeasts, and protozoa. This includes beige-like protein (BGL), which is an isoform of neurobeachin but does not bind to the regulatory subunit of protein kinase (Repetto et al. 2018; Wang et al. 2000; Dyomin et al. 2002; Gaudet et al. 2011) (Fig. 11.4).

11.5 Neurobeachin, an Autism Candidate Protein

Neurobeachin is located on chromosome q13 and recently was identified as a nonfamilial autism spectrum disorder-related protein. It was first recognized in a boy from a family with negative history of ASD or any other psychiatric or developmental disorders as balanced de novo translocation t(5;13)(q12.1;q13.2) and in three other patients as a monoallelic deletion. In patients with ASD, the NBEA gene is found to be either deleted or translocated (the breakpoint is located in intron 2 distal to the promoter), resulting in the absence of its expression. In a number of studies on ASD patients using genome-wide assessment (GWAS) and whole-genome comparative microarray, multiple other structural variations and autism-specific copy number variants (CNVs), including inversion and duplication, were also identified in some cases but with no known, precise causative mechanism (Odent et al. 2021; Creemers et al. 2014; Castermans et al. 2003; Marshall et al. 2008; Christian et al. 2008).

11.6 Neurobeachin Studies in Animal Models

To understand the role played by NBEA in autism pathogenesis, few ASD-related features were studied using animal models. In a study on mice, the NBEA-deficient mice showed some behavioral alterations, such as changes in memory and learning, self-grooming, social responses, and fear reactions. The observed symptoms coincide with an increase in long-term potentiation (LTP) in their CA1 region. The noticed changes in memory and learning and hippocampal LTP are associated with decreased expression of the immediate-early gene zif268 in dorsomedial striatum and CA1 region of the hippocampus, and also increased CREB phosphorylation and increased hippocampal BDNF expression. Such changes in NBEA-deficient mice could underlie the ASD symptoms in NBEA mutant individuals. In another in vivo study also done in mice, data indicated that mice lacking one allele of the NBEA gene exhibited unusual and specific cell excitability changes, which may contribute to the behavioral abnormalities in NBEA-deficient mice and can be related to ASD symptoms in patients. In another study, two independent mouse models have demonstrated a role in neurotransmitter release and synaptic functioning. Rugose (RG) in *Drosophila* is a homolog of the mammalian, including human, NBEA gene. It encodes an A-kinase anchor protein (DAKAP 550) and interacts with the epidermal growth factor receptors (EGFR), and Notch-mediated signaling pathways. Protein-protein interaction with NOTCH 1 is most relevant for ASD pathogenesis because NOTCH signaling is essential for neural development. Data from a functional study of the larval neuromuscular junctions revealed abnormal neurodevelopmental synaptic physiology. Additionally, RG mutant adult Drosophila showed unusual social behavior, diminished acclimatization, changes of motion, and overactivity resembling human ASD. Furthermore, the NBEA homologue in *C. elegans*, SEL-2, was identified as a negative regulator of Notch activity. A separate study has shown that NBEA acts as an important regulator in the postsynaptic neurons of zebrafish and is required for electrical and chemical synapse formation. It also showed a correlation to abnormal behavior (Nuytens et al. 2013a; Muellerleile et al. 2020; Wise et al. 2015; Miller et al. 2015).

11.7 Functions of Neurobeachin

The functions of neurobeachin have not been fully understood. It is characterized by its high binding affinity to the regulatory unit type II of protein kinase A (R II PKA) and targeting the cell membrane. NBEA deficiency or its absence can disturb protein kinase A (PKA)-mediated phosphorylation. NBEA has been shown to regulate the nucleus transcriptional process. It plays a not fully known role in spine formation, which includes small actin-rich protrusions from dendrites where most excitatory synapses are located, and it has an influence on actin distribution. It has been shown that deletion of the NBEA gene in cultured neurons from knockout mice and its deletion in cortical tissue from heterozygous mice lead to reduced numbers of spinous synapses and change the miniature postsynaptic currents (mEPSCs).

Neurobeachin has also been proved to target the postsynaptic neurotransmitter receptor in other species such as Drosophila and zebrafish. In addition, a novel interaction between NBEA, which is a nucleus transcriptional regulator, and NOTCH1, which is an essential protein-coding gene for neural development, was identified as the most relevant pathogenesis for ASD. NBEA haploinsufficiency was found to affect the morphological structure of dense granules in blood platelets leading to insufficient secretion regulation, a possible endophenotype in autism. In addition, it can affect receptor trafficking and synaptic structure (Repetto et al. 2018; Miller et al. 2015; Tuand et al. 2016; Niesmann et al. 2011; Nuytens et al. 2013b).

11.8 Lysosomal Trafficking Regulator Protein (LYST)

LYST is the first protein to be discovered in the BEACH family. It is a large cytosolic protein of 3801 amino acids (430 kDa). LYST gene contains PH-BEACH domain, ConA-like lectin domain, and WD40 repeats. The human LYST gene or CHS1 is located on chromosome 1 (1q42-43) and is believed to be related to material trafficking into lysosomes. Lysosomes help to recycle processes within the cells. LYST gene disruption, including nonsense and missense mutations, deletions, and insertions, leads to Chediak-Higashi syndrome (CHS), which is a rare, autosomal recessive condition that affects various body systems. Patients with CHS are characterized by severe immunodeficiency and frequent infections; hypopigmentation of the hair, skin, and eyes (albinism); poor blood coagulation leading to easy bruising; and prolonged bleeding time in addition to neurologic problems, such as neuropathies and ataxia, which accelerate with age (Cullinane et al. 2013; Ward et al. 2002; Ajitkumar et al. 2021).

11.9 WDFY3 and WDFY4

WD and FYVE zinc finger domain-containing protein 3 or autophagy-linked FYVE (Alfy) is a large scaffold protein that belongs to the human BDCP family. It is known to be the only protein in BDCP family that contains zinc finger domain FYVE, which is a domain also found in several other human proteins and exhibits a selective autophagy function, especially under unusual conditions such as starvation. It plays a vital role in mitochondrial homeostasis as well. It is localized near organelle membranes, making it easier to interact directly with the PtdIns(3)P phospholipids. The PH-like domain presence binds other phosphorylated inositides, but normally not PtdIns(3)P. In addition to the classical presence of ConA-like lectin and BEACH domains, the WD repeats found in the C-terminal are thought to be responsible for the co-localization of WDFY3. The human WDFY3 gene is located on chromosome 4 (4q21.23) and is expressed in developing as well as in the adult central nervous system. Its mutations are recently linked in animal models to conditions of decreased intellectual abilities, neurodevelopmental delay, familial microcephaly, and psychiatric conditions like attention-deficit hyperactivity disorder (ADHD) and ASD with macrocephaly (Napoli et al. 2018; Isakson et al. 2013).

WD and FYVE zinc finger domain-containing protein 4 (WDFY4) is a large protein of 3184 amino acids belonging to the human BDCP family. Like other proteins of this family, it contains multiple functional domains, including WD40 and BEACH. However, despite its name and unlike WDFY3, WDFY4 does not encode the FYVE domain. WDFY4 gene is located on chromosome 10 (10q11.23) and is highly expressed in immune tissues like lymph nodes, tonsils, thymus gland, and spleen. Studies strongly suggest the relation between WDFY4 mutation and the autoimmune disease systemic lupus erythematosus (SLE) pathogenesis; however, its exact function is not fully identified (Cullinane et al. 2013; Yuan et al. 2018).

11.10 Neurobeachin-Like 1 and Neurobeachin-Like 2

Based on the presence of the BEACH domain, neurobeachin-like 1 (NBEAL1) and neurobeachin-like 2 (NBEAL2) as BEACH domain-containing proteins (BDCPs) have been identified as mammalian homologues of NBEA. NBEAL1 is a typical large BDCP with a total of 2694 amino acids. The human NBEAL1 gene is located on chromosome 2 (2q33-2q34); in addition to a ConA-like lectin, a PH-like, BEACH, and WD domain, it contains a vacuolar-targeting peptide motif ILPK, which suggests the possible protein localization in the lysosome. However, this has to be confirmed by cellular localization studies. Biopsies from different grades of glioma patients showed upregulation of the NBEAL1 gene, especially in the lower grade gliomas, which suggests their possible correlation. Studies also suggested the correlation of NBEAL1 to several other tumors like ovarian serous adenocarcinoma and metastasis of specific mammary gland breast cancer (Volders et al. 2011; Chen et al. 2004).

NBEAL2 is a protein with 2754 amino acids, which belongs to the same family and like other members in this protein family, it contains a BEACH domain and multiple WD40 repeats. Human NBEAL2 is a gene with previously unknown functions until its involvement in granule development was noticed. It is located in chromosome 3 (3p21-3q31) and was found to be related to thrombopoiesis and thought to interfere with megakaryocyte alpha-granule biogenesis. Its mutation is identified as the cause of gray platelet syndrome (GPS), which is a rare congenital autosomal recessive disorder caused by the decrease or complete absence of alpha-granules in blood platelets, leading to mild-to-moderate bleeding tendency, thrombocytopenia, and a marked reduction or lack of platelet alpha-granules and the proteins contained in them. Many patients tend to develop myelofibrosis later in life. Unlike neurobeachin, NBEAL2 is highly expressed in blood cells, mainly megakaryocytes (MKs) and granulocytes, and NBEAL2 expression levels increase during granulocyte maturation. It has low expression in the brain tissues (Albers et al. 2011; Fabbro et al. 2011).

Other members of this family are also connected to different disorders when mutated. LPS-responsive and beige-like anchor gene (LRBA) is generally involved in the immune response and cell apoptosis and proliferation and is required for several pathways such as regulation of EGFR and PKA pathways. LRBA deficiency in humans was found to be related to some immunodeficiency conditions. It is observed to be more expressed in several cancer types, e.g., breast cancer related to estrogen and p53 mutation, melanoma, and gastric cancer (Bratanič et al. 2017; Wang et al. 2004). The neutral sphingomyelinase activation-associated factor (NSMAF) protein or factor associated with N-Smase activation (FAN) is a unique member of the BDCPs. NSMAF is comparatively not very large, and it consists of 948 amino acids only. It contains BEACH, and WD repeat domains, but not the ConA-like lectin domain. It is the only member with a membrane-associated GRAM domain instead of the membrane-associated PH-like domain found in a number of other BDCPs. For NSMAF to function, it has to bind phospholipids like PtdIns(4,5)P and be localized in the plasma membrane. It is required for TNF-mediated activation of neutral sphingomyelinase, and it is believed to have a role in regulating cellular inflammatory responses induced by TNF (Haubert et al. 2007). A WD repeat domain 81 (WDR81) is a transmembrane protein belonging to BDPCs, and similarly, the WD repeat domains precede the BEACH domain, but unlike most other BDCPs, WDR81 does not contain a PH-like or ConA-like lectin domain. It is expressed mainly in the corpus callosum and cerebellum, particularly in the cerebellar Purkinje cells. Mutation of WDR81 is linked with posture and gait abnormalities in humans. It has been demonstrated to be associated with neurological disorders. This needs further studies to be well identified (Cullinane et al. 2013; Gulsuner et al. 2011; Wang et al. 2021).

11.11 Conclusions

The NBEA is one of the several ASD candidate genes, which has been identified in a patient with a de novo chromosomal translocation. In animal models, NBEA gene loss was found to mimic autism as the defect affects signaling at neuronal junctions (synapses). The NBEA gene encodes a large multi-domain scaffolding protein that functions in neuronal post-Golgi membrane trafficking. NBEA possesses several domains that mediate protein-protein interactions. NBEA protein plays a diverse biological role. NBEA deficiency affects regulated secretion [negative regulator of secretion of large dense-core vesicles (LDCVs)], receptor trafficking, synaptic architecture [synaptic transmission is a fundamental step in brain function], and PKA-mediated phosphorylation [PKA is virtually a universal cellular component in eukaryotes, where diminished PKA-mediated phosphorylation of proteins thus leads to abnormalities in cellular signaling].

As a putative regulator of membrane protein trafficking to synaptic contacts, NBEA presumed function is consistent with the “excitatory-inhibitory imbalance” model of autism. While NBEA seems to be relevant to ASD social behavior, further investigation is warranted to understand how alterations in NBEA function might contribute to the pathogenesis of ASD.

References

Ajitkumar A, Yarrarapu SNS, Ramphul K (2021) Chediak Higashi syndrome. In: StatPearls. StatPearls Publishing. Copyright © 2021, StatPearls Publishing LLC, Treasure Island

Albers CA et al (2011) Exome sequencing identifies NBEAL2 as the causative gene for gray platelet syndrome. Nat Genet 43(8):735–737

Alexiou A et al (2018) Proteins commonly linked to autism spectrum disorder and Alzheimer's disease. Curr Protein Pept Sci 19(9):876–880

Baio J et al (2018) Prevalence of autism spectrum disorder among children aged 8 years—autism and developmental disabilities monitoring network, 11 sites, United States, 2014. MMWR Surveill Summ 67(6):1–23

Barbosa MD et al (1996) Identification of the homologous beige and Chediak-Higashi syndrome genes. Nature 382(6588):262–265

Begovac I et al (2009) The loss of criteria for mental retardation and regression of symptoms of childhood autism during a five-year follow-up—a case report. Acta Clin Croat 48(4):445–450

Bratanič N et al (2017) Multifocal gastric adenocarcinoma in a patient with LRBA deficiency. Orphanet J Rare Dis 12(1):131

Castermans D et al (2003) The neurobeachin gene is disrupted by a translocation in a patient with idiopathic autism. J Med Genet 40(5):352

Chaste P, Leboyer M (2012) Autism risk factors: genes, environment, and gene-environment interactions. Dialogues Clin Neurosci 14(3):281–292

Chen J et al (2004) Identification and characterization of NBEAL1, a novel human neurobeachin-like 1 protein gene from fetal brain, which is up regulated in glioma. Brain Res Mol Brain Res 125(1–2):147–155

Christian SL et al (2008) Novel submicroscopic chromosomal abnormalities detected in autism spectrum disorder. Biol Psychiatry 63(12):1111–1117

Creemers JWM, Nuytens K, Tuand K (2014) Neurobeachin gene in autism. In: Patel VB, Preedy VR, Martin CR (eds) Comprehensive guide to autism. Springer, New York, pp 825–844

Cullinane AR, Schäffer AA, Huizing M (2013) The BEACH is hot: a LYST of emerging roles for BEACH-domain containing proteins in human disease. Traffic (Copenhagen, Denmark) 14(7): 749–766

Dyomin VG et al (2002) BCL8 is a novel, evolutionarily conserved human gene family encoding proteins with presumptive protein kinase A anchoring function. Genomics 80(2):158–165

Engström W, Ward A, Moorwood K (2010) The role of scaffold proteins in JNK signalling. Cell Prolif 43(1):56–66

Fabbro S et al (2011) Homozygosity mapping with SNP arrays confirms 3p21 as a recessive locus for gray platelet syndrome and narrows the interval significantly. Blood 117(12):3430–3434

Freitag CM (2007) The genetics of autistic disorders and its clinical relevance: a review of the literature. Mol Psychiatry 12(1):2–22

Garbett D, Bretscher A (2014) The surprising dynamics of scaffolding proteins. Mol Biol Cell 25(16):2315–2319

Gaudet P et al (2011) Phylogenetic-based propagation of functional annotations within the Gene Ontology consortium. Brief Bioinform 12(5):449–462

Greenblatt EJ, Spradling AC (2018) Fragile X mental retardation 1 gene enhances the translation of large autism-related proteins. Science 361(6403):709–712

Gulsuner S et al (2011) Homozygosity mapping and targeted genomic sequencing reveal the gene responsible for cerebellar hypoplasia and quadrupedal locomotion in a consanguineous kindred. Genome Res 21(12):1995–2003

Hata Y, Iida J (2009) Scaffold protein. In: Binder MD, Hirokawa N, Windhorst U (eds) Encyclopedia of neuroscience. Springer, Berlin, Heidelberg, pp 3613–3616

Haubert D et al (2007) PtdIns(4,5)P-restricted plasma membrane localization of FAN is involved in TNF-induced actin reorganization. EMBO J 26(14):3308–3321

Hodges H, Fealko C, Soares N (2020) Autism spectrum disorder: definition, epidemiology, causes, and clinical evaluation. Transl Pediatr 9(Suppl 1):S55–s65
Howlin P et al (2004) Adult outcome for children with autism. J Child Psychol Psychiatry 45(2): 212–229
Isakson P, Holland P, Simonsen A (2013) The role of ALFY in selective autophagy. Cell Death Differ 20(1):12–20
Levy SE, Mandell DS, Schultz RT (2009) Autism. Lancet (London, England) 374(9701): 1627–1638
Locasale JW, Shaw AS, Chakraborty AK (2007) Scaffold proteins confer diverse regulatory properties to protein kinase cascades. Proc Natl Acad Sci 104(33):13307–13312
Marshall CR et al (2008) Structural variation of chromosomes in autism spectrum disorder. Am J Hum Genet 82(2):477–488
Miller AC et al (2015) Neurobeachin is required postsynaptically for electrical and chemical synapse formation. Curr Biol 25(1):16–28
Muellerleile J et al (2020) Enhanced LTP of population spikes in the dentate gyrus of mice haploinsufficient for neurobeachin. Sci Rep 10(1):16058
Napoli E et al (2018) Beyond autophagy: a novel role for autism-linked Wdfy3 in brain mitophagy. Sci Rep 8(1):11348
Niesmann K et al (2011) Dendritic spine formation and synaptic function require neurobeachin. Nat Commun 2(1):557
Nuytens K et al (2013a) Haploinsufficiency of the autism candidate gene Neurobeachin induces autism-like behaviors and affects cellular and molecular processes of synaptic plasticity in mice. Neurobiol Dis 51:144–151
Nuytens K et al (2013b) Platelets of mice heterozygous for neurobeachin, a candidate gene for autism spectrum disorder, display protein changes related to aberrant protein kinase A activity. Mol Autism 4(1):43
Odent P et al (2021) Spectrum of social alterations in the Neurobeachin haploinsufficiency mouse model of autism. Brain Res Bull 167:11–21
Park HR et al (2016) A short review on the current understanding of autism spectrum disorders. Exp Neurobiol 25(1):1–13
Repetto D et al (2018) Molecular dissection of neurobeachin function at excitatory synapses. Front Synaptic Neurosci 10:28
Shaw AS, Filbert EL (2009) Scaffold proteins and immune-cell signalling. Nat Rev Immunol 9(1): 47–56
Shen L et al (2018) iTRAQ-based proteomic analysis reveals protein profile in plasma from children with autism. Proteomics Clin Appl 12(3):e1700085
Styles M et al (2020) Risk factors, diagnosis, prognosis and treatment of autism. Front Biosci (Landmark Ed) 25:1682–1717
Teh OK et al (2015) BEACH-domain proteins act together in a cascade to mediate vacuolar protein trafficking and disease resistance in Arabidopsis. Mol Plant 8(3):389–398
Tuand K et al (2016) Nuclear localization of the autism candidate gene neurobeachin and functional interaction with the NOTCH1 intracellular domain indicate a role in regulating transcription. PLoS One 11(3):e0151954
Volders K, Nuytens K, Creemers JWM (2011) The autism candidate gene neurobeachin encodes a scaffolding protein implicated in membrane trafficking and signaling. Curr Mol Med 11(3): 204–217
Wang X et al (2000) Neurobeachin: a protein kinase A-anchoring, beige/Chediak-higashi protein homolog implicated in neuronal membrane traffic. J Neurosci 20(23):8551–8565
Wang J-W et al (2004) Deregulated expression of LRBA facilitates cancer cell growth. Oncogene 23(23):4089–4097
Wang W et al (2017) Striatopallidal dysfunction underlies repetitive behavior in Shank3-deficient model of autism. J Clin Invest 127(5):1978–1990

Wang M et al (2021) WDR81 regulates adult hippocampal neurogenesis through endosomal SARA-TGFβ signaling. Mol Psychiatry 26(2):694–709

Ward DM, Shiflett SL, Kaplan J (2002) Chediak-Higashi syndrome: a clinical and molecular view of a rare lysosomal storage disorder. Curr Mol Med 2(5):469–477

Wieczorek M et al (2017) Major histocompatibility complex (MHC) class I and MHC class II proteins: conformational plasticity in antigen presentation. Front Immunol 8:292

Wise A et al (2015) Drosophila mutants of the autism candidate gene neurobeachin (rugose) exhibit neuro-developmental disorders, aberrant synaptic properties, altered locomotion, and impaired adult social behavior and activity patterns. J Neurogenet 29(2–3):135–143

Yuan Q et al (2018) WDFY4 is involved in symptoms of systemic lupus erythematosus by modulating B cell fate via noncanonical autophagy. J Immunol 201(9):2570–2578

Regulatory Role of ADGRL3, PARK2, and CNTNAP2 in Neurodevelopmental Disorders

12

Vidya Murugesan and Senthilkumar Rajagopal

Abstract

Adhesion G protein-coupled receptors (AGPCRs) are a 33-member subfamily of Class B GPCRs that control many physiological processes and are implicated in disease. AGPCRs play a role in cell-cell adhesion and neuron guidance via different proteins present in the surface of the cells and play a role in the development of glutamatergic synapses in the cortex. The most crucial function of Parkin RBR E3 ubiquitin protein ligase (PARK2) gene is unknown; moreover, the encoded protein is a component of a multiprotein E3 ubiquitin ligase complex that mediates the targeting of substrate proteins for proteasomal degradation. Mutations in the gene are known to cause Parkinson disease and autosomal recessive juvenile Parkinson disease. The researchers have shown that the Contactin-associated protein-like 2 (CNTNAP2) gene is associated with different symptoms of autism spectrum disorders (ASDs) and other neurodevelopmental disorders. The CNTNAP2 gene, coding for the cell adhesion glycoprotein Caspr2, is thought to be one of the major susceptibility genes for ASD. A large number of rare heterozygous missense CNTNAP2 variants have been identified in ASD patients. However, the intricate biochemical and molecular machinery contributing to the neurological disorders is still unknown. Here, we discuss the regulatory role of these proteins in neurodevelopmental disorders (NDDs).

V. Murugesan
Department of Chemistry and Biochemistry, Ramaiah College of Arts, Science and Commerce, Bengaluru, Karnataka, India

S. Rajagopal (✉)
Department of Biotechnology, School of Applied Sciences, REVA University, Bengaluru, Karnataka, India

M. W. Qoronfleh et al. (eds.), *Proteins Associated with Neurodevelopmental Disorders*, Nutritional Neurosciences, https://doi.org/10.1007/978-981-15-9781-7_12

Keywords

Attention-deficit/hyperactivity disorder · ADHD · Orphan receptor · Unc-5 receptor · Netrin · ADGRL3 · PARK2 · CNTNAP2 and neurodevelopmental disorders

Abbreviations

ADGRL3	Adhesion G protein-coupled receptor L3
ADHD	Attention-deficit/hyperactivity disorder (ADHD)
aGPCRs	Adhesion G protein-coupled receptors
ARJP	Autosomal recessive juvenile parkinsonism
ASD	Autism spectrum disorder
CNTNAP2	Contactin-associated protein-like 2
DAT	Dopamine transporter
ECDs	Extracellular domains
ECR47	Evolutionarily conserved region 47
FLRT3	Fibronectin leucine-rich transmembrane protein 3
LPHN3	Latrophilin 3
PARK2	Parkin RBR E3 ubiquitin protein ligase
PD	Parkinson's disease
SNPs	Single-nucleotide polymorphisms
UNC5	Unc-5 netrin receptor
VGKC	Voltage-gated potassium channel complex

12.1 Latrophilin 3/Adhesion G Protein-Coupled Receptor L3 (LPHN3/ADGRL3)

The G protein-coupled receptor (GPCR) superfamily is the biggest group of cell membrane receptors. The seven transmembrane proteins transfer the external signals internally by connections between diverse stimuli like peptides, metabolites, light, hormones, ions, proteins, and N-terminal extracellular domains (ECDs). In humans, 33 members of the adhesion G protein-coupled receptor (aGPCR) family are present. Although the bulk of these is orphan receptors with unknown activities, many studies have shown that some members of this family play crucial roles in neurodevelopment, myelination, organogenesis, cancer progression, and angiogenesis. Significantly, human diseases have been related to mutations in various aGPCRs (Folts et al. 2019; Maraschi et al. 2014; Hu et al. 2016).

Attention-deficit/hyperactivity disorder (ADHD) disruptive behavior comorbidity, long-term prognosis, severity, and response to the treatment are predicted by variants in the ADGRL3 (LPHN3) gene (Acosta et al. 2016). The gene coding for latrophilin 3 (additionally known as adhesion G protein-coupled receptor L3 or

ADGRL3 or LPHN3) has been linked to ADHD vulnerability in independent ADHD samples (Bruxel et al. 2021) from human and animal studies. It was also demonstrated via fine mapping of a genetic linkage region for ADHD. ADGRL3 gene is expressed strongly in the caudate nucleus, amygdala, cerebral cortex, and cerebellum (Arcos-Burgos et al. 2010). During neurodevelopment, ADGRL3 and its ligands appear to play a crucial role in defining the connection rates between the primary neurons in the cortex (O'Sullivan et al. 2014) as well as neurotransmitter exocytosis and synaptic function. ADGRL3, also known as latrophilin 3 (LPHN3), is found in both the pre- and postsynaptic terminals of interneuron connections, suggesting that it might play a major role in the development and/or function of the synapse (Ribasés et al. 2011). As a result, changes in ADGRL3 expression might disrupt the proper establishment and maintenance of neural circuits, resulting in neurodevelopmental disorders (NDDs) like ADHD.

The creation of a trimeric complex with Unc-5 netrin receptor (UNC5) and fibronectin leucine-rich transmembrane protein 3 (FLRT3) by ADGRL3 mediates some of its actions at synaptic terminals. Both glutamatergic synapse formation and transcellular adhesion are aided by the above complex (Jackson et al. 2015). In mouse, zebrafish, and Drosophila, silencing or disruption of the ADGRL3 orthologue expression has consistently enhanced locomotor activity across species (Orsini et al. 2016; van der Voet et al. 2016), implying that this gene's function has been remarkably consistent throughout evolution. ADGRL3.1 and ADGRL3.2 are zebrafish paralogues of human ADGRL3, with ADGRL3.1 showing more particular expression patterns throughout embryonic development (Lange et al. 2012).

A prevalent ADGRL3 haplotype was connected to ADHD susceptibility in humans, a finding that was reproduced in both childhood and aged ADHD populations (Ribasés et al. 2011; Hwang et al. 2015; Kappel et al. 2017). An analysis of brain tissue transcriptomes in mice deficient in ADGRL3 reveals that gene expression for calcium signaling proteins and cell adhesion molecules is altered at distinct developmental time points, which in turn could influence neuronal function and structure (Martinez et al. 2016). ADGRL3 comprises an ultraconserved motif in the evolutionarily conserved region 47 (ECR47) that works as a transcriptional enhancer, according to an extensive investigation that included in silico, in vitro, and in vivo tests (Martinez et al. 2016). The authors also found that an ADHD risk haplotype (rs17226398, rs56038622, and rs2271338) lowered the enhancer activity in astrocytoma ad neuroblastoma cell lines by 40%. The rs2271338 risk allele interferes with the binding of the YY1 transcription factor to ECR47, which is critical for the function and development of the central nervous system. The haplotype causes the binding location of a crucial neurodevelopmental transcription factor to be disrupted.

Additionally, brain expression data indicate that ADGRL3 has maximum expression across infant and fetal stages and relatively high expression levels throughout life, suggesting that the gene is necessary for proper brain function (Martinez et al. 2016). YY1 knockdown, on the other hand, had no effect on ADGRL3 expression in differentiated cells, implying that ECR47 is only active during the developmental stages when the expression of ADGRL3 is higher (Martinez et al. 2016).

ADGRL3 interactions with other genes could also make an individual more prone to ADHD. ADGRL3 interacts with several genes that span a section on chromosome 11. Single-nucleotide polymorphisms (SNPs) in the 11q cluster interact with ADGRL3 SNPs to double the risk of ADHD and enhance the severity of the illness (Bruxel et al. 2015; Acosta et al. 2011; Puentes-Rozo et al. 2019). The genes present in the cascade play a vital role in brain development, confirming the neurological significance of ADHD.

12.2 Parkin RBR E3 Ubiquitin Protein Ligase (PARK2)

Parkin, a 465-amino acid protein, is a member of a multiprotein E3 ubiquitin ligase complex and targets the substrate proteins for degradation of proteasomes. It is essential for mitochondrial homeostasis and is encoded by the PARK2 gene. Mutations in PARK2 gene situated on 6q26 chromosome have been linked to Parkinson's disease, although structural changes have been reported in patients suffering from neurodevelopmental abnormalities, implying a widespread pathological effect in the brain's neurodegenerative and neurodevelopmental brain processes (Conceição et al. 2017). PARK2 gene is a neurodevelopmental gene that was first discovered as one of the reasons of early-onset Parkinson disease (Kitada et al. 1998) and has been linked to autism spectrum disorder (Glessner et al. 2009), schizophrenia (Xu et al. 2008), and attention-deficit/hyperactivity disorder (ADHD) (Jarick et al. 2014). According to Glessner et al., seven patients with ASD were shown to have a chromosome 6 copy number loss involving the PARK2 gene area (Glessner et al. 2009).

Parkin has a variety of substrates, demonstrating that it is a multifunctional protein engaged in various intracellular activities, including apoptosis regulation, management of mitochondrial integrity, and regulation of transcription (Charan and LaVoie 2015). Wild-type Parkin might affect cardiac health (Piquereau et al. 2013), Alzheimer's disease (Burns et al. 2009), cancer risk (Hu et al. 2016), multiple sclerosis (Witte et al. 2009), autism (Glessner et al. 2009), inclusion body myositis (Rosen et al. 2006), and leprosy (Mira et al. 2004). In addition, Parkin modulates a wide range of biological functions in both non-neuronal and neuronal cells (Charan and LaVoie 2015).

Parkinson's disease (PD) is a neurodegenerative movement illness due to the death of dopamine-producing neurons in the substantia nigra pars compacta. Damaged mitochondria may play a crucial role in PD pathophysiology, according to studies correlating PD to abnormalities in the electron transport chain (Venderova and Park 2012). PINK1 (PTEN-induced putative kinase protein 1 or PARK6) and Parkin (PARK2), two recessive PD genes, have provided solid insight into the role of damaged mitochondria in PD pathophysiology (Valente et al. 2004). PINK1 is the only protein kinase reported to have a mitochondrial targeting domain, while Parkin is a cytosolic E3 ubiquitin ligase. The two proteins are implicated in a similar pathway that promotes selective autophagy (mitophagy) of depolarized mitochondria and regulates mitochondrial quality control (Narendra et al. 2012).

Table 12.1 List of synaptic proteins which interact with Parkin

Binding member for Parkin	Impact on physiological processes	References
CDCrel-1/SEPT5_v1	Control of neurotransmitter release	Zhang et al. (2000)
Synphilin-1	Regulation of ubiquitination of α-synuclein	Chung et al. (2001)
Synaptotagmin XI	Docking and vesicle budding	Huynh et al. (2003)
GluK2	Regulation of kainate receptor currents	Maraschi et al. (2014)
DAT	Regulation of reuptake of dopamine	Jiang et al. (2004)

Parkin appears to play a role in cytoskeletal integrity, cell survival, and cell mitosis, among others (Moore 2006).

Patients suffering from NDDs like intellectual disability (ID), developmental delay (DD), autism spectrum disorder (ASD), and attention-deficit/hyperactivity disorder (ADHD) exhibit structural genetic changes in PARK2, known as copy number variants (CNVs) (Glessner et al. 2009; Jarick et al. 2014; Scheuerle and Wilson 2011; Mariani et al. 2013; Roberts et al. 2014). Mutations in the PARK2 gene reported in patients with autosomal recessive juvenile Parkinsonism (ARJP) might result in dysregulation of dopaminergic and glutamatergic synapses, leading to dopaminergic neuronal malfunction and death (Sassone et al. 2017).

The CDCrel-1 turnover, a protein that interacts with synaptic vesicles and governs their dynamics, is regulated by wild-type Parkin by interacting with ubiquitinate. Parkin mutations raise the amount of CDCrel-1, preventing neurotransmitter release (Zhang et al. 2000). Synphilin-1, a synaptic vesicle-binding protein whose physiological role is unknown, is likewise ubiquitinated by Parkin (Chung et al. 2001) and synaptotagmin XI, a presynaptic protein engaged in synaptic vesicle production and docking interactions to this protein (Huynh et al. 2003). Parkin has been shown to interact with proteins involved in synaptic vesicle release, implying that presynaptic Parkin might control dopamine release. Parkin associates to and ubiquitinates dopamine transporter (DAT), raising the DAT expression on the plasma membrane and promoting dopamine absorption (Jiang et al. 2004) (Table 12.1).

12.3 Contactin-Associated Protein-Like 2 (CNTNAP2)

The molecular pathways that govern central glutamatergic synapses are arising as common substrates in the etiology of mental diseases. In mice, Contactin-associated protein-like 2 (CNTNAP2), which is encoded by CNTNAP2, is critical for dendritic spine formation and produces disease-related abnormalities in its absence. Exon deletions, copy number variations, single-nucleotide variants, truncations, and polymorphisms in the CNTNAP2 gene have been linked to epilepsy, language difficulties, intellectual property, autism, and schizophrenia (Varea et al. 2015).

Table 12.2 Genes that have been associated with attention-deficit/hyperactivity disorder (ADHD) and autism spectrum disorder (ASD)

Name of the genes	Associated disorder	Function of the gene	Phenotypical characteristics
CNTNAP2	ASD	Neuron-glia adhesion	Alteration of the dopaminergic system
ADGRL3	ADHD	Transcellular adhesion	Dysfunction of the dopaminergic system
PARK2	ASD ADHD	Mitochondrial quality control, E3 Ub protein ligase	Mitochondrial dysfunction

CNTNAP2 belongs to the neurexin family and is made of a 24-exon transcript that codes for the CASPR2 protein, which is involved in various neuronal activities such as dendritic arborization, neuronal migration, and synaptic transmission. Neurexins are cell adhesion proteins that play an essential role in synapse generation and synaptic property modulation. CASPR2 is responsible for the clustering of voltage-gated potassium channels and conduction of axon potentials at the juxtaparanodes in myelinated axons of both the spinal cord and the central nervous system (Varea et al. 2015; Flaherty et al. 2017). The strong expression of the protein in the Broca's area and other perisylvian regions is consistent with its novel role in social communication and normal language development (Bakkaloglu et al. 2008; Abrahams et al. 2007). In human neurodevelopmental impairments like autism, epilepsy, and intellectual disability, mutations in the CNTNAP2 gene coding CASPR2 have been reported. CASPR2, on the other hand, has been demonstrated to have a role in the localization of the voltage-gated potassium channel complex (VGKC), which comprises TAG-1, Kv1.1, and Kv1.2. This complex was identified in the node of Ranvier, the axon beginning segment, and the synapse, all of which are important for action potential propagation (Saint-Martin et al. 2018).

Axonal development was hindered, and synaptic abnormalities were identified in CNTNAP2 deletion neurons, suggesting that these factors may play a role in autism (Canali et al. 2018). Furthermore, mice with CNTNAP2 deletion exhibited stereotypic tendencies and communicative and social abnormalities, which are the main signs and symptoms of autism (Brumback et al. 2018; Scott et al. 2017). Thus, CNTNAP2 was deemed to be one of the most high-risk genes for ASD. The gene CNTNAP2 was one of the first to be linked to autism and epilepsy in Amish children (Strauss et al. 2006). Reduced presynaptic gamma-aminobutyric acid (GABA) and enhanced dopamine release in Cntnap4 knockout mice have been associated with severe, highly penetrant, recurring, and perseverative movements observed in human autism spectrum disorder patients (Li et al. 2018). Table 12.2 shows the overview of the genes that are associated strongly with ADHD and ASD.

12.4 Conclusions

ADGRL3 has putative roles in neuronal migration and synapse function. Various polymorphisms in ADGRL3 have been linked to an increased risk of attention-deficit/hyperactivity disorder (ADHD) in human studies. Impaired functioning of CNTNAP2 causes autism-related alterations in social interactions, stereotypic behavior, and sensory processing. Here, the authors have revealed present evidences for the contributions of ADGRL3, PARK2, and CNTNAP2 in NDDs such as ASD, Parkinson's diseases, and ADHA. PARK2 might be a pathological factor for NDDs. Essential functions of the above mentioned genes associated with NDDs might be important in the clinical disease presentation, and they act as suitable targets for therapeutic intervention.

References

Abrahams BS, Tentler D, Perederiy JV, Oldham MC, Coppola G, Geschwind DH (2007) Genome-wide analyses of human perisylvian cerebral cortical patterning. Proc Natl Acad Sci U S A 104(45):17849–17854

Acosta MT, Vélez JI, Bustamante ML, Balog JZ, Arcos-Burgos M, Muenke M (2011) A two-locus genetic interaction between LPHN3 and 11q predicts ADHD severity and long-term outcome. Transl Psychiatry 1(7):e17

Acosta MT, Swanson J, Stehli A, Molina BSG, MTA Team, Martinez AF et al (2016) ADGRL3 (LPHN3) variants are associated with a refined phenotype of ADHD in the MTA study. Mol Genet Genomic Med 4(5):540–547

Arcos-Burgos M, Jain M, Acosta MT, Shively S, Stanescu H, Wallis D et al (2010) A common variant of the latrophilin 3 gene, LPHN3, confers susceptibility to ADHD and predicts effectiveness of stimulant medication. Mol Psychiatry 15(11):1053–1066

Bakkaloglu B, O'Roak BJ, Louvi A, Gupta AR, Abelson JF, Morgan TM et al (2008) Molecular cytogenetic analysis and resequencing of contactin associated protein-like 2 in autism spectrum disorders. Am J Hum Genet 82(1):165–173

Brumback AC, Ellwood IT, Kjaerby C, Iafrati J, Robinson S, Lee AT et al (2018) Identifying specific prefrontal neurons that contribute to autism-associated abnormalities in physiology and social behavior. Mol Psychiatry 23(10):2078–2089

Bruxel EM, Salatino-Oliveira A, Akutagava-Martins GC, Tovo-Rodrigues L, Genro JP, Zeni CP et al (2015) LPHN3 and attention-deficit/hyperactivity disorder: a susceptibility and pharmacogenetic study. Genes Brain Behav 14(5):419–427

Bruxel EM, Moreira-Maia CR, Akutagava-Martins GC, Quinn TP, Klein M, Franke B et al (2021) Meta-analysis and systematic review of ADGRL3 (LPHN3) polymorphisms in ADHD susceptibility. Mol Psychiatry 26:2277–2285

Burns MP, Zhang L, Rebeck GW, Querfurth HW, Moussa CEH (2009) Parkin promotes intracellular Abeta1-42 clearance. Hum Mol Genet 18(17):3206–3216

Canali G, Garcia M, Hivert B, Pinatel D, Goullancourt A, Oguievetskaia K et al (2018) Genetic variants in autism-related CNTNAP2 impair axonal growth of cortical neurons. Hum Mol Genet 27(11):1941–1954

Charan RA, LaVoie MJ (2015) Pathologic and therapeutic implications for the cell biology of Parkin. Mol Cell Neurosci 66(Pt A):62–71

Chung KK, Zhang Y, Lim KL, Tanaka Y, Huang H, Gao J et al (2001) Parkin ubiquitinates the alpha-synuclein-interacting protein, synphilin-1: implications for Lewy-body formation in Parkinson disease. Nat Med 7(10):1144–1150

Conceição IC, Rama MM, Oliveira B, Café C, Almeida J, Mouga S et al (2017) Definition of a putative pathological region in PARK2 associated with autism spectrum disorder through in silico analysis of its functional structure. Psychiatr Genet 27(2):54–61
Flaherty E, Deranieh RM, Artimovich E, Lee IS, Siegel AJ, Levy DL et al (2017) Patient-derived hiPSC neurons with heterozygous CNTNAP2 deletions display altered neuronal gene expression and network activity. NPJ Schizophr 3(1):35
Folts CJ, Giera S, Li T, Piao X (2019) Adhesion G protein-coupled receptors as drug targets for neurological diseases. Trends Pharmacol Sci 40(4):278–293
Glessner JT, Wang K, Cai G, Korvatska O, Kim CE, Wood S et al (2009) Autism genome-wide copy number variation reveals ubiquitin and neuronal genes. Nature 459(7246):569–573
Hu HH, Kannengiesser C, Lesage S, André J, Mourah S, Michel L et al (2016) PARKIN inactivation links Parkinson's disease to melanoma. J Natl Cancer Inst 108(3)
Huynh DP, Scoles DR, Nguyen D, Pulst SM (2003) The autosomal recessive juvenile Parkinson disease gene product, Parkin, interacts with and ubiquitinates synaptotagmin XI. Hum Mol Genet 12(20):2587–2597
Hwang IW, Lim MH, Kwon HJ, Jin HJ (2015) Association of LPHN3 rs6551665 A/G polymorphism with attention deficit and hyperactivity disorder in Korean children. Gene 566(1):68–73
Jackson VA, del Toro D, Carrasquero M, Roversi P, Harlos K, Klein R et al (2015) Structural basis of latrophilin-FLRT interaction. Structure (London, England: 1993) 23(4):774–781
Jarick I, Volckmar AL, Pütter C, Pechlivanis S, Nguyen TT, Dauvermann MR et al (2014) Genome-wide analysis of rare copy number variations reveals PARK2 as a candidate gene for attention-deficit/hyperactivity disorder. Mol Psychiatry 19(1):115–121
Jiang H, Jiang Q, Feng J (2004) Parkin increases dopamine uptake by enhancing the cell surface expression of dopamine transporter. J Biol Chem 279(52):54380–54386
Kappel DB, Schuch JB, Rovaris DL, da Silva BS, Cupertino RB, Winkler C et al (2017) Further replication of the synergistic interaction between LPHN3 and the NTAD gene cluster on ADHD and its clinical course throughout adulthood. Prog Neuropsychopharmacol Biol Psychiatry 79 (Pt B):120–127
Kitada T, Asakawa S, Hattori N, Matsumine H, Yamamura Y, Minoshima S et al (1998) Mutations in the parkin gene cause autosomal recessive juvenile parkinsonism. Nature 392(6676):605–608
Lange M, Norton W, Coolen M, Chaminade M, Merker S, Proft F et al (2012) The ADHD-susceptibility gene lphn3.1 modulates dopaminergic neuron formation and locomotor activity during zebrafish development. Mol Psychiatry 17(9):946–954
Li C, Zheng Z, Ha P, Chen X, Jiang W, Sun S et al (2018) Neurexin superfamily cell membrane receptor contactin-associated protein like-4 (Cntnap4) is involved in neural EGFL-like 1 (Nell-1)-responsive osteogenesis. J Bone Miner Res 33(10):1813–1825
Maraschi A, Ciammola A, Folci A, Sassone F, Ronzitti G, Cappelletti G et al (2014) Parkin regulates kainate receptors by interacting with the GluK2 subunit. Nat Commun 5(1):5182
Mariani M, Crosti F, Redaelli S, Fossati C, Piras R, Biondi A et al (2013) Partial duplication of the PARK2 gene in a child with developmental delay and her normal mother: a second report. Am J Med Genet B Neuropsychiatr Genet 162B(5):485–486
Martinez AF, Abe Y, Hong S, Molyneux K, Yarnell D, Löhr H et al (2016) An ultraconserved brain-specific enhancer within ADGRL3 (LPHN3) underpins attention-deficit/hyperactivity disorder susceptibility. Biol Psychiatry 80(12):943–954
Mira MT, Alcaïs A, Nguyen VT, Moraes MO, Di Flumeri C, Vu HT et al (2004) Susceptibility to leprosy is associated with PARK2 and PACRG. Nature 427(6975):636–640
Moore DJ (2006) Parkin: a multifaceted ubiquitin ligase. Biochem Soc Trans 34(Pt 5):749–753
Narendra D, Walker JE, Youle R (2012) Mitochondrial quality control mediated by PINK1 and Parkin: links to parkinsonism. Cold Spring Harb Perspect Biol 4(11):a011338
O'Sullivan ML, Martini F, von Daake S, Comoletti D, Ghosh A (2014) LPHN3, a presynaptic adhesion-GPCR implicated in ADHD, regulates the strength of neocortical layer 2/3 synaptic input to layer 5. Neural Dev 9:7

Orsini CA, Setlow B, DeJesus M, Galaviz S, Loesch K, Ioerger T et al (2016) Behavioral and transcriptomic profiling of mice null for Lphn3, a gene implicated in ADHD and addiction. Mol Genet Genomic Med 4(3):322–343

Piquereau J, Godin R, Deschênes S, Bessi VL, Mofarrahi M, Hussain SN et al (2013) Protective role of PARK2/Parkin in sepsis-induced cardiac contractile and mitochondrial dysfunction. Autophagy 9(11):1837–1851

Puentes-Rozo PJ, Acosta-López JE, Cervantes-Henríquez ML, Martínez-Banfi ML, Mejia-Segura-E, Sánchez-Rojas M et al (2019) Genetic variation underpinning ADHD risk in a Caribbean community. Cells 8(8):907

Ribasés M, Ramos-Quiroga JA, Sánchez-Mora C, Bosch R, Richarte V, Palomar G et al (2011) Contribution of LPHN3 to the genetic susceptibility to ADHD in adulthood: a replication study. Genes Brain Behav 10(2):149–157

Roberts JL, Hovanes K, Dasouki M, Manzardo AM, Butler MG (2014) Chromosomal microarray analysis of consecutive individuals with autism spectrum disorders or learning disability presenting for genetic services. Gene 535(1):70–78

Rosen KM, Veereshwarayya V, Moussa CE, Fu Q, Goldberg MS, Schlossmacher MG et al (2006) Parkin protects against mitochondrial toxins and beta-amyloid accumulation in skeletal muscle cells. J Biol Chem 281(18):12809–12816

Saint-Martin M, Joubert B, Pellier-Monnin V, Pascual O, Noraz N, Honnorat J (2018) Contactin-associated protein-like 2, a protein of the neurexin family involved in several human diseases. Eur J Neurosci 48(3):1906–1923

Sassone J, Serratto G, Valtorta F, Silani V, Passafaro M, Ciammola A (2017) The synaptic function of Parkin. Brain 140(9):2265–2272

Scheuerle A, Wilson K (2011) PARK2 copy number aberrations in two children presenting with autism spectrum disorder: further support of an association and possible evidence for a new microdeletion/microduplication syndrome. Am J Med Genet B Neuropsychiatr Genet 156B(4): 413–420

Scott R, Sánchez-Aguilera A, van Elst K, Lim L, Dehorter N, Bae SE et al (2017) Loss of Cntnap2 causes axonal excitability deficits, developmental delay in cortical myelination, and abnormal stereotyped motor behavior. Cereb Cortex 29(2):586–597

Strauss K, Puffenberger E, Huentelman M, Gottlieb S, Dobrin S, Parod J et al (2006) Recessive symptomatic focal epilepsy and mutant contactin-associated protein-like 2. N Engl J Med 354: 1370–1377

Valente EM, Abou-Sleiman PM, Caputo V, Muqit MM, Harvey K, Gispert S et al (2004) Hereditary early-onset Parkinson's disease caused by mutations in PINK1. Science (New York, NY) 304(5674):1158–1160

van der Voet M, Harich B, Franke B, Schenck A (2016) ADHD-associated dopamine transporter, latrophilin and neurofibromin share a dopamine-related locomotor signature in Drosophila. Mol Psychiatry 21(4):565–573

Varea O, Martin-de-Saavedra MD, Kopeikina KJ, Schürmann B, Fleming HJ, Fawcett-Patel JM et al (2015) Synaptic abnormalities and cytoplasmic glutamate receptor aggregates in contactin associated protein-like 2/Caspr2 knockout neurons. Proc Natl Acad Sci U S A 112(19): 6176–6181

Venderova K, Park DS (2012) Programmed cell death in Parkinson's disease. Cold Spring Harb Perspect Med 2(8):a009365

Witte ME, Bol JG, Gerritsen WH, van der Valk P, Drukarch B, van Horssen J et al (2009) Parkinson's disease-associated Parkin colocalizes with Alzheimer's disease and multiple sclerosis brain lesions. Neurobiol Dis 36(3):445–452

Xu B, Roos JL, Levy S, van Rensburg EJ, Gogos JA, Karayiorgou M (2008) Strong association of de novo copy number mutations with sporadic schizophrenia. Nat Genet 40(7):880–885

Zhang Y, Gao J, Chung KK, Huang H, Dawson VL, Dawson TM (2000) Parkin functions as an E2-dependent ubiquitin- protein ligase and promotes the degradation of the synaptic vesicle-associated protein, CDCrel-1. Proc Natl Acad Sci U S A 97(24):13354–13359

Essential Role of nSR100 and CPEB4 Proteins During the Development of the Nervous System

13

GaddeVenkata Swarnalatha and Senthilkumar Rajagopal

Abstract

mRNA localization and transport are the gene regulation mechanisms of post-transcription. Activation and repression of protein synthesis are mediated by cytoplasmic polyadenylation element-binding (CPEB) proteins. CPEB proteins are important in the development of the nervous system. At the same time, the disruption of CPEB proteins leads to diseases like autism spectrum disorder and brain cancer. CPEB also provides targets for recovery from brain-related syndromes. In the nervous system of vertebrates, alternate splicing (AS) yields complexity in transcriptomes. A Ser/Arg repeat of neural-specific nSR100 has been reported to be important in neurite outgrowth, cortex formation, and axon formation in the corpus callosum. Altered AS, through microexons, is an emerging class for the regulation of different interactions of proteins in neuron development and some disorders related to neurons. Thus, both CPEB and nSR100 are essential for the proper development of the nervous system.

Keywords

Alternate splicing · CPEB · nSR100 · Autism spectrum disorder · Intersectin 1 · Microexons · Neuroguidin

G. Swarnalatha
Department of Biochemistry, Rayalaseema University, Kurnool, Andhra Pradesh, India

S. Rajagopal (✉)
Department of Biotechnology, REVA University, Bengaluru, Karnataka, India

M. W. Qoronfleh et al. (eds.), *Proteins Associated with Neurodevelopmental Disorders*, Nutritional Neurosciences, https://doi.org/10.1007/978-981-15-9781-7_13

Abbreviations

AS	Alternative splicing
ASD	Autism spectrum disorder
CPEB	Cytoplasmic polyadenylation element binding
CREB	cAMP-response element-binding protein
Itsn1	Intersectin 1
mRNA	Messenger RNA
NDUFV2	NADH dehydrogenase [ubiquinone] flavoprotein 2
Ngd	Neuroguidin
Orb2	The Drosophila protein encoded by the oo18 RNA-binding (orb) gene
PABP	Poly A-binding protein
PAP	Poly A polymerase
PARN	Poly A ribonuclease
PKA	Protein kinase A
Rbfox	RNA-binding fox-1 homolog 1
RNA	Ribonucleic acid
UTR	Untranslated region

13.1 Introduction

Coordinated regulations of genes at multiple layers of the brain are essential for the normal function and development of the nervous system. Maturation stages of neurons lead to progression into specific subtype formation during the neurogenesis development; perturbations of these cellular or molecular mechanisms affect brain development and may lead to neurodevelopmental problems/disorders (Gao et al. 2013). The exocrine and endocrine cues are used by the subventricular neurons of the cortex for establishing their cortical layers. The intrinsic and extrinsic factors also play an important role in the development of neuronal projections. The neurons attain their morphology and function due to subcellular processes, vesicular transport, and cytoskeleton remodeling. The diversity of transcriptomics and proteomics due to posttranscriptional regulation is prominent during the development of the nervous system (Norris and Calarco 2012; Zheng and Black 2013).

The production of multiple transcripts by choosing different splice sites from a single gene is called alternative splicing (AS). The cis-acting and trans-acting factors concert and cognate for productive spliceosomes (Chen and Manley 2009; Braunschweig et al. 2013). Enhancers and silencers are binding proteins of RNA that regulate the AS. Thus, AS is the main factor with respect to complexity evolution towards the function and development of the nervous system (Barbosa-Morais et al. 2012). Comparing to all other tissues, the AS patterns of the brain are complex and they are also involved in synaptic plasticity, a complex process of neurons (Lipscombe 2005; Ule and Darnell 2006). In the vertebrates, a differential splicing mechanism, namely alternative cassette exon inclusion, is found which is exclusive for brain (Barbosa-Morais et al. 2012; Merkin et al. 2012). Thus, there exists an AS conserved function along with species-specific neuronal behavior and developmental characters. Much information is not available on the protein factors that are actually important for the complexity of AS for the nervous system development. Nova, Rbfox, and Ptbp are the splicing regulators of neurons. Nova proteins are autoantigens that have control on inhibitory synapse, and knockout of these proteins results in defects of neuro-muscle junction and cortical migration (Yano et al. 2010; Ruggiu et al. 2009). Cerebellar development is disrupted in Rbfox1 and Rbfox2 mutant mice (Gehman et al. 2012). Few other neuronal genes have also been identified in vivo. One of the enriched proteins of neurons is the splicing factor nSR100. nSR100 is a 100 kDa protein with Ser/Arg repeats (Raj et al. 2011). As it is a splicing factor, it binds near the 3′ splice site at UGC enhancer elements. The decrease in nSR100 expression results in increased activity of neurons. Regulation experiments of nSR100 in cell culture reported around 50% of brain-specific inclusion patterns in the data of transcriptomic profiling (Raj et al. 2014). nSR100 knockdown impairs branching of trigeminal ganglia and also prevents the differentiation of cortical neuronal progenitors (Raj et al. 2011; Calarco et al. 2009).

One of the classes of cassette exons is microexons and they are located in the disordered intrinsic regions. They exhibit alternative versions of proteins. About one-third of the splicing events of neurons are regulated by microexons of the brain. In some diseases like autism, these are misregulated that further leads to decreased expression of nSR100.

The nervous system functioning depends on the neurons' ability to store, transmit, and perceive the information that is encoded in chemical and electrical signals. Stimulus to the response is due to alteration in RNA and protein distribution, and quality and number change in synaptic membrane receptors. The localized mRNA's translation is activated by polyadenylation, which is the reason for both the neuronal and intracellular brain architecture. Cytoplasmic polyadenylation elements: CPE at 3′ untranslated region are the critical regulatory elements for cytoplasmic polyadenylation. Cytoplasmic polyadenylation element-binding (CPEB) proteins are a family of proteins involved in protein synthesis activation and repression. The CPEB proteins mainly act by mediating the transcripts' polyadenylation and deadenylation in untranslated 3′ region. The CPEB proteins are complexes of ribonucleoproteins involved in the transportation of mRNA and subcellular localizations. The functions of CPEB proteins are similar in vertebrates and invertebrates as they are conserved proteins. The role of CPEB proteins is crucial in the case of neural development, memory, cell division, and learning process of developing the nervous system. Improper functioning of the CPEB proteins leads to several disorders like autism and brain cancer. On the other side, the regulation of CPEB genes has a positive effect on brain recovery functions in Huntington's disease and fragile X syndrome patients. Thus, CPEB genes act as important targets towards gene therapy.

13.2 Regulation Through CPEB Proteins

In the oocytes of *Xenopus laevis*, the CPE-mediated mRNA regulation was described initially. In CPE of 3′UTRs, the other important factors like symplekin, maskin, poly (A) ribonuclease (PARN), and polymerase (PAP) and binding protein (PABP) are present and regulate the 3′UTR region. In the RNA complex, bound PAPB decreases when the poly (A) is reduced by the prevailing activity of PARN over PAP (Kim and Richter 2006). The assembly of the translation initiation complex is prevented due to the binding of maskin to eIF4E. For promoting translation, an activation signal is required. Phosphorylation of CPEB proteins changes the RNA complex composition. PAP activity increases when PARN is dissociated, leading to poly (A) tail elongation and increased PAPB binding to mRNA. Hence, maskin (a CPEB-associated factor that interacts with eIF4E) affinity decreases to eIF4E, allowing translation. There may be variations in different cell types and organisms with respect to proteins like neuroguidin (Ngd) which is present instead of maskin in Drosophila neurons (Udagawa et al. 2012).

13.3 Properties of CPEB Proteins

There are two subfamilies for CPEB proteins; one is Drosophila Orb protein, which belongs to CPEB1 subfamily and the other one is CPEB1 of mammals, including humans. Oogenesis and embryonic development are the two translation regulation stages for these proteins. Recent findings state that they are also found in synapses and are further involved in the functioning and development of the nervous system. Orb 2 and $CPEB_{2–4}$ belong to $CPEB_2$ subfamily, which also regulates the translation with varied mechanisms (Pai et al. 2013). Postsynaptic localized density protein is CPEB3, whose expression site is in the brain, and the function is towards memory and learning (Kozlov et al. 2021). The memory function of CPEB proteins is conserved in phylogenetically distant species throughout the evolution. Around 40% of translational control in mammalian, *Xenopus*, and human mRNAs is through CPE mediation (Piqué et al. 2008). The subfamilies of CPEB proteins have multiple transcript targets, which might not be identical but overlap each other (Stepien et al. 2016). The CPEB function efficiency depends on the CPE position on mRNA at 3′UTR.

The CPEB proteins have two domains: one is the RNA-binding domain of RRM type and the other is zinc finger domain. At N-terminus, polyglutamine- or polyalanine-rich domains are present in some of the CPEB family because of which these are also known as prion-like proteins. The important function of prions is to show varied functions with stable confirmations (Shorter and Lindquist 2005). They also offer resistance to proteases and chemical agents.

13.4 CPEB Proteins in the Development of the Nervous System

The monomeric forms of CPEB proteins are mainly involved in neurogenesis, and they show the functions as a part of RNP complex. These proteins transport the mRNA that is present locally. Dendrite-RNP complex is regulated by $CPEB_1$ protein. The $CPEB_1$ protein interacts with 11 transcripts involved in long-term potentiation, memory formation, and synapse morphogenesis. Research states that the $CPEB_1$ binds to mRNAs of β-catenin, which leads to neuronal growth cone localization (Ohashi and Shiina 2020).

CPEBs play an essential role in the neuroblastic asymmetric division in Drosophila through mRNA localization. Mutant CPEB protein embryos show disrupted asymmetric division in neuroblasts. There is a disturbance in synapse and a neuromuscular connection of the central nervous system by Brat mislocalization due to the deletion of CPEB proteins (Santana and Casas-Tintó 2017). $CPEB_1$ proteins increase the level of translation along with β-catenin localization. The NDUFV2 mRNA is also regulated by mRNA; as a result, the brain ATP level is reduced which in turn leads to impaired growth and branching of dendrites in knockout $CPEB_1$ mice (Oruganty-Das et al. 2012).

13.5 CPEB Role in Memory

The main phenomenon for memory and learning is the plasticity of the synapse. Information storage is key for short-term and long-term memory. Long-term memory is only regulated at the molecular level. Synaptic plasticity is classified in terms of depression and potentiation. More information on memory formation is studied from *Aplysia,* a sea slug that has a simple organization, and *Drosophila.* cAMP, protein kinase A (PKA), CREB1, CREB2, MAPK (Mek1/2 in mice), and CPEB play a major role in the nervous system of *Aplysia* (Kandel et al. 2014). Short-term synaptic transmission is mediated by serotonin in Aplysia, and five serotonin applications to sensory neurons result in long-term transmission. Repetitive stimulus forms mRNA that is useful for long-term memory. The mRNAs are transported to synapses through the neuronal cell body. The synapses of long-term memory undergo structural and molecular changes (Sudhakaran and Ramaswami 2017). The main problem for maintaining long-term memory is how the memory stays for long if the protein decays. To explain the above scenario, the prion theory came into existence where it explained the nontoxic conformation of prions. As discussed earlier, CPEB proteins, which are prion-like proteins, are potentially implicated. mRNA transport, regulation of synaptic tagging, and regulation of mRNA translation are the main functions of CPEB proteins (Tompa and Friedrich 1998).

A silent mRNA is transported by CPEB along with kinesin and dynein proteins of microtubules. After the stimulation of synapse, various enzymes activate the CPEBs, and they act as translational repressors. Phosphorylation of CPEB activates them, and then induces mRNA polyadenylation and protein synthesis finally. $CPEB_4$ is present in the nuclei of neurons and protects it during the conditions of stress (Kozlov et al. 2021).

13.6 CPEB Proteins in Disease

A polygenic disease, namely autism spectrum disorder (ASD), is correlated with impaired mRNA splicing of $CPEB_4$. The condition is characterized by disability in the development and inability to maintain social interaction and repetitive behaviors. Though the levels of $CPEB_4$ mRNA are normal in ASD, its protein is decreased in the brain cells which is due to the splicing variant with decreased fourth microexon. The 24-nucleotide microexon has been placed in $CPEB_4$ mRNAs that shorten the poly A tail and finally reduce the translation which in turn results in an impaired ratio of $CPEB_4$ splicing variants that leads to decreased expression. Further, the misregulation targets other genes of ASD (Parras et al. 2018).

In cell differentiation and proliferation, the CPEB proteins are involved. So CPEB proteins play a role in tumor development and form a potential target for gene therapy. Brain gliomas are influenced by $CPEB_1$ and $CPEB_4$ gene expression. $CPEB_1$ gene expression is reduced in glioblastoma multiform brain tumor (Yin et al. 2014). In FMR1 knockout mice with fragile X syndrome, the pathological processes of the syndrome are reduced due to the mutations of the CPEB1 gene.

13.7 nSR100 Role in the Development of the Nervous System

nSR100 is a 100 kDa protein with neural-specific Ser/Arg. Cell culture experiments reported that around 30–50% of nSR100 shows the transcriptome profiling inclusive patterns of the brain (Raj et al. 2011). Neural outgrowth and ganglia branching are impaired in nSR100 Neuro 2a cell knockdown of zebrafish (Calarco et al. 2009). In mice, the knockdown of in utero nSR100 prevented the cortical neuro-progenitor's differentiation. The hearing and balance effects of the inner-ear hair cells are defective in nSR100-mapped mouse mutation (Nakano et al. 2012). The neurodevelopmental phenotypes lead to altered phenotypes of some other knockout splicing factor in nSr100-deficient mice. The genes related to functions of neural development are present in nSR-regulated exons. These exons are mainly involved in the regulation of protein-protein interactions and present primarily in the disordered regions of surface-accessible proteins (Ellis et al. 2012). The inclusion of microexons of 3–27 nt is promoted by nSR100 (Irimia et al. 2014). These microexons are concentrated in the domains of protein-protein or protein-lipid interactions. The interactions promote the increased expression of nSR100 (Irimia et al. 2014). The proper functioning and maturation of the neurons depend on the protein interaction regulations by nSR100. Studies on the deletion of an exon Srrm4 of nSR100 in mice resulted in the loss of protein. The postnatal survival of mutant animals requires nSR100. The defective layering of the cortex, diaphragm neurite outgrowth impairment, and impaired crossing of callosal axons are the effects of nSR100 loss in mice. There are several microexons and alternative cassette exons that are regulated by neural-enriched nSR100 than other AS events. The exons play a crucial role in the neuronal maturation as they are conserved to form the proteins for maturation. The mouse neurite growth of primary neurons is promoted by Unc13b gene, which is an nSR-regulated 6 nt microexon. The mutant phenotype of $nSR100^{\Delta 7-8/\Delta 7-8}$ mice is rescued from the effect of neuritogenesis by Unc 13b transcript expression with microexon.

nSR100 deficit might lead to phenotype alteration of other knockout splicing factors. Ptbp2, which is expressed in neurons and skeletal and cardiac muscle, is a splicing regulator which on loss causes neonatal lethality. The mice are paralyzed during birth with Ptbp2 lacking (Licatalosi et al. 2012; Li et al. 2014). The expression of Ptbp2 is promoted by nSR100 where the Ptbp2 nonsense transcript decay is prevented by alternative exon inclusion. nS100 knockout mice show a neurodevelopmental aberration with axon midline crossing defect in the corpus callosum (Paul et al. 2007; Donahoo and Richards 2009). Alternative Slit2 exon regulation also depends on nSR100, where axon controlling complementary mechanism is observed. nSR100 loss shows effects on the neuron distribution and axon growth. In early brain development, nSR loss occurs and the neural progenitors fail to accomplish asymmetric division. Thus, premature postmitotic fate is formed. This states that nSR100 is an important regulator for the timing of neurogenesis. There are several proteins that interact and are involved in functions like vesicle recycling and trafficking, namely Itsn1, Ppfia2, Rims2, Dnm2, Nbea, Abi1, Ptprd, and Vav2. These are all regulated by microexons. Out of the 72 nSR100-activated microexons, 65 exons are frame preserving and can insert around nine amino acids to corresponding proteins.

13.8 nSR in Pathology

About 76% of microexons affected by the loss of nSR100 are affected in vivo and they are conserved in humans. ASD is subjected to 46% loss of these exons in brain cortices (Irimia et al. 2014). ASD results from reduced nSR expression levels. Epilepsy and schizophrenia are also linked with the misregulation of microexons (Ovadia and Shifman 2011; Rusconi et al. 2015).

13.9 nSR100 and CPEB in Autism

Risk genetic variants of ASD are the genetic contributors in common that have minimal effect size. In idiopathic ASD, the environmental factors perturb the development of neurons. It is also necessary to investigate the risk gene regulators in ASD during the process of neurodevelopment. Low nSR100 levels of mutant mice show important autism features including altered neuronal and synaptic transmission activity, hypersensitivity, and socialization deficits. The autistic phenotypes are due to disruption of alternative splicing regulatory network. The above condition is due to few molecular mechanisms that increase the activity of neurons where it finally disrupts the network in human brains with autism. In the neurons of nSR100, mutants with microexon splicing reflect those of primary neurons at an active state. Hence with respect to neuronal activity, there is a decrease in nSR100 levels. Thus, a change in the activity of the brain is noticed in ASD. It is clear from this that, in healthy neurons, the microexon splicing program is activated by the expression of nSR100. Misregulated behavior is found in autistic brains due to the loss of nSR100 protein expression. Experiments on mice with reduced nSR100 levels genetically generate the wild-type neurons with splicing of microexons and cause autistic-like behavior (Quesnel-Vallières et al. 2016). The neurodevelopmental function is normal with proper nSR100 activity-dependent dynamic modulation (Irimia et al. 2014). In the case of ASD, transcriptomic profiling disrupts a common target. Therefore, there is a linkage between genetic alterations and neurological disorders. In epileptic seizures or stalled GABAergic neuron maturation, depolarization of neurons occurs, and this results in reduced nSR100 that exhibits ASD-related deficits in neurodevelopment functions.

As discussed earlier, the CPEB proteins of 1–4 are involved in the development of embryo and the plasticity of the synapse by regulating and modulating their mRNAs and their poly (A) tails. The CPEB4 proteins bind to ASD risk gene transcripts. In the brains of ASD patients, the isoform transcripts of CPEB4 show imbalances due to the reduced specific neuron microexon. Also, it causes 9% reduction in the poly (A) tail length. In the case of high-confidence ASD risk genes, that percentage is further increased, corresponding to reduced ASD risk gene expression and their protein products. ASD-like electrophysiological, behavioral, and neuroanatomical phenotypes are induced due to changes in the expression and polyadenylation of ASD risk genes in mice with an imbalance in the CPEB4 transcript isoforms (Parras et al. 2018). Thus, it clearly states that an essential regulator for ASD risk genes is CPEB4.

13.10 Conclusions

CPEB proteins are essential in ontogeny. They maintain the target mRNA's maintenance, formation, transport, localization, and cell polarity. They are also involved in target mRNA's activation or translational repression. CPEB proteins are majorly associated with neurogenesis and neuron functions. Long-term memory of the brain is due to prion-like stable conformation of these proteins. In the pathology of the nervous system also, the CPEB proteins are involved. Disturbances of CPEB proteins cause few other pathological processes like tumor invasion, carcinogenesis, and angiogenesis. Thus, CPEB proteins also have interactions with the oncological process. Further, CPEB proteins play a role in metabolic and liver diseases. Similarly, nSR100 also plays an important role in the development of the nervous system. Further investigations on CPEB and nSR100 are important to understand the various molecular mechanisms that are essential for the proper functioning of the nervous system and its correlation to other parts of the body and eventually find ways to treat various diseases.

References

Barbosa-Morais NL et al (2012) The evolutionary landscape of alternative splicing in vertebrate species. Science 338(6114):1587–1593

Braunschweig U et al (2013) Dynamic integration of splicing within gene regulatory pathways. Cell 152(6):1252–1269

Calarco JA et al (2009) Regulation of vertebrate the nervous system alternative splicing and development by an SR-related protein. Cell 138(5):898–810

Chen M, Manley JL (2009) Mechanisms of alternative splicing regulation: insights from molecular and genomics approaches. Nat Rev Mol Cell Biol 10(11):741–754

Donahoo A-L, Richards L (2009) Understanding the mechanisms of callosal development through the use of transgenic mouse models. Semin Pediatr Neurol 16(3):127–142

Ellis JD et al (2012) Tissue-specific alternative splicing remodels protein–protein interaction networks. Mol Cell 46(6):884–892

Gao P et al (2013) Lineage-dependent circuit assembly in the neocortex. Development 140(13): 2645–2655

Gehman LT et al (2012) The splicing regulator Rbfox2 is required for both cerebellar development and mature motor function. Genes Dev 26(5):445–460

Irimia M et al (2014) A highly conserved program of neuronal microexons is misregulated in autistic brains. Cell 159(7):1511–1523

Kandel ER, Dudai Y, Mayford MR (2014) The molecular and systems biology of memory. Cell 157(1):163–186

Kim JH, Richter JD (2006) Opposing polymerase-deadenylase activities regulate cytoplasmic polyadenylation. Mol Cell 24(2):173–183

Kozlov E et al (2021) The role of CPEB family proteins in the nervous system function in the norm and pathology. Cell Biosci 11:64

Li Q et al (2014) The splicing regulator PTBP controls a program of embryonic splicing required for neuronal maturation. eLife 3:e01201

Licatalosi DD et al (2012) Ptbp2 represses adult-specific splicing to regulate the generation of neuronal precursors in the embryonic brain. Genes Dev 26(14):1626–1642

Lipscombe D (2005) Neuronal proteins custom designed by alternative splicing. Curr Opin Neurobiol 15(3):358–363

Merkin J et al (2012) Evolutionary dynamics of gene and isoform regulation in mammalian tissues. Science 338(6114):1593–1599

Nakano Y et al (2012) A mutation in the Srrm4 gene causes alternative splicing defects and deafness in the Bronx waltzer mouse. PLoS Genet 8(10):e1002966

Norris AD, Calarco JA (2012) Emerging roles of alternative pre-mRNA splicing regulation in neuronal development and function. Front Neurosci 6:122

Ohashi R, Shiina N (2020) Cataloguing and selection of mRNAs localized to dendrites in neurons and regulated by RNA-binding proteins in RNA granules. Biomolecules 10(2):167

Oruganty-Das A et al (2012) Translational control of mitochondrial energy production mediates neuron morphogenesis. Cell Metab 16(6):789–800

Ovadia G, Shifman S (2011) The genetic variation of RELN expression in schizophrenia and bipolar disorder. PLoS One 6(5):e19955

Pai TP et al (2013) Drosophila Orb protein in two mushroom body output neurons is necessary for long-term memory formation. Proc Natl Acad Sci U S A 110(19):7898–7903

Parras A et al (2018) Autism-like phenotype and risk gene mRNA deadenylation by CPEB4 mis-splicing. Nature 560(7719):441–446

Paul LK et al (2007) Agenesis of the corpus callosum: genetic, developmental and functional aspects of connectivity. Nat Rev Neurosci 8(4):287–299

Piqué M et al (2008) A combinatorial code for CPE-mediated translational control. Cell 132(3): 434–448

Quesnel-Vallières M et al (2016) Misregulation of an activity-dependent splicing network as a common mechanism underlying autism spectrum disorders. Mol Cell 64(6):1023–1034

Raj B et al (2011) Cross-regulation between an alternative splicing activator and a transcription repressor controls neurogenesis. Mol Cell 43(5):843–850

Raj B et al (2014) A global regulatory mechanism for activating an exon network required for neurogenesis. Mol Cell 56(1):90–103

Ruggiu M et al (2009) Rescuing Z+ agrin splicing in Nova null mice restores synapse formation and unmasks a physiologic defect in motor neuron firing. Proc Natl Acad Sci 106(9):3513–3518

Rusconi F et al (2015) LSD1 neurospecific alternative splicing controls neuronal excitability in mouse models of epilepsy. Cereb Cortex 25(9):2729–2740

Santana E, Casas-Tintó S (2017) Orb2 as modulator of Brat and their role at the neuromuscular junction. J Neurogenet 31(14):181–188

Shorter J, Lindquist S (2005) Prions as adaptive conduits of memory and inheritance. Nat Rev Genet 6(6):432–450

Stepien BK et al (2016) RNA-binding profiles of Drosophila CPEB proteins Orb and Orb2. Proc Natl Acad Sci 113(45):E7030–E7038

Sudhakaran IP, Ramaswami M (2017) Long-term memory consolidation: the role of RNA-binding proteins with prion-like domains. RNA Biol 14(5):568–586

Tompa P, Friedrich P (1998) Prion proteins as memory molecules: a hypothesis. Neuroscience 86(4):1037–1043

Udagawa T et al (2012) Bidirectional control of mRNA translation and synaptic plasticity by the cytoplasmic polyadenylation complex. Mol Cell 47(2):253–266

Ule J, Darnell RB (2006) RNA binding proteins and the regulation of neuronal synaptic plasticity. Curr Opin Neurobiol 16(1):102–110

Yano M et al (2010) Nova2 regulates neuronal migration through an RNA switch in disabled-1 signaling. Neuron 66(6):848–858

Yin J et al (2014) CPEB1 modulates differentiation of glioma stem cells via downregulation of HES1 and SIRT1 expression. Oncotarget 5(16):6756–6769

Zheng S, Black DL (2013) Alternative pre-mRNA splicing in neurons: growing up and extending its reach. Trends Genet 29(8):442–448

Current Trends of Stem Cells in Neurodegenerative Diseases

14

Christos Tsagkaris, Dimitrios V. Moysidis, Andreas S. Papazoglou, Andleeb Khan, Stavros Papadakos, Anna Maria Louka, Dorothy Martha Scordilis, Anastasiia Shkodina, Kyriakoula Varmpompiti, Gaber El-Saber Batiha, and Athanasios Alexiou

C. Tsagkaris
Novel Global Community Educational Foundation, Hebersham, NSW, Australia

D. V. Moysidis · A. S. Papazoglou
Faculty of Medicine, Aristotle University of Thessaloniki, Thessaloniki, Greece

A. Khan
Department of Pharmacology and Toxicology, College of Pharmacy, Jazan University, Jazan, Saudi Arabia

S. Papadakos
Department of Pathology, University of Athens, Athens, Greece

A. M. Louka
Faculty of Medicine, University of Thessaly, Larissa, Greece

D. M. Scordilis
National Institutes of Health, Bethesda, MD, USA

A. Shkodina
Ukrainian Medical and Stomatological Academy, Poltava, Ukraine

K. Varmpompiti
Department of Neurosciences, Kings College Hospital, London, UK

G. E.-S. Batiha
Department of Pharmacology and Toxicology, Faculty of Veterinary Medicine, Damanhour University, Damanhour City, Egypt

A. Alexiou (✉)
Novel Global Community Educational Foundation, Hebersham, NSW, Australia

AFNP Med Austria, Wien, Austria
e-mail: alexiou@ngcef.net

M. W. Qoronfleh et al. (eds.), *Proteins Associated with Neurodevelopmental Disorders*, Nutritional Neurosciences, https://doi.org/10.1007/978-981-15-9781-7_14

Abstract

Currently, there is extensive interest in stem cell technology. While the management of neurodegenerative diseases is still an open problem for the patients and the public health systems, this book chapter aims to provide information on neurodegeneration and their possible treatment by stem cells. The process of extension of stem cell research into translational and clinical therapies is the main aim of this book chapter. The authors have explained various neurodegenerative diseases and their burden on society with an emphasis on the usage of stem cells. Different types of stem cells used in therapy from bench to bedside have also been discussed. Finally, a descriptive account of the current applications of stem cell therapy in Alzheimer's disease (AD), amyotrophic lateral sclerosis (ALS), Parkinson's disease (PD), Huntington's disease (HD), and autism spectrum disorder (ASD) has been given. The book chapter is distinct in its compilation to provide updated and recent advances of stem cell therapy for the treatment of neurodegenerative diseases.

Keywords

Alzheimer's disease · Amyotrophic lateral sclerosis · Autism spectrum disorder · Clinical trials · Huntington's disease · Neurodegeneration · Parkinson's disease · Stem cells

Abbreviations

AD	Alzheimer's disease
ALS	Amyotrophic lateral sclerosis
ASDs	Autism spectrum disorders
BBB	Blood-brain barrier
CT	Computed tomography
ESCs	Embryonic stem cells
HD	Huntington's disease
iPSCs	Induced pluripotent stem cells
MRI	Magnetic resonance imaging
MSCs	Mesenchymal stem cells
NSCs	Neural stem cells
PD	Parkinson's disease
PET	Positron-emission tomography
SMA	Smooth-muscle antibody

14.1 Introduction

Neurodegenerative disease is a general term for several diseases that mainly affect neurons in the human brain. Neurons are the building blocks of the nervous system, which includes the brain and spinal cord. Neurons usually do not spontaneously reproduce and do not replace themselves, and therefore when they are damaged or die, they cannot be replaced by the body. Acute neurodegeneration can result from an immediate attack, such as a stroke or injury, leading to a loss of neurons in the lesion area. Chronic neurodegeneration can develop over a long period and results in a generalized loss of neuronal populations (Gitler et al. 2017). Examples of neurodegenerative diseases include Parkinson's, Alzheimer's, Huntington's, and amyotrophic lateral sclerosis (*Neurodegenerative Diseases | MedlinePlus* n.d., accessed in 2021; *Amyotrophic Lateral Sclerosis (ALS) Fact Sheet | National Institute of Neurological Disorders and Stroke* n.d., accessed in 2021; Winner et al. 2011). Neurodegenerative diseases are incurable and debilitating conditions that lead to progressive degeneration and/or eventually death of nerve cells. Degenerative nerve diseases can be severe or life-threatening, depending upon their type. Most of them however have no cure.

Treatment can help improve and slow the progression of symptoms, relieve pain, and increase mobility. As neurons deteriorate, a person may first experience relatively mild symptoms like problems with coordination or remembering names. But as vast numbers of neurons continue to die, the symptoms gradually get worse. In some cases, patients lose the ability to walk, think clearly, and generally function in the world. After all, many of these diseases are fatal. Degenerative nerve diseases affect many of your body's activities, such as balance, movement disorders, speech, mental function, respiration, and heart function. Dementias are responsible for the greater incidence of the disease, with Alzheimer's accounting for about 60–70% of cases (Roberts and Knopman 2013; Erkkinen et al. 2018).

This book chapter focuses on different neurodegenerative diseases and their prevention by stem cell therapy. The chapter is an elaborative description of the clinical studies using stem cells as a line of treatment for different neurodegenerative diseases.

14.1.1 Alzheimer's Disease (AD)

Alzheimer's disease (AD) is a progressive, neurodegenerative disease of the brain that slowly erodes memory and thinking skills and eventually causes an inability to perform simple tasks. It is the most common cause of dementia, accounting for about 50–70% of all dementia cases. Alzheimer's disease is a chronic neurodegenerative disease of the central nervous system (CNS). It is characterized in its mild form by gradual loss of memory and limitation of the brain's other mental functions. It is pathologically characterized by the deposition of two pathological proteins in the brain: the amyloid-beta protein (Aβ) and the tau protein. A buildup of these proteins leads to CNS dysfunction and subsequent death of nerve cells. The disease slowly affects nerve cells in all areas of the cerebral cortex and some surrounding structures, making it difficult for a person to regulate emotions, recognize mistakes and patterns, coordinate movement, and remember (*Dementia Pathology: Dementia, Alzheimer Disease, Vascular Dementia* 2019; Mayeux and Stern 2012; Leyns and Holtzman 2017). The risk of developing AD increases exponentially with advanced age. It is a disease of the "elderly," although sporadic cases might occur under 65. The most significant risk for Alzheimer's disease is advanced age, with the risk continuing to increase as we age (Petersen et al. 2014; Edwards et al. 2019; Silva et al. 2019).

Although the main etiology of the disease is unknown, several risk factors are known to affect the development (Bekris et al. 2010). The condition can continue for many years without symptoms, referring to the preclinical or pre-symptomatic disease stage.

Genetic factors seem to play an important role. Simultaneously, environmental factors, including vascular disease (e.g., diabetes, hypertension, dyslipidemia) and lifestyle (e.g., education, occupation, mental, social activities, physical activity, diet), seem to influence the likelihood of developing the disease. Changes and lesions in the brain can take more than 20 years before symptoms develop (Ballard et al. 2011; Nazarko 2019). Eventually, a person with AD loses memory and many other mental functions. According to the World Health Organization, 10% of the population aged 65 years and over have AD, and over 14 million Americans will develop AD by 2060 (2020 Alzheimer's disease facts and figures 2020).

There is no etiological treatment for AD (Ballard et al. 2011; Alexiou et al. 2017). Drug therapy aims to slow the progression of the disease and treat the symptoms associated with the disease. The benefit of the drugs used to treat Alzheimer's disease is usually small. Therefore, patients and their families may not notice any benefit. Patients and their families should discuss with their physicians whether medication can help improve behavior or functional abilities. It should also be discussed whether medications should be prescribed early in the disease. Cholinesterase inhibitors and NMDA antagonists are often prescribed to treat Alzheimer's disease. The commonly used medicines include galantamine, donepezil, rivastigmine, and memantine (Giacobini and Gold 2013; Mendiola-Precoma et al. 2016).

14.1.2 Parkinson's Disease (PD)

Parkinson's disease (PD) is a progressive, degenerative disease of the central nervous system. As part of this degenerative process, some of the brain's nerve cells, primarily responsible for programming and coordination of movement, lose their functionality in adulthood (Winner et al. 2011; Erkkinen et al. 2018). This disturbance results in the gradual reduction of the individual's mobility. The programming and harmony in the execution of a movement are controlled by a complex system of structures in the brain (primary ganglia-cortex). The substance that plays a significant role in the communication of these structures is dopamine. This substance is produced by specialized nerve cells in an area at the brain's base, called a black substance. Over the years and under the influence of genetic and environmental factors, these cells' functions may degenerate, reducing dopamine production levels in the brain. Low dopamine levels and disruption of the primary ganglia and cerebral cortex connections lead to decreased harmony of movement (Erkkinen et al. 2018; Tarakad and Jankovic 2020). Movement is controlled by neurons in the motor cortex and the basal ganglia, along with pyramidal and other afferent neuronal tracts. The initiation of movement is mediated by a dopamine-dependent neural circuit involving the basal ganglia, the cortex, and the substantia nigra. In healthy people, these messages are transmitted smoothly. However, in patients with PD, the messages are blocked and not transmitted correctly due to a lack of dopamine (Marino et al. 2019). In people with PD, 70–80% of dopamine-producing cells have degenerated and died. The degeneration mainly occurs in a small area of the brain called the substantia nigra. When there is dopamine deficiency, nerve cells do not function properly and cannot transmit messages to the brain, resulting in PD symptoms. Although dopamine is the primary neurotransmitter affected, PD can be disrupted by other neurotransmitters as well. The above mechanism partly explains why simple dopamine replacement therapy does not produce the expected results (Tysnes and Storstein 2017; Balestrino and Schapira 2020).

Disorders of other neurotransmitters might also explain the numerous non-motor symptoms in PD. It is unclear as to why dopamine-producing cells deplete so quickly. Multiple factors are generally thought to be involved, and modern research focuses on aging, viruses, and genetic and environmental factors (Balestrino and Schapira 2020).

It is also not clear why certain people develop Parkinson's disease and not others. The causes of the disease are still unknown. Systematic research is currently undertaken by many teams worldwide to determine the cause and possible association of the disease with environmental and genetic factors (Tysnes and Storstein 2017).

14.1.3 Amyotrophic Lateral Sclerosis (ALS)

Amyotrophic lateral sclerosis (ALS) is a progressive neurodegenerative disease that affects nerve cells in the brain and the spinal cord. ALS is the most common motor neuron disease in adults (Pupillo et al. 2014). Motor neuron diseases cause selective loss of nerve cells that connect the brain to the muscles. ALS affects both the upper and lower motor neurons throughout the brain and spinal cord. Motor neurons travel from the brain to the spinal cord, and from there, the spinal cord transmits neuronal signals to the muscles throughout the body. The progressive degeneration of motor neurons in ALS eventually leads to their death. When motor neurons die, the brain's ability to move and control muscle movement is lost. With the muscles' voluntary action being gradually affected, patients in the advanced stages of the disease can become paralyzed completely (Martin et al. 2017; Masrori and Van Damme 2020). ALS most often affects people between the ages of 40 and 60; however, persons younger and older than the age range could also develop the disease. It is shown that men seem to become affected more often than women (Talbott et al. 2016).

The etiology of ALS is still unknown. An essential step in answering this question was made in 1993 when scientists from the National Institute of Neurological Disorders and Stroke (NINDS) discovered that mutations in the gene produced by the enzyme SOD1 were linked to some cases of familial ALS (Majoor-Krakauer et al. 2003). Although the mechanism behind the mutations in the SOD1 gene resulting in degeneration of motor neurons remains unclear, there is growing evidence that the mutated SOD1 protein might become toxic (Štětkářová and Ehler 2021).

14.1.4 Huntington's Disease (HD)

Huntington's disease (HD) is caused by neuronal degeneration at the basal ganglia, which is the area responsible for movement and coordination. The neural structures and circuits responsible for thought, perception, emotion, and memory are also affected, probably due to the connections of the basal ganglia with frontal lobes (Roos 2010; Erkkinen et al. 2018). HD is nowadays recognized as one of the most common genetic disorders. More than 1/4 of a million Americans suffer from HD or are at risk of inheriting the disease from an ailing parent. HD is characterized by substantial variability in its expression, even within the same families. The identification and localization of the responsible HTT gene facilitated the determination of the people who will develop the disease by DNA analysis. The documentation of family history is essential by means of genetic testing for HD (Nguyen and Weydt 2018). Brain imaging modalities, such as computed tomography (CT) and magnetic resonance imaging (MRI), can detect atrophy in specific areas of the brain, especially in the caudate nucleus and putamen. Other studies have yielded generalized atrophy of the brain (Negi et al. 2014; Fazio et al. 2018). A DNA mutation analysis can be clinically valuable to predict HD in a person at risk, set a prenatal diagnosis in high-risk pregnancy, or confirm the disease after being suspected by relative symptoms. The symptoms include physical abnormalities (e.g., involuntary movements, anxiety, loss of balance, strange walking, poor coordination, dysarthria), cognitive changes (e.g., memory loss, miscalculation, disorganization), and emotional and behavioral disorders (e.g., depression, apathy, paranoia, anger, withdrawal, anxiety) (Nguyen and Weydt 2018).

HD is an inherited, fatal degenerative brain disorder. No treatment is available yet; however, Xenazine and Deutetrabenazine have been approved as symptomatic treatments for HD-associated chorea (Bachoud-Lévi et al. 2019). HD gradually reduces the affected person's ability to walk and talk. Eventually, the patient becomes entirely dependent on others for their care. HD profoundly affects the lives of the entire family emotionally, socially, and financially. The early symptoms of HD can affect the person's cognitive ability or mobility and include depression, mood swings, memory loss, awkwardness, involuntary contractions, and lack of coordination. Later on, as the disease progresses, the patient's concentration and short-term memory decrease, while the head, torso, and involuntary movements of the limbs increase. Speech, walking ability, and swallowing coordination also continue to degenerate. Finally, the patients are not capable of taking care of themselves. Death occurs from complications such as drowning, infection, or heart failure (Mestre et al. 2009; Frank 2014).

HD becomes clinically apparent between 30 and 50 years, although it can begin from early childhood. Those children who develop the juvenile form of the disease rarely live to adulthood (Erkkinen et al. 2018). The prevalence of HD is similar between women and men and transcends all ethnic and racial boundaries. Every person who carries the HTT gene will develop the disease. Each child who suffers from HD has an inevitable 50% chance of fatally inheriting the gene. The HD gene was isolated in 1993. The genetic test, which was developed, can determine whether a person carries the HD gene accurately. The test, however, could not predict when the symptoms would start (Nopoulos 2016).

14.1.5 Autism Spectrum and Neurodevelopmental Disorders

Autism spectrum disorders (ASDs) consist of a heterogeneous group of neurodevelopmental disorders. The principal features are stereotypical behaviors and problematic social communication and reciprocal interaction (Sharma et al. 2018). ASDs have become more prevalent over the past two decades, although their increasing prevalence could be partly attributed to the elevated level of awareness among physicians, educators, and parents. In the United States, ASDs affect up to 1 in 88 children (1 in 54 males and 1 in 252 females) older than 8 years (*2020 Community Report on Autism* 2020). Globally, it is estimated that 1 in 160 children suffer from ASDs. Up to 75% of patients with ASDs have psychiatric comorbidities, including but not limited to depression, bipolar disorder, and attention-deficit–hyperactivity disorder (ADHD) (Antshel et al. 2013).

ASDs are multifactorial disorders in terms of etiology. Risk factors implicated in their pathogenesis include impaired immune responses, neuroinflammation, neurotransmission abnormalities, dysfunction of the mitochondria, oxidative stress, and environmental stressors/toxins. Associated genetic deficits and disorders include fragile X syndrome, tuberous sclerosis, epilepsy, and Down syndrome (Lacivita et al. 2017; Eissa et al. 2018).

The International Classification of Diseases (ICD-10) has reserved the F.84.0 code for ASDs, classifying them as "pervasive developmental disorders" (*F84.0—Autistic disorder | ICD-10-CM* n.d., accessed in 2021). The Diagnostic and Statistical Manual of Mental Disorders (DSM-5) of the American Psychiatric Association has set the major criteria for the diagnosis of ASDs: persistent deficits in social interaction, restricted stereotype behavioral patterns, symptoms with an onset in the early developmental age, considerable functional impairment of everyday activity, and failure to better interpret these symptoms using other intellectual deficits (*Autism Diagnosis Criteria: DSM-5 | Autism Speaks* n.d., accessed in 2021). Disorders of the autism spectrum include but are not limited to autistic disorder, Rett's and

Asperger's syndromes, disintegrative childhood disorder, and pervasive developmental disorder (PDD). Despite the common perception, individuals without significant intellectual disability (high-functioning autism) are usually capable of attending college, graduating, and living independently, although they typically struggle with social interaction. In many cases, the failure of individuals with autism to live independently or graduate is due to increased social demands and lack of adequate psychosocial support (Eissa et al. 2018; Sharma et al. 2018).

The treatment of ASDs is mainly symptomatic and includes various pharmacological and non-pharmacological modalities. Approved pharmacological agents such as serotonin reuptake inhibitors (fluoxetine), tricyclic antidepressants (imipramine), anticonvulsants (lamotrigine), atypical antipsychotics (clozapine), and inhibitors of acetylcholinesterase (rivastigmine) are used to mitigate the behavioral symptoms of ASDs (Leskovec et al. 2008; Accordino et al. 2016). Educational enhancement therapies with a focus on speech, language, auditory ability, and social integration are used in combination with various models of psychotherapy. Ongoing research for curative treatments focuses on anti-inflammatory agents, novel psychotropic agents, and food supplementation (Eissa et al. 2018).

14.2 Stem Cells

Stem cells constitute primate cells which are precursors of all the cells in the human body and, under suitable conditions, can be transformed into each human cell. Stem cells could potentially heal all organs that contain cells into which they can be transformed (Imamura and Inoue 2012).

14.2.1 Stem Cell Sources

The primary sources of stem cells are umbilical cord blood and tissue, placenta, deciduous teeth, adipose tissue, and bone marrow (Wislet-Gendebien et al. 2012a).

The **embryonic stem cells (ESC)**, as their name implies, are derived from embryos. They are developed from eggs that have been fertilized in vitro at fertilization clinics and then made available for research purposes with donors' consent (Girlovanu et al. 2015). Embryonic stem cells are not derived from fertilized eggs in a woman's body. Adult stem cells are thought to be undifferentiated cells found among differentiated cells in tissues or organs. Adult stem cells can be renewed and differentiated, while they can generate some of the most important cells of a tissue, too. An adult stem cell's primary role in a living organism is to maintain and repair the tissue in which it is located. Adult stem cells usually produce the cell types of the tissue in which they live. For example, the bone marrow's hematopoietic adult stem cells usually give rise to many types of blood cells (Goya et al. 2018; Zakrzewski et al. 2019).

Moreover, in recent years, experiments have shown that stem cells originating from one tissue could generate a completely different tissue type (Liu et al. 2020).

The issue remains an area of significant discussion within the research community. The above controversy underlines adult stem cell studies' challenges, and it seems that additional research is needed to understand their potential as future therapies entirely (Lim et al. 2013). In contrast with embryonic stem cells, which are defined by their origin, the origin of adult stem cells in specific mature tissues is still under investigation. Stem cells are distinguished from other cell types by two characteristics. First, they are undifferentiated cells capable of regenerating themselves through cell division (sometimes after long periods of inactivity). Second, under certain physiological or experimental conditions, they could be stimulated to generate specific organ or tissue cells with specific functions. They are characterized by the remarkable ability to evolve into many different cell types of the body during early life and growth. When a stem cell is divided, each new cell that occurs has the potential to either remain a stem cell or become another type of cell with a more specialized function, such as a muscle cell, a red blood cell, or a brain cell. Besides, stem cells can serve as a kind of internal repair system in many tissues, being continuously divided for the reconstitution of other cells, as long as the person or animal is still alive (Sivakumar et al. 2015; Liu and Deng 2016).

Another possible application of stem cells is their usage in medical treatments. Nowadays, there is an increasing demand for donated organs, which are needed to replace diseased or damaged organs. Unfortunately, the number of people who require a transplant far exceeds the number of organs available for transplantation (Sivakumar et al. 2015).

The **multipotent stem cells** are a renewable source of replacement cells or tissues to treat many diseases, conditions, and disabilities, including PD, spinal cord injury, burns, heart disease, stroke, and ALS (Hermann and Storch 2013). In some specific organs, such as the intestine and the bone marrow, stem cells undergo continuous division to repair and replace degenerated or damaged tissue. However, in other organs like the pancreas and the heart, stem cells divide only under specific conditions (Kurtzberg 2017).

The hematopoietic stem cells (HSC) are responsible for creating blood cells: red blood cells, B-lymphocytes, T-lymphocytes, neutrophils, basophils, eosinophils, monocytes, and macrophages (Lim et al. 2013).

The mesenchymal stem cells (MSC) can create various cell types: bone cells, cartilage cells, fat cells, and other types of connective tissue cells, such as those in tendons (Table 14.1).

Neural stem cells (NSC) can create three main cell types: nerve cells and two classes of non-neuronal cells, astrocytes, and oligodendrocytes, as presented in Table 14.2.

The epithelial stem cells differentiate into many types of cells, including but not limited to absorptive cells, goblet cells, and Paneth cells (Girlovanu et al. 2015).

The epidermal stem cells are located in the basal layer of the skin and at the base of hair follicles. Epidermal stem cells create keratinocytes, which eventually migrate towards the skin's surface to form a protective layer (and eventually shed off) (Lenkiewicz 2019) (Fig. 14.1).

Table 14.1 An overview of the literature related to mesenchymal stem cells in neurodegenerative disease research

Albani et al. (2013)	Literature review	Hydrogel-based nanocomposites and mesenchymal stem cells: a promising synergistic strategy for neurodegenerative disorder therapy
Chen et al. (2018)	Literature review	Mesenchymal stem cell-mediated immunomodulation in cell therapy of neurodegenerative diseases
Fričová et al. (2020)	Literature review	Challenges and translational considerations of mesenchymal stem/stromal cell therapy for Parkinson's disease
Huang et al. (2012)	Literature review	Mesenchymal stem cells as therapeutic agents and potential targeted gene delivery vehicles for brain diseases
Lo Furno et al. (2018)	Literature review	Functional role of mesenchymal stem cells in the treatment of chronic neurodegenerative diseases
Nery et al. (2013)	Classification, flow cytometry	Human mesenchymal stem cells: From immunophenotyping by flow cytometry to clinical applications
Ng (2014)	Literature review	Progress of mesenchymal stem cell therapy for neural and retinal diseases
Olson et al. (2012)	Literature review, develop a siRNA delivery system	Genetically engineered mesenchymal stem cells as a proposed therapeutic for Huntington's disease
Peng et al. (2013)	Literature review	Mesenchymal stem cells: a revolution in therapeutic strategies of age-related diseases
Shariati et al. (2020)	Literature review	Mesenchymal stromal cells (MSCs) for neurodegenerative disease: A promising frontier
Staff et al. (2019)	Literature review	Mesenchymal stromal cell therapies for neurodegenerative diseases

Recently, cell replacement therapy was proven to help relieve symptoms or even reverse the progression of neurological disorders, where neither pharmacological interventions nor other treatments were sufficient or available (Shin et al. 2011; Abdullah et al. 2012). Thus, various stem cells have been transplanted into the injured brain, including mesenchymal stem cells (MSCs), to release or stimulate the release of nutrients (Huang et al. 2012). MSCs are of utmost importance for treatment strategies because of the simplicity in their isolation process (Lescaudron et al. 2012). MSCs are defined by their surface marker expression pattern and can be easily extracted from the patient's bone marrow or adipose tissue. They can also be replanted into the same patient to repair injured or degenerated tissues (Kan et al. 2007).

It has been shown that the MSCs and genetically modified MSCs offer therapeutic benefits in brain diseases, such as strokes, neurodegenerative diseases, and brain stem gliomas (Tanna and Sachan 2014). They are considered a promising treatment for PD due to their neuro-rehabilitation properties and constitute a promising

Table 14.2 An overview of the regenerative medicine for neurodegenerative diseases and relevant research

Ahani-Nahayati et al. (2021)	Literature review	Stem cell in neurodegenerative disorders; an emerging strategy
De Filippis and Binda (2012)	Literature review	Concise review: Self-renewal in the central nervous system: Neural stem cells from an embryo to adult
Díaz (2019)	Literature review	Regenerative medicine: Could Parkinson's be the first neurodegenerative disease to be cured?
Harris et al. (2020)	Literature review	Emerging regenerative medicine and tissue engineering strategies for Parkinson's disease
Hermann and Storch (2013)	Literature review	Induced neural stem cells (iNSCs) in neurodegenerative diseases
Kim et al. (2013)	Cell culture experiments	Neural stem cell-based treatment for neurodegenerative diseases
Kittappa et al. (2012)	Literature review	The role of eNSCs in neurodegenerative disease
Relaño-Ginés et al. (2014)	Literature review	Prion diseases and adult neurogenesis: How do prions counteract the brain's endogenous repair machinery?
Struzyna et al. (2017)	Original research report	Anatomically Inspired Three-dimensional Micro-tissue Engineered Neural Networks for Nervous System Reconstruction, Modulation, and Modeling
Van Den Berge et al. (2013)	Literature review	Resident adult neural stem cells in Parkinson's disease—The brain's own repair system?
Wislet-Gendebien et al. (2012a)	Literature review	Adult bone marrow: Which stem cells for cellular therapy protocols in neurodegenerative disorders?
Ziemka-Nalecz (2012)	Literature review	Endogenous neurogenesis induced by ischemic brain injury or neurodegenerative diseases in adults

therapeutic tool for reducing Aβ deposits in patients with AD (Kim et al. 2013). The application on transferring agents provides benefits over other methods. Emerging evidence suggests that the implantation of MSCs in the corpus striatum might delay the loss of median neurons in HD.

MSCs have also been proven to be superior to NSCs, as their isolation and application are well established and could be received from various adult tissues (Lescaudron et al. 2012). In addition, they are highly interactive with their microenvironment and can share proteins, RNA, and even mitochondria with damaged tissue (Wislet-Gendebien et al. 2012b). Viral vectors have been used to construct MSCs, which overexpress cell receptors on their cell surface. On the other hand, nonviral vectors might be more suitable for the controlled release of genes into cells or tissues. A significant obstacle to the effective implementation of MSC therapy is the inability to deliver these cells under minimally invasive conditions (Dey et al. 2010).

Magnetic resonance imaging (MRI) is mainly used to detect stem cells, and when correlated with positron-emission tomography (PET) imaging, it can visualize metabolic events. Several cytometry techniques are being developed to visualize stem cell transplantation into damaged tissue (Nery et al. 2013). In chronic neuronal

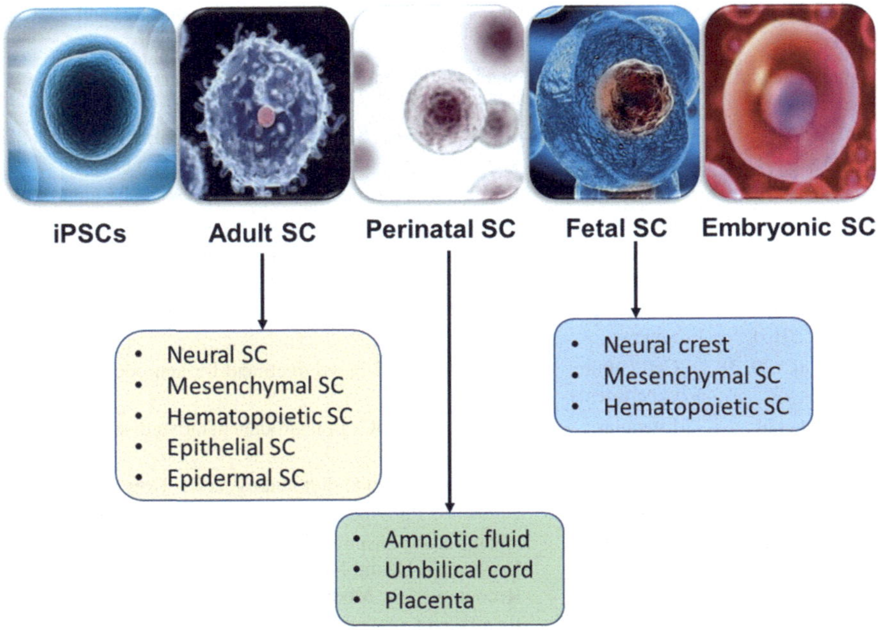

Fig. 14.1 Classification of stem cells based on the source of origin

degeneration, nanocomposite gels might be a powerful tool for achieving a controlled and long-lasting drug delivery in close contact with affected areas (Albani et al. 2013). Severe side effects were not observed or reported in the use of MSCs. Few studies have shown contradictory results on the immunomodulatory properties of MSCs, which can be explained by the heterogeneous stem cell population (Lo Furno et al. 2018).

Significant challenges need to be overcome before MSC treatment can be adopted in everyday clinical practice. These include (1) poor MSC in vivo retention; (2) poor MSC implantation, viability, and in vivo function; (3) unclear mechanisms of action; and (4) lack of standardized, controlled studies. MSCs do not seem to be a viable alternative to NSCs as a strategy for neural tissue replacement. There is no precise definition of the mechanisms behind stem cell therapies for brain diseases yet (Peng et al. 2013).

Nevertheless, MSCs provide a promising platform able to produce neural regulation agents. The importance of enhancing MSCs' targeting capacity for future treatment of brain diseases has been highlighted (Joyce et al. 2010). It is also imperative that the surgical and transport strategies need to be improved, and safety issues should be thoroughly studied. It has been noted that rodent MSC cultures are often contaminated with hematopoietic elements and do not sufficiently reflect human stem cell biology (Ramakrishnan et al. 2013). Further investigation is needed to study the details of cell therapy. Additional clinical studies, combined with genetic and pharmacological approaches for increasing cell survival and uptake of endogenous repair mechanisms, would reveal the stem cell source's potential. It is expected that the infusion and transplantation of preserved, genetically allogenic, or homologous MSCs for the treatment of multiple human neurological diseases would be improved in the upcoming years. Therefore, future research should focus on neurodegenerative diseases.

Cell replacement therapy and gene transfer in the diseased, degenerated, or injured brain provide the basis for developing new therapeutic strategies for human neurological diseases (De Filippis and Binda 2012). The endogenous neural stem cells (eNSCs) can allow therapeutic interventions in neurodegenerative diseases, especially in cases where the damaged brain tissue is located close to the NSC niches, such as the striatum in PD, close to the subventricular zone (SVZ). The NSCs are present in the SVZ of patients suffering from PD. The mobilization of the SVZ endogenous NSCs to reconstitute the dopamine striatum constitutes an attractive approach strategy for treating motor symptoms in patients with PD without the ethical and immune problems of NSCs and fatal brain transplantation (Alexiou et al. 2020). It has been observed that the proliferation of NSCs in SVZ was not statistically significantly different between patients and controls.

In conclusion, dopamine deficiency reported to be caused by various toxins in different animal models had a variable effect on the proliferation of NSCs in SVZ (De Filippis and Binda 2012; Yuan and Shaner 2013).

Immortalized NSCs have emerged as a highly effective tool for genetic manipulation and ex vivo gene transfer to the central nervous system. The cells were genetically modified in vitro, survived, integrated into host tissues, and differentiated into neurons and glial cells after transplantation into the intact or damaged brain in vivo. Clonally generated immortalized cell lines of human NSCs, as produced by introduction of oncogenes, have advantages for cell therapy and gene therapy. The advantages include (1) homogeneity of NSCs since they are produced from a single clone, (2) potential to be extended to a large number of cells in vitro, and (3) stabilization of the expression of therapeutic genes to be easily achieved (Ramakrishnan et al. 2013).

The development of innovative methods for creating NSCs in vitro has been one of the main goals of researchers after discovering active neurogenesis in the central nervous system of mammals (Compagnucci et al. 2014). The new technology for production of the induced NSCs (iNSCs) from easily accessible cell sources, such as skin fibroblasts, opens up a new field in personalized medicine for neurodegenerative diseases. iNSCs appear to provide some advantages over induced neurons (iNS) and possibly iPSCs in replacing and modeling neurodegenerative disease cells (Hermann and Storch 2013). Although the available studies convincingly showed the similarities between iNSCs and primary NSCs, many open debates and questions remain regarding the following aspects: (1) Are retroviral vectors completely silent in iNSCs? (2) Does the iNSC conversion process lead to genetic changes? (3) Are the iNSC conversion states stable, and does it lead to a stable change of the epigenome to NSC?

There is no doubt that the replacement of endangered or damaged cells can alleviate the symptoms of the disease. However, it does not necessarily stop the progression of the disease itself (Kittappa et al. 2012; Ziemka-Nalecz 2012). Evidence from several research groups has underlined that extensive eNSCs are present throughout the adult brain and across the whole spinal cord (Wislet-Gendebien et al. 2012b). Hence, as reported in the models of degenerative disease and acute attacks, the treatments that increase the numbers of cells in living tissues also provide neuroprotection and rescue of neurons from death. The determination of the presence of eNSCs depends mainly on two experimental approaches. The first is the identification of biomarkers for the immediate detection of eNSCs. The second approach is to extract the cells from the primary tissue and grow them under conditions that support survival and self-renewal (Lescaudron et al. 2012).

Regarding cell therapies, the most obvious question is the source utilized for cell replacement (Kittappa et al. 2012). Should the emphasis be kept on embryonic or adult stem cells? What is the optimal source to avoid immunocompatibility, inefficient cell proliferation, and cancer induction and minimize ethical issues? Understanding the endogenous configuration of the neurogenesis system is expected to bolster the development of effective therapies based on nerve stem cells (Relaño-Ginés et al. 2014) (Table 14.3).

Due to the lack of appropriate disease models and a sufficient number of brain biopsy specimens, the true etiology and pathology of many neurological diseases remain unknown (Kunkanjanawan et al. 2011). It has been proposed that the induced pluripotent stem cells (iPSCs) could become a new tool for the production of patient-specific multipotent stem cells to be used as a model of genetic neurological diseases (Mohamet 2014). Human PSCs, including iPSCs and embryonic stem cells (ESCs), can renew themselves, grow indefinitely, and retain the ability to form all kinds of cells in the body. The iPSCs are induced pluripotent stem cells derived from fibroblasts through forced expression of critical multivalent transcription factors of human embryonic stem cells (hESCs) (Oct4, c-myc, SOX2, and KLF4) (Abdullah et al. 2012). The two types of multivalent cells exhibit quite similar properties, such as self-renewal, differentiation, and same cell surface antigens and gene expression profile. However, studies comparing hESCs with iPSCs have indicated that they are genetically and epigenetically similar but not identical (Schwartz et al. 2012). Modeling neurodegenerative diseases requires a differentiation process of iPSCs into specific neuronal cell types (Compagnucci et al. 2014). One of the limitations of the iPSCs' human technologies is that it is time consuming to generate neurons from multivalent stem cells. Both the formation of iPSCs and the subsequent differentiation into neuronal cells require 1–2 months. Although iPSC technology is still in its infancy and faces several problems and obstacles, it has a great potential for identifying therapeutic targets for treating neurodegenerative diseases (Gao et al. 2013). Compared to ESCs, the therapeutic use of iPSCs is considered to be ethically more profitable because it does not involve the destruction of human embryos. The iPSCs seem to be potential sources for cell therapy because they can be differentiated into NSCs and MSCs to replace damaged cells. The unlimited possibilities of iPSCs to differentiate might allow us to model AD and ALS and provide possible treatment without HD. The iPSC technology has emerged as a revolutionary tool in medical research and clinical therapy (De Filippis and Binda 2012; Schwartz et al. 2015).

Table 14.3 Pluripotent stem cells in neurodegenerative disease—an overview of the available literature

Abdullah et al. (2012)	Literature review	The path from skin to brain: generation of functional neurons from fibroblasts
Chang et al. (2017)	Original research report	Combining Induced Pluripotent Stem Cells and Genome Editing Technologies for Clinical Applications
Chang et al. (2018)	Literature review	Induced Pluripotent Stem Cells: A Powerful Neurodegenerative Disease Modeling Tool for Mechanism Study and Drug Discovery
Compagnucci et al. (2014)	Literature review	In vitro neurogenesis: Development and functional implications of iPSC technology
Gao et al. (2013)	Literature review	Potential therapeutic applications of differentiated induced pluripotent stem cells (iPSCs) in the treatment of neurodegenerative diseases
Imaizumi and Okano (2014)	Literature review, electron microscopic observation, analysis of postmortem brain tissue, expression analysis	Modeling human neurological disorders with induced pluripotent stem cells
Imamura and Inoue (2012)	Literature review	Research on neurodegenerative diseases using induced pluripotent stem cells
Jongkamonwiwat and Noisa (2013)	Reports review	Biomedical and clinical promises of human pluripotent stem cells for neurological disorders
Jung et al. (2012)	Literature review	Human-induced pluripotent stem cells and neurodegenerative disease: Prospects for novel therapies
Kunkanjanawan et al. (2011)	Literature review	Modeling neurological disorders by human induced pluripotent stem cells
Liu and Deng (2016)	Original research report	Reverse Engineering Human Neurodegenerative Disease Using Pluripotent Stem Cell Technology
Mohamet (2014)	Literature review, cluster analysis	Familial Alzheimer's disease modeling using induced pluripotent stem cell technology
Pen and Jensen (2017)	Literature review	Current status of treating neurodegenerative disease with induced pluripotent stem cells
Siller et al. (2013)	Literature review	Modelling human disease with pluripotent stem cells
Yuan and Shaner (2013)	Method	Bioengineered stem cells in neural development and neurodegeneration research

Nowadays, scientists can isolate a patient biopsy and, through iPSC technology, grow cells, stimulate pluripotency, and differentiate the resulting iPSCs into the specific cell type affected by the disease (Siller et al. 2013). iPSC technology could also be applied to understand the molecular mechanisms involved in cancer and oncogenicity (Siller et al. 2013). iPSC technology has potential applications in (1) disease modeling, (2) drug screening, and (3) stem cell transplantation therapy (Compagnucci et al. 2014). The first and most crucial step in building cellular models is to create iPSC-derived patient cell lines (Jung et al. 2012). Smooth-muscle antibody (SMA) is one of the first human neurodegenerative diseases characterized by the iPSC in vitro model of the disease.

Furthermore, PD is one of the diseases in which cell therapy with iPSCs is effectively applied. It is also possible to create iPSC models of patients and study how each individual could develop the disease (Yuan and Shaner 2013; Payne et al. 2015). The use of iPSCs offers improved clinical cell models that are expected to significantly reduce the time and costs needed to develop new therapies, thereby increasing the number of new drugs available on the market for neurodegenerative diseases. Nowadays, there are multiple methods to derive the cell types that researchers are interested in; the method of selection depends mainly on the research question and its effectiveness. Current research shows that disease-specific iPSC technology can accurately reflect and image conditions before the onset of clinical disease, or in many cases, during the early stages of the disease (Imaizumi and Okano 2014). Dramatic progress in elucidating pathogenetic mechanisms is expected in the following years, with the assistance of iPSC technology, whole-genome analysis, and noninvasive imaging technology. Although hPSCs have been proven to be functional in vitro and have been able to treat phenotypes in diseased mice, there are still several issues which are needed to be resolved before this technology is applied to everyday clinical practice, such as the purity of the transplanted cells, the transplant sites, and the oncogenesis (Shin et al. 2011; *Transplantation of Neural Stem Cell-Derived Neurons for Parkinson's Disease* 2017). There is an imperative need to improve our knowledge of the mechanisms that control neurogenesis in vivo to effectively guide cell modeling and possible therapeutic applications of iPSC technology. While iPSC technology is a powerful technique that allows scientists to investigate the process and stratification of degeneration in neurological diseases and discover new drug therapies and strategies (Kunkanjanawan et al. 2011), researchers should examine the following (Siller et al. 2013): (1) what should be defined as a high-quality iPSC, (2) the standardization of methods necessary to confirm pluripotency, and (3) the determination of the minimum number of clones, subjects, and controls that are needed for genetic variation.

14.3 Clinical Trials for Stem Cell Therapies

In the following part of the chapter, the authors provide a twofold account of clinical trials for stem cell therapies in neurodegenerative diseases. First, an overview of stem cells investigated in clinical trials is provided. In addition, the ongoing and projected clinical trials in the context of particular neurodegenerative diseases, such as AD, PD, and HD, are discussed.

14.3.1 Human Embryonic Stem Cells

Even though scientists were able to extract ESCs from mice in the 1980s, it was not until 1998 when a research team from the University of Wisconsin-Madison isolated and kept the human fetal embryonic stem cells alive in cell cultures (Kim et al. 2013). This time interval was necessary to develop the required techniques for tracing embryonic stem cells. This can be partly attributed to the fact that adult stem cells are inherent, indistinguishable in shape, size, and function. They also reside deeply in tissues in countable populations, hence complicating their tracing and isolation. Despite the common belief that human fetal stem cells could contribute as treatment options in many catastrophic diseases, basic and clinical research is still in its infancy. The first clinical trials have just started to be published. The National Institutes of Health (USA) funded the first ESC research study in 2002 (*Embryonic Stem Cells* n.d. | *stemcells.nih.gov*, accessed in 2021). From then onwards, biotechnology companies rely on these foundations to develop human stem cell therapies. There are currently two active clinical trials utilizing human embryonic stem cell-based therapies conducted by a biotechnology company called ACT (*Sub-retinal Transplantation of hESC Derived RPE(MA09-hRPE)Cells in Patients With Stargardt's Macular Dystrophy* 2020). The company has started enrolling patients for Phase I as follows: (1) The first clinical trial was conducted to document the safety of human fetal ESCs isolated from the retina for the treatment of Stargardt disease (SMD), an inherited form of macular degeneration. (2) The second clinical trial was concerned with the safety of human fetal retinal-derived ESCs to treat patients with age-related macular degeneration. In January 2012, researchers published a preliminary report about the first two patients treated with human-derived embryonic stem cells (Schwartz et al. 2012).

Pfizer has launched a study in collaboration with the University College London to test the treatment of human embryonic germinal cell-derived stem cells for acute wet age-related macular degeneration. The results of this study are expected to be announced in due time (Coffey 2019). A third clinical trial with hESCs was stopped on November 14, 2011. The trial was conducted by the biotechnology company Geron. Four patients with a spinal cord injury were included in the clinical trial treated with hESCs (Scott and Magnus 2014). Oligodendrocyte precursor cells derived from hESCs were injected directly at the site of spinal cord injury. On November 14, Geron announced that it was shutting down its stem cell programs to focus on cancer programs. The vulnerability of patients, in combination with the complexity of the transplantation procedure, has contributed to its continuation (Scott and Magnus 2014).

14.3.2 Induced Pluripotent Stem Cells

In late 2007, scientists reported that they could reprogram adult human skin cell iPSCs to behave like ESCs (Baker 2007). From the first reports until now, researchers have rapidly improved the iPSC creation techniques, providing a valuable way to differentiate cells whose developmental fate has been determined. In July 2013, Japan's health minister approved the first clinical trial with iPSCs, in an attempt to cure age-related macular degeneration, a form of blindness (Sipp 2013).

14.3.3 Bone Marrow and Umbilical Cord Stem Cells

Bone marrow contains hematopoietic stem cells that have been used for decades to treat blood cancers and other blood disorders (Tigue et al. 2007). Umbilical cord blood is another source of hematopoietic stem cells used in therapeutics (Lo Furno et al. 2018). Such stem cells are either deposited to a global hematopoietic stem cell blood or stored in private blood banks. The therapeutic potential of the latter practice is controversial and limited, although private banks make stem cells constantly available to the owners (Gluckman et al. 2011; Iriberri 2011). More than 2100 clinical studies are currently investigating the therapeutic potential of these cells (Chivu-Economescu and Rubach 2016). About 42 trials registered in Clinicaltrials.gov are currently investigating bone marrow stem cells in neurodegenerative disorders, particularly AD, PD, and ALS (*Search of: Bone Marrow Stem Cells | Neurodegenerative Diseases* 2021), while 19 registered trials are studying umbilical cord stem cells in this context (*Search of: Umbilical cord stem cells | Neurodegenerative Diseases* 2021). The majority of the latter studies also focus on AD, PD, and ALS, although fewer are related to juvenile and hereditary neurogenerative disorders such as hereditary cerebellar ataxia.

14.3.4 Human Spinal Cord Stem Cells

A biotechnology company called Neuralstem is conducting a clinical trial using human stem cells to treat spinal ALS. The company received FDA approval to conduct a Phase I trial and began enrolling patients in January 2010. About 12 participants received transplants in the lumbar region, and in March 2012, a second group of participants received microinjections in the cervical region (Feldman et al. 2014; Riley et al. 2014). Results are expected to be announced in due time.

14.3.5 Human Mesenchymal Stem Cells

Osiris Therapeutics conducts three separate Phase II clinical trials with derivatives of adult mesenchymal cells. These clinical trials are concerned with (1) protection of islet pancreatic β-cells in children and adults diagnosed with type 1 diabetes, (2) repair of heart tissue after a heart attack, (3) repair of lung tissue in patients with chronic obstructive pulmonary disease (COPD) and particularly with regard to neurodegenerative diseases, and (4) use of MSCs for the generation of brain-derived neurotrophic factor (BDNF) in patients with HD (Nolta 2016). The main challenges regarding the latter consist of the safety of MSC transplantation, in terms of procedures and in terms of immune compromising and histocompatibility.

14.4 Clinical Trials in the Context of Neurodegenerative Diseases

14.4.1 Clinical Trials for Alzheimer's Disease

Going back in time, efforts to utilize stem cells in the therapy of Alzheimer's disease have been done from the 2000s decade. The first attempts were aiming at the mobilization of endogenous bone marrow-derived stem cells. Tsai KJ et al. using the granulocyte colony-stimulating factor (G-CSF), a therapeutic regimen widely prescribed in medical oncology to reverse the neutropenia caused by chemotherapy, documented a positive effect on treated mice model. Their mental status did not deteriorate and the neuronal loss was reversed by hematopoietic stem cells (Tsai et al. 2007). These results have been verified by several studies in the course of time (Guo et al. 2020). Some researchers taking a few steps further enlighten the cellular processes of chemotaxis of BM-SC in the damaged area. Wu CC et al. reported that BM-MSCs are recruited in the brain with CXCR4/SDF-1-depended chemotaxis (Wu et al. 2017), which can give more specific treatments (Shin et al. 2011). From the RADAR, a systematic review in order to record the side effects of G-CSF, a causal link between the drug and carcinogenesis, was not established (Tigue et al. 2007). Given all the above, a Phase II clinical trial of filgrastim has been completed, but their results are pending (*To Evaluate the Efficacy and Safety/Tolerability Profiles of G-CSF in Subjects With Mild to Moderate Alzheimer's Disease—Full Text View—ClinicalTrials.gov* 2018). Researchers from Nature Cell Co. Ltd explore in a Phase IIB clinical trial the role of mesenchymal stem cells derived from the patient's adipose tissue in the treatment of Alzheimer's (*Study to Evaluate the Safety and Efficacy of AstroStem in Treatment of Alzheimer's Disease—Full Text View—ClinicalTrials.gov* 2020). Researchers from John Wayne Cancer Institute still recruit patients with mild Alzheimer's dementia in Phase II clinical trial with allogenic mesenchymal stem cells (*Allogeneic Human Mesenchymal Stem Cells for Alzheimer's Disease* 2016). Finally, researchers from the University of Miami have designed a Phase I clinical study to certify the safety of manifold infusions of allogenic mesenchymal stem cells (*Alzheimer's Disease Stem Cells Multiple Infusions* 2019).

14.4.2 Clinical Trials for Parkinson's Disease

The body of clinical trials has just started to grow. There is only one clinical trial which has finished patient recruitment. Researchers give increasing doses of allogenic bone marrow-derived mesenchymal stem cells in order to certify the safety of the method (*Allogeneic Bone Marrow-Derived Mesenchymal Stem Cell Therapy for Idiopathic Parkinson's Disease* 2015). At the same center, the University of Texas Health Science Center at Houston, the next stage of the aforementioned study is conducted, a Phase IIb clinical trial. It still recruits participants (*Phase IIa Randomized Placebo Controlled Trial: Mesenchymal Stem Cells as a Disease-modifying Therapy for iPD* 2020). Two trials grasp the attention of the scientific community. Independent researchers from two centers, the NeuroGeneration which is a biotechnology corporation and the Shanghai East Hospital, have planned to stereotactically implant progenitor cells. The first team will implant neural progenitor cells and the other will transplant human amniotic epithelial stem cells (hAESCs). Both trials are designed as Phase I (*Transplantation of Neural Stem Cell-Derived Neurons for Parkinson's Disease* 2017; *Stereotactic Transplantation of hAESCs for Parkinson's Disease* 2018).

14.4.3 Clinical Trials for Amyotrophic Lateral Sclerosis (ALS)

There is a multitude of different treatment approaches that are being investigated with clinical trials. Since 2007, researchers from Peking University have conducted a Phase II clinical trial about the effect of granulocyte colony-stimulating factor (G-CSF) on the natural history of ALS. Preliminary results have been published, but no further update has come to the attention of the authors to date *[Basic and clinical researches on amyotrophic lateral sclerosis/motor neuron disease]—PubMed* 2009). Nabavi SM et al. documented that both intravenous administration and injection of mesenchymal cell derived from the patient's bone marrow in the spinal canal are safe (Nabavi et al. 2019). Researchers from the University Hospital "Dr. José Eleuterio González" in Mexico conducted a Phase II/III with bone marrow-derived hematopoietic stem cells injected through lumbar puncture directly in cerebrospinal fluid. Data are not published yet (*Effect of Intrathecal Administration of Hematopoietic Stem Cells in Patients With Amyotrophic Lateral Sclerosis (ALS)* 2013). Another clinical trial enrolled patients in Phase II study with autologous adipose tissue-derived mesenchymal cells injected intrathecally (*Intrathecal Autologous Adipose-derived Mesenchymal Stromal Cells for Amyotrophic Lateral Sclerosis (ALS)* 2017). A more straightforward approach is designed by Q Therapeutics, Inc. In a Phase I/II study, the researchers would check the safety and efficacy of progenitor glial cells implanted surgically in the patients' spinal cord. The study is not enrolling patients at the moment (*Study to Investigate the Safety of the Transplantation (by Injection) of Human Glial Restricted Progenitor Cells (hGRPs; Q-Cells®) Into Subjects With Amyotrophic Lateral Sclerosis (ALS)* 2015). A hallmark trial for the study of stem cell-based therapies in neurodegenerative diseases is the Neurologic Stem Cell Treatment Study (NEST). Researchers estimate to enroll 300 patients to applicate bone marrow stem cell-derived treatments locally through the nasal canal and systemically (*Neurologic Stem Cell Treatment Study* 2016).

14.4.4 Clinical Trials for Huntington's Disease

There are only a few clinical trials on HD. The most mature trial is a Phase II/III clinical trial designed to inject the subjects with Cellavita-HD, a stem cell-based biologic therapy, intravenously (*Clinical Extension Study for Safety and Efficacy Evaluation of Cellavita-HD Administration in Huntington's Patients* 2020). Encouraging perspectives are emerging from the basic research. It is easily perceived that transplantation of allogenic tissues raises serious concerns, as these biologic products become immunogenic when they are injected into the circulation or implanted locally in tissues. Immunosuppression should be implemented to achieve successful implantation, which comes with severe complications. Difficult-to-treat infections such as bacterial meningitis and encephalitis are common adverse events of lumbar puncture, especially in immunocompromised patients, and the need for regular infusions exacerbates those risks. To address those concerns, Wu Z. et al. succeeded to transform stromal cells of the brain parenchyma into functional neurons. Researchers infected glial cells with a genetically manipulated adenovirus vector, inducing the coexpression of NeuroD1 and Dlx2, two critical transcription factors (Wu et al. 2020). Clinical trials in humans are expected.

14.4.5 Clinical Trials for Autism Spectrum Neurodevelopmental Disorders

According to the Clinicaltrials.gov records, 19 clinical trials are investigating the use of stem cells in the context of autism spectrum disorders, of which 6 studies have been completed. Of the six completed studies, one is a Phase I trial, and five are Phase II trials. Out of the ongoing ten studies, only two are Phase II trials. Most of the studies investigate autologous transplantation of bone marrow or umbilical stem cells (*Stem cells | Autism Spectrum Disorder* n.d., accessed in 2021).

One Phase I trial has been completed in 2012, and its results have been published. In this study, 15 participants with autism spectrum disorders underwent infusion of autologous umbilical cord blood. The principal outcome measure was the improvement of language ability and behavioral improvement over 55 weeks. The most encouraging results were related to motor capacity within 6 months, which was significantly improved compared to the placebo group (Mauron 2012).

Clinical trials for autism spectrum disorders face several challenges, given the high cost of autologous stem cell storage and its controversial efficacy. The use of stem cells from global umbilical cord banks or bone marrow donors appears quite intricate. Future directions for research include mesenchymal, neural, and fetal stem cells. Moreover, research focused on particular ASDs might yield more encouraging disease-specific results (Siniscalco et al. 2012, 2014; Yuan and Shaner 2013; Bradstreet et al. 2014).

14.5 Conclusions

There is a tremendous societal burden due to neurodegenerative disorders owing to their devastating sequelae and lack of effective therapies. Till date, only stem cell is the potential therapy which can offer "cure" for neurodegenerative disorders. The capability of stem cells to cross BBB and migrate to brain has made them a versatile treatment modality. The usage of stem cells in clinical trials has showed promising effects on people suffering from these dreaded diseases. Further, more research is needed to understand the mechanisms of action underlying the efficacy of stem cells in the context of neurodegeneration.

References

[Basic and clinical researches on amyotrophic lateral sclerosis/motor neuron disease]—PubMed (2009). https://pubmed.ncbi.nlm.nih.gov/19727208/. Accessed 1 Apr 2021

2020 Alzheimer's disease facts and figures (2020) Alzheimers Dement 16(3):391–460. https://doi.org/10.1002/alz.12068. Wiley

2020 Community Report on Autism (2020) *NCBDDD | CDC*. https://www.cdc.gov/ncbddd/autism/addm-community-report/index.html. Accessed 2 Apr 2021

Abdullah AI, Pollock A, Sun T (2012) The path from skin to brain: generation of functional neurons from fibroblasts. Molecul Neurobiol 45(3):586–595. https://doi.org/10.1007/s12035-012-8277-6. Humana Press Inc.

Accordino RE et al (2016) Psychopharmacological interventions in autism spectrum disorder. Expert Opin Pharmacother 17(7):937–952. https://doi.org/10.1517/14656566.2016.1154536. Taylor and Francis Ltd

Ahani-Nahayati M et al (2021) Stem cell in neurodegenerative disorders; an emerging strategy. Int J Dev Neurosci. https://doi.org/10.1002/jdn.10101. Wiley

Albani D et al (2013) Hydrogel-based nanocomposites and mesenchymal stem cells: a promising synergistic strategy for neurodegenerative disorders therapy. Scientific World J. https://doi.org/10.1155/2013/270260. ScientificWorld Ltd

Alexiou A et al (2017) A Bayesian model for the prediction and early diagnosis of Alzheimer's disease. Front Aging Neurosci 9:77. https://doi.org/10.3389/fnagi.2017.00077. Frontiers Research Foundation

Alexiou A et al (2020) Social and ethical impact of advanced artificial and biological enhancements. Open Public Health J 13(1):62–68. https://doi.org/10.2174/1874944502013010062. Bentham Science Publishers Ltd

Allogeneic Bone Marrow-Derived Mesenchymal Stem Cell Therapy for Idiopathic Parkinson's Disease (2015). *ClinicalTrials.gov*. https://clinicaltrials.gov/ct2/show/NCT02611167?term=stem+cells&recrs=e&cond=Parkinson+Disease&draw=2&rank=1. Accessed 1 Apr 2021

Allogeneic Human Mesenchymal Stem Cells for Alzheimer's Disease (2016). *ClinicalTrials.gov*. https://clinicaltrials.gov/ct2/show/NCT02833792?term=stem+cells&cond=Alzheimer+Disease&draw=2&rank=3. Accessed 1 Apr 2021

Alzheimer's Disease Stem Cells Multiple Infusions (2019). *ClinicalTrials.gov*. https://clinicaltrials.gov/ct2/show/NCT04040348?term=stem+cells&cond=Alzheimer+Disease&draw=2&rank=4. Accessed 1 Apr 2021

Amyotrophic Lateral Sclerosis (ALS) Fact Sheet | National Institute of Neurological Disorders and Stroke (n.d.). https://www.ninds.nih.gov/Disorders/Patient-Caregiver-Education/Fact-Sheets/Amyotrophic-Lateral-Sclerosis-ALS-Fact-Sheet. Accessed 31 Mar 2021

Antshel KM, Zhang-James Y, Faraone SV (2013) The comorbidity of ADHD and autism spectrum disorder. Expert Rev Neurother 13(10):1117–1128. https://doi.org/10.1586/14737175.2013.840417

Autism Diagnosis Criteria: DSM-5 | Autism Speaks (n.d.). https://www.autismspeaks.org/autism-diagnosis-criteria-dsm-5. Accessed 2 Apr 2021

Bachoud-Lévi AC et al (2019) International guidelines for the treatment of Huntington's disease. Front Neurol. https://doi.org/10.3389/fneur.2019.00710. Frontiers Media S.A.

Baker M (2007) From skin cell to stem cell. Nat Rep Stem Cells. https://doi.org/10.1038/stemcells.2007.6. Springer Science and Business Media LLC

Balestrino R, Schapira AHV (2020) Parkinson disease. Eur J Neurol 27(1):27–42. https://doi.org/10.1111/ene.14108. Blackwell Publishing Ltd

Ballard C et al (2011) Alzheimer's disease. Lancet 377(9770):1019–1031. https://doi.org/10.1016/S0140-6736(10)61349-9. Elsevier BV

Bekris LM et al (2010) Genetics of Alzheimer disease. J Geriatr Psychiatry Neurol 23(4):213–227. https://doi.org/10.1177/0891988710383571. NIH Public Access

Bradstreet JJ et al (2014) Efficacy of fetal stem cell transplantation in autism spectrum disorders: an open-labeled pilot study. Cell Transplant 23(Suppl 1):S105–S112. https://doi.org/10.3727/096368914X684916. Sage

Chang C-Y et al (2017) Combining induced pluripotent stem cells and genome editing technologies for clinical applications. Cell Transplant. https://doi.org/10.3727/096368917x695119. Sage

Chang CY et al (2018) Induced pluripotent stem cells: a powerful neurodegenerative disease modeling tool for mechanism study and drug discovery. Cell Transplant 27(11):1588–1602. https://doi.org/10.1177/0963689718775406. Sage

Chen X, Wang S, Cao W (2018) Mesenchymal stem cell-mediated immunomodulation in cell therapy of neurodegenerative diseases. Cellular Immunology 326:8–14. https://doi.org/10.1016/j.cellimm.2017.06.006. Academic

Chivu-Economescu M, Rubach M (2016) Hematopoietic stem cells therapies. Curr Stem Cell Res Ther 12(2):124–133. https://doi.org/10.2174/1574888x10666151026114241. Bentham Science Publishers Ltd

Clinical Extension Study for Safety and Efficacy Evaluation of Cellavita-HD Administration in Huntington's Patients (2020). *ClinicalTrials.gov*. https://clinicaltrials.gov/ct2/show/NCT04219241?term=stem+cells&cond=Huntington+Disease&draw=2&rank=3. Accessed 1 Apr 2021

Coffey P (2019) The London project to cure blindness at 10 years: have we found a cure? Acta Ophthalmol 97(S263):j.1755-3768.2019.1004. https://doi.org/10.1111/j.1755-3768.2019.1004. Wiley

Compagnucci C et al (2014) In vitro neurogenesis: development and functional implications of iPSC technology. Cell Mol Life Sci 71(9):1623–1639. https://doi.org/10.1007/s00018-013-1511-1. Birkhauser Verlag AG

De Filippis L, Binda E (2012) Concise review: self-renewal in the central nervous system: neural stem cells from embryo to adult. Stem Cells Transl Med 1(4):298–308. https://doi.org/10.5966/sctm.2011-0045. Wiley

Dementia Pathology: Dementia, Alzheimer Disease, Vascular Dementia (2019). https://emedicine.medscape.com/article/2003174-overview. Accessed 1 Apr 2021

Dey ND et al (2010) Genetically engineered mesenchymal stem cells reduce behavioral deficits in the YAC 128 mouse model of Huntington's disease. Behav Brain Res 214(2):193–200. https://doi.org/10.1016/j.bbr.2010.05.023

Díaz ML (2019) Regenerative medicine: could Parkinson's be the first neurodegenerative disease to be cured? Future Sci OA 5(9):FSO418. https://doi.org/10.2144/fsoa-2019-0035. Future Medicine Ltd

Edwards GA et al (2019) Modifiable risk factors for Alzheimer's disease. Front Aging Neurosci. https://doi.org/10.3389/fnagi.2019.00146. Frontiers Media S.A.

Effect of Intrathecal Administration of Hematopoietic Stem Cells in Patients With Amyotrophic Lateral Sclerosis (ALS) (2013). *ClinicalTrials.gov*. https://clinicaltrials.gov/ct2/show/NCT01933321?term=stem+cells&cond=Amyotrophic+Lateral+Sclerosis&draw=2. Accessed 1 Apr 2021

Eissa N et al (2018) Current enlightenment about etiology and pharmacological treatment of autism spectrum disorder. Front Neurosci 12:304. https://doi.org/10.3389/fnins.2018.00304. Frontiers Media SA

Embryonic Stem Cells | stemcells.nih.gov (n.d.) https://stemcells.nih.gov/info/Regenerative_Medicine/2006Chapter1.htm. Accessed 2 Apr 2021

Erkkinen MG, Kim MO, Geschwind MD (2018) Clinical neurology and epidemiology of the major neurodegenerative diseases. Cold Spring Harb Perspect Biol 10(4):a033118. https://doi.org/10.1101/cshperspect.a033118. Cold Spring Harbor Laboratory Press

F84.0—Autistic disorder | ICD-10-CM (n.d.). https://www.unboundmedicine.com/icd/view/ICD-10-CM/875604/all/F84_0___Autistic_disorder. Accessed 2 Apr 2021

Fazio P et al (2018) Novel imaging biomarkers for Huntington's disease and other hereditary Choreas. Curr Neurol Neurosci Rep 18(12):85. https://doi.org/10.1007/s11910-018-0890-y. Current Medicine Group LLC

Feldman EL et al (2014) Intraspinal neural stem cell transplantation in amyotrophic lateral sclerosis: phase 1 trial outcomes. Ann Neurol 75(3):363–373. https://doi.org/10.1002/ana.24113. Wiley

Frank S (2014) Treatment of Huntington's disease. Neurotherapeutics 11(1):153–160. https://doi.org/10.1007/s13311-013-0244-z. Springer

Fričová D, Korchak JA, Zubair AC (2020) Challenges and translational considerations of mesenchymal stem/stromal cell therapy for Parkinson's disease. NPJ Regen Med 5(1):1–10. https://doi.org/10.1038/s41536-020-00106-y. Nature Research

Gao A et al (2013) Potential therapeutic applications of differentiated induced pluripotent stem cells (iPSCs) in the treatment of neurodegenerative diseases. Neuroscience 228:47–59. https://doi.org/10.1016/j.neuroscience.2012.09.076

Giacobini E, Gold G (2013) Alzheimer disease therapy—moving from amyloid-β to tau. Nat Rev Neurol 9(12):677–686. https://doi.org/10.1038/nrneurol.2013.223

Girlovanu M et al (2015) Stem cells—biological update and cell therapy progress. Med Pharm Rep 88(3):265–271. https://doi.org/10.15386/cjmed-483. Clujul Medical

Gitler AD, Dhillon P, Shorter J (2017) Neurodegenerative disease: models, mechanisms, and a new hope. Dis Model Mech 10(5):499–502. https://doi.org/10.1242/dmm.030205. Company of Biologists Ltd

Gluckman E et al (2011) Family-directed umbilical cord blood banking. Haematologica 96(11):1700–1707. https://doi.org/10.3324/haematol.2011.047050

Goya RG et al (2018) Rejuvenation by cell reprogramming: a new horizon in gerontology. Stem Cell Res Ther 9(1):349. https://doi.org/10.1186/s13287-018-1075-y. BioMed Central

Guo X et al (2020) Reparative effects of stem cell factor and granulocyte colony-stimulating factor in Aged APP/PS1 mice. Aging Dis 11(6):1423–1443. https://doi.org/10.14336/AD.2020.0201. International Society on Aging and Disease

Harris JP et al (2020) Emerging regenerative medicine and tissue engineering strategies for Parkinson's disease. npj Parkinsons Dis 6:1–14. https://doi.org/10.1038/s41531-019-0105-5. Nature Research

Hermann A, Storch A (2013) Induced neural stem cells (iNSCs) in neurodegenerative diseases. J Neural Transm (Vienna). https://doi.org/10.1007/s00702-013-1042-9

Huang B, Tabata Y, Gao JQ (2012) Mesenchymal stem cells as therapeutic agents and potential targeted gene delivery vehicle for brain diseases. J Control Release 162(2):464–473. https://doi.org/10.1016/j.jconrel.2012.07.034

Imaizumi Y, Okano H (2014) Modeling human neurological disorders with induced pluripotent stem cells. J Neurochem 129(3):388–399. https://doi.org/10.1111/jnc.12625. Blackwell Publishing Ltd

Imamura K, Inoue H (2012) Research on neurodegenerative diseases using induced pluripotent stem cells. Psychogeriatrics 12(2):115–119. https://doi.org/10.1111/j.1479-8301.2011.00394.x

Intrathecal Autologous Adipose-derived Mesenchymal Stromal Cells for Amyotrophic Lateral Sclerosis (ALS) (2017). *ClinicalTrials.gov*. https://clinicaltrials.gov/ct2/show/NCT03268603?term=stem+cells&cond=Amyotrophic+Lateral+Sclerosis&draw=2. Accessed 1 Apr 2021

Iriberri A (2011) Spanish haematologists warn against storing umbilical cord stem cells in private banks. BMJ 343:d7499–d7499. https://doi.org/10.1136/bmj.d7499

Jongkamonwiwat N, Noisa P (2013) Biomedical and clinical promises of human pluripotent stem cells for neurological disorders. Biomed Res Int. https://doi.org/10.1155/2013/656531

Joyce N et al (2010) Mesenchymal stem cells for the treatment of neurodegenerative disease. Regen Med 5(6):933–946. https://doi.org/10.2217/rme.10.72

Jung YW et al (2012) Human induced pluripotent stem cells and neurodegenerative disease: prospects for novel therapies. Curr Opin Neurol 25(2):125–130. https://doi.org/10.1097/WCO.0b013e3283518226

Kan I, Melamed E, Offen D (2007) Autotransplantation of bone marrow-derived stem cells as a therapy for neurodegenerative diseases. Handb Exp Pharmacol 180:219–242. https://doi.org/10.1007/978-3-540-68976-8_10. Springer Science and Business Media, LLC

Kim SU, Lee HJ, Kim YB (2013) Neural stem cell-based treatment for neurodegenerative diseases. Neuropathology 33(5):491–504. https://doi.org/10.1111/neup.12020

Kittappa R, Bornstein SR, Androutsellis-Theotokis A (2012) The role of eNSCs in neurodegenerative disease. Mol Neurobiol 46(3):555–562. https://doi.org/10.1007/s12035-012-8303-8. Humana Press Inc.

Kunkanjanawan T, Noisa P, Parnpai R (2011) Modeling neurological disorders by human induced pluripotent stem cells. J Biomed Biotechnol. https://doi.org/10.1155/2011/350131

Kurtzberg J (2017) A history of cord blood banking and transplantation. Stem Cells Transl Med 6(5):1309–1311. https://doi.org/10.1002/sctm.17-0075. Wiley-Blackwell

Lacivita E et al (2017) Targets for drug therapy for autism spectrum disorder: challenges and future directions. J Med Chem 60(22):9114–9141. https://doi.org/10.1021/acs.jmedchem.7b00965. American Chemical Society

Lenkiewicz AM (2019) Epidermal stem cells. In: Advances in experimental medicine and biology. Springer, pp 239–259. https://doi.org/10.1007/978-3-030-31206-0_12

Lescaudron L et al (2012) Assessing the potential clinical utility of transplantations of neural and mesenchymal stem cells for treating neurodegenerative diseases. Methods Mol Biol 879:147–164. https://doi.org/10.1007/978-1-61779-815-3_10

Leskovec TJ, Rowles BM, Findling RL (2008) Pharmacological treatment options for autism spectrum disorders in children and adolescents. Harv Rev Psychiatry 16(2):97–112. https://doi.org/10.1080/10673220802075852

Leyns CEG, Holtzman DM (2017) Glial contributions to neurodegeneration in tauopathies. Mol Neurodegen. https://doi.org/10.1186/s13024-017-0192-x. BioMed Central Ltd

Lim WF et al (2013) Hematopoietic cell differentiation from embryonic and induced pluripotent stem cells. Stem Cell Res Ther 4(3):71. https://doi.org/10.1186/scrt222

Liu Y, Deng W (2016) Reverse engineering human neurodegenerative disease using pluripotent stem cell technology. Brain Res 1638:30–41. https://doi.org/10.1016/j.brainres.2015.09.023. Elsevier BV

Liu G et al (2020) Advances in pluripotent stem cells: history, mechanisms, technologies, and applications. Stem Cell Rev Rep 16(1):3–32. https://doi.org/10.1007/s12015-019-09935-x. Springer

Lo Furno D, Mannino G, Giuffrida R (2018) Functional role of mesenchymal stem cells in the treatment of chronic neurodegenerative diseases. J Cell Physiol 233(5):3982–3999. https://doi.org/10.1002/jcp.26192. Wiley-Liss Inc.

Majoor-Krakauer D, Willems PJ, Hofman A (2003) Genetic epidemiology of amyotrophic lateral sclerosis. Clin Genet 63(2):83–101. https://doi.org/10.1046/j.0009-9163.2002.00001.x

Marino BLB et al (2019) Parkinson's disease: a review from pathophysiology to treatment. Mini Rev Med Chem 20(9):754–767. https://doi.org/10.2174/1389557519666191104110908. Bentham Science Publishers Ltd

Martin S, Al Khleifat A, Al-Chalabi A (2017) What causes amyotrophic lateral sclerosis? F1000Research. https://doi.org/10.12688/f1000research.10476.1. Faculty of 1000 Ltd

Masrori P, Van Damme P (2020) Amyotrophic lateral sclerosis: a clinical review. Eur J Neurol 27(10):1918–1929. https://doi.org/10.1111/ene.14393. Blackwell Publishing Ltd

Mauron A (2012) Autism, stem cells, and magical powder. Rev Med Suisse 8(354):1795. https://pubmed.ncbi.nlm.nih.gov/23097920/. Accessed 2 Apr 2021

Mayeux R, Stern Y (2012) Epidemiology of Alzheimer disease. Cold Spring Harb Perspect Med 2(8). https://doi.org/10.1101/cshperspect.a006239. Cold Spring Harbor Laboratory Press

Mendiola-Precoma J et al (2016) Therapies for prevention and treatment of Alzheimer's disease. BioMed Res Int. https://doi.org/10.1155/2016/2589276. Hindawi Publishing Corporation

Mestre T et al (2009) Therapeutic interventions for symptomatic treatment in Huntington's disease. Cochrane Database Syst Rev. https://doi.org/10.1002/14651858.CD006456.pub2. Wiley

Mohamet L (2014) Familial Alzheimer's disease modelling using induced pluripotent stem cell technology. World J Stem Cells 6(2):239. https://doi.org/10.4252/wjsc.v6.i2.239. Baishideng Publishing Group Inc.

Nabavi SM et al (2019) Safety, feasibility of intravenous and intrathecal injection of autologous bone marrow derived mesenchymal stromal cells in patients with amyotrophic lateral sclerosis: an open label phase I clinical trial. Cell J 20(4):592–598. https://doi.org/10.22074/cellj.2019.5370. Royan Institute (ACECR)

Nazarko L (2019) Dementia: prevalence and pathophysiology. Br J Healthc Assist 13(6):266–270. https://doi.org/10.12968/bjha.2019.13.6.266. Mark Allen Group

Negi RS, Manchanda KL, Sanga S (2014) Imaging of Huntington's disease. Med J Armed Forces India 70(4):386–388. https://doi.org/10.1016/j.mjafi.2012.08.002

Nery AA et al (2013) Human mesenchymal stem cells: from immunophenotyping by flow cytometry to clinical applications. Cytometry A 83(1):48–61. https://doi.org/10.1002/cyto.a.22205

Neurodegenerative Diseases | MedlinePlus (n.d.). https://medlineplus.gov/degenerativenervediseases.html. Accessed 31 Mar 2021

Neurologic Stem Cell Treatment Study (2016). *ClinicalTrials.gov*. https://clinicaltrials.gov/ct2/show/NCT02795052?term=stem+cells&cond=Amyotrophic+Lateral+Sclerosis&draw=2. Accessed 1 Apr 2021

Ng TK (2014) Progress of mesenchymal stem cell therapy for neural and retinal diseases. World J Stem Cells 6(2):111. https://doi.org/10.4252/wjsc.v6.i2.111. Baishideng Publishing Group Inc.

Nguyen HHP, Weydt P (2018) Huntington disease. Med Genet 246–251. https://doi.org/10.1007/s11825-018-0190-6. Springer

Nolta JA (2016) "Next-generation" mesenchymal stem or stromal cells for the in vivo delivery of bioactive factors: progressing toward the clinic. Transfusion 15S–17S. https://doi.org/10.1111/trf.13564. Blackwell Publishing Inc.

Nopoulos PC (2016) Huntington disease: a single-gene degenerative disorder of the striatum. Dialog Clin Neurosci 18(1):91–98. https://doi.org/10.31887/dcns.2016.18.1/pnopoulos. Les Laboratoires Seriver

Olson SD et al (2012) Genetically engineered mesenchymal stem cells as a proposed therapeutic for Huntington's disease. Mol Neurobiol 45(1):87–98. https://doi.org/10.1007/s12035-011-8219-8. Humana Press Inc.

Payne NL et al (2015) Application of human induced pluripotent stem cells for modeling and treating neurodegenerative diseases. New Biotechnol 32(1):212–228. https://doi.org/10.1016/j.nbt.2014.05.001. Elsevier

Pen AE, Jensen UB (2017) Current status of treating neurodegenerative disease with induced pluripotent stem cells. Acta Neurol Scand 135(1):57–72. https://doi.org/10.1111/ane.12545. Blackwell Publishing Ltd

Peng Y et al (2013) Mesenchymal stem cells: a revolution in therapeutic strategies of age-related diseases. Ageing Res Rev 12(1):103–115. https://doi.org/10.1016/j.arr.2012.04.005

Petersen RC et al (2014) Mild cognitive impairment: a concept in evolution. J Intern Med 275(3): 214–228. https://doi.org/10.1111/joim.12190

Phase IIa Randomized Placebo Controlled Trial: Mesenchymal Stem Cells as a Disease-modifying Therapy for iPD (2020). *ClinicalTrials.gov*. https://clinicaltrials.gov/ct2/show/NCT04506073?term=stem+cells&cond=Parkinson+Disease&draw=2&rank=9. Accessed 1 Apr 2021

Pupillo E et al (2014) Long-term survival in amyotrophic lateral sclerosis: a population-based study. Ann Neurol 75(2):287–297. https://doi.org/10.1002/ana.24096. Wiley

Ramakrishnan A, Torok-Storb B, Pillai MM (2013) Primary marrow-derived stromal cells: isolation and manipulation. Methods Mol Biol 1035:75–101. https://doi.org/10.1007/978-1-62703-508-8_8. Humana Press Inc.

Relaño-Ginés A, Lehmann S, Crozet C (2014) Prion diseases and adult neurogenesis: how do prions counteract the brain's endogenous repair machinery? Prion 8(3):240–246. https://doi.org/10.4161/pri.29021. Landes Bioscience

Riley J et al (2014) Intraspinal stem cell transplantation in amyotrophic lateral sclerosis: a phase I trial, cervical microinjection, and final surgical safety outcomes. Neurosurgery 74(1):77–87. https://doi.org/10.1227/NEU.0000000000000156

Roberts R, Knopman DS (2013) 'Classification and epidemiology of MCI. Clin Geriatr Med 29(4): 753–772. https://doi.org/10.1016/j.cger.2013.07.003. NIH Public Access

Roos RAC (2010) Huntington's disease: a clinical review. Orphanet J Rare Dis. https://doi.org/10.1186/1750-1172-5-40

Schwartz SD et al (2012) Embryonic stem cell trials for macular degeneration: a preliminary report. Lancet 379(9817):713–720. https://doi.org/10.1016/S0140-6736(12)60028-2. Elsevier BV

Schwartz SD et al (2015) Human embryonic stem cell-derived retinal pigment epithelium in patients with age-related macular degeneration and Stargardt's macular dystrophy: follow-up of two open-label phase 1/2 studies. Lancet 385(9967):509–516. https://doi.org/10.1016/S0140-6736(14)61376-3. Lancet Publishing Group

Scott CT, Magnus D (2014) Wrongful termination: lessons from the Geron clinical trial. Stem CellS Transl Med 3(12):1398–1401. https://doi.org/10.5966/sctm.2014-0147. Wiley

Search of: Bone Marrow Stem Cells | Neurodegenerative Diseases (2021). *ClinicalTrials.gov*. https://clinicaltrials.gov/ct2/results?recrs=&cond=Neurodegenerative+Diseases&term=bone+marrow+stem+cells&cntry=&state=&city=&dist=. Accessed 2 Apr 2021

Search of: umbilical cord stem cells | Neurodegenerative Diseases (2021). *ClinicalTrials.gov*. https://clinicaltrials.gov/ct2/results?cond=Neurodegenerative+Diseases&term=Umbilical+cord+stem+cells&cntry=&state=&city=&dist=&Search=Search. Accessed 2 Apr 2021

Shariati A et al (2020) Mesenchymal stromal cells (MSCs) for neurodegenerative disease: a promising frontier. Eur J Cell Biol. https://doi.org/10.1016/j.ejcb.2020.151097. Elsevier GmbH

Sharma SR, Gonda X, Tarazi FI (2018) Autism spectrum disorder: classification, diagnosis and therapy. Pharmacol Ther 190:91–104. https://doi.org/10.1016/j.pharmthera.2018.05.007. Elsevier Inc.

Shin JW et al (2011) Combined effects of hematopoietic progenitor cell mobilization from bone marrow by granulocyte colony stimulating factor and AMD3100 and chemotaxis into the brain using stromal cell-derived factor-1α in an Alzheimer's disease mouse model. Stem Cells 29(7): 1075–1089. https://doi.org/10.1002/stem.659

Siller R et al (2013) Modelling human disease with pluripotent stem cells. Curr Gene Ther 13(2): 99–110. https://doi.org/10.2174/1566523211313020004. Bentham Science Publishers Ltd

Silva MVF et al (2019) Alzheimer's disease: risk factors and potentially protective measures. J Biomed Sci 26(1). https://doi.org/10.1186/s12929-019-0524-y. BioMed Central Ltd

Siniscalco D et al (2012) Autism spectrum disorders: is mesenchymal stem cell personalized therapy the future? J Biomed Biotechnol. https://doi.org/10.1155/2012/480289

Siniscalco D et al (2014) Mesenchymal stem cells in treating autism: novel insights. World J Stem Cells 6(2):173–178. https://doi.org/10.4252/wjsc.v6.i2.173

Sipp D (2013) *Pilot clinical study into iPS cell therapy for eye disease starts in Japan | RIKEN.* https://www.riken.jp/en/news_pubs/research_news/pr/2013/20130730_1/. Accessed 2 Apr 2021

Sivakumar M et al (2015) Stem cells: an insight into the therapeutic aspects from medical and dental perspectives. J Pharm Bioallied Sci 7:S361–S371. https://doi.org/10.4103/0975-7406.163453. Medknow Publications

Staff NP, Jones DT, Singer W (2019) Mesenchymal stromal cell therapies for neurodegenerative diseases. Mayo Clin Proc 94(5):892–905. https://doi.org/10.1016/j.mayocp.2019.01.001. Elsevier Ltd

Stem cells | Autism Spectrum Disorder (n.d.). *ClinicalTrials.gov*. https://clinicaltrials.gov/ct2/results?term=stem+cells&cond=Autism+Spectrum+Disorder&Search=Apply&age_v=&gndr=&type=&rslt=. Accessed 2 Apr 2021

Stereotactic Transplantation of hAESCs for Parkinson's Disease (2018). *ClinicalTrials.gov*. https://clinicaltrials.gov/ct2/show/NCT04414813?term=stem+cells&cond=Parkinson+Disease&draw=2&rank=14. Accessed 1 Apr 2021

Štětkářová I, Ehler E (2021) Diagnostics of amyotrophic lateral sclerosis: up to date. Diagnostics (Basel, Switzerland) 11(2):231. https://doi.org/10.3390/diagnostics11020231. MDPI AG

Struzyna LA et al (2017) Anatomically inspired three-dimensional micro-tissue engineered neural networks for nervous system reconstruction, modulation, and modeling. J Vis Exp 2017(123). https://doi.org/10.3791/55609

Study to Evaluate the Safety and Efficacy of AstroStem in Treatment of Alzheimer's Disease—Full Text View—ClinicalTrials.gov (2020). *Clinicaltrials.gov*. https://clinicaltrials.gov/ct2/show/NCT04482413?term=stem+cells&cond=Alzheimer+Disease&draw=2. Accessed 1 Apr 2021

Study to Investigate the Safety of the Transplantation (by Injection) of Human Glial Restricted Progenitor Cells (hGRPs; Q-Cells®) Into Subjects With Amyotrophic Lateral Sclerosis (ALS) (2015). *ClinicalTrials.gov*. https://clinicaltrials.gov/ct2/show/NCT02478450?term=stem+cells&cond=Amyotrophic+Lateral+Sclerosis&draw=2. Accessed 1 Apr 2021

Sub-retinal Transplantation of hESC Derived RPE(MA09-hRPE)Cells in Patients With Stargardt's Macular Dystrophy (2020). *ClinicalTrials.gov*. https://www.clinicaltrials.gov/ct2/show/NCT01345006?term=ACT+and+human+embryonic. Accessed 2 Apr 2021

Talbott EO, Malek AM, Lacomis D (2016) The epidemiology of amyotrophic lateral sclerosis. Handb Clin Neurol 138:225–238. https://doi.org/10.1016/B978-0-12-802973-2.00013-6. Elsevier BV

Tanna T, Sachan V (2014) Mesenchymal stem cells: potential in treatment of neurodegenerative diseases. Curr Stem Cell Res Ther 9(6):513–521. https://doi.org/10.2174/1574888x0966140923101110. Bentham Science Publishers Ltd

Tarakad A, Jankovic J (2020) Recent advances in understanding and treatment of Parkinson's disease. Faculty Rev 9:6. https://doi.org/10.12703/b/9-6. Faculty Opinions Ltd

Tigue CC et al (2007) Granulocyte-colony stimulating factor administration to healthy individuals and persons with chronic neutropenia or cancer: an overview of safety considerations from the Research on Adverse Drug Events and Reports project. Bone Marrow Transplant 40(3):185–192. https://doi.org/10.1038/sj.bmt.1705722

To Evaluate the Efficacy and Safety/Tolerability Profiles of G-CSF in Subjects With Mild to Moderate Alzheimer's Disease—Full Text View—ClinicalTrials.gov (2018). *Clinicaltrials.gov*. https://clinicaltrials.gov/ct2/show/NCT03656042?term=stem+cells&cond=Alzheimer+Disease&draw=2. Accessed 1 Apr 2021

Transplantation of Neural Stem Cell-Derived Neurons for Parkinson's Disease (2017). *ClinicalTrials.gov*. https://clinicaltrials.gov/ct2/show/NCT03309514?term=stem+cells&cond=Parkinson+Disease&draw=2&rank=12. Accessed 1 Apr 2021

Tsai KJ, Tsai YC, Shen CKJ (2007) G-CSF rescues the memory impairment of animal models of Alzheimer's disease. J Exp Med 204(6):1273–1280. https://doi.org/10.1084/jem.20062481

Tysnes OB, Storstein A (2017) Epidemiology of Parkinson's disease. J Neural Transm 124(8): 901–905. https://doi.org/10.1007/s00702-017-1686-y. Springer Wien

Van Den Berge SA, Van Strien ME, Hol EM (2013) Resident adult neural stem cells in Parkinson's disease—the brain's own repair system? Eur J Pharmacol 719(1–3):117–127. https://doi.org/10.1016/j.ejphar.2013.04.058. Elsevier BV

Winner B, Kohl Z, Gage FH (2011) Neurodegenerative disease and adult neurogenesis. Eur J Neurosci 33(6):1139–1151. https://doi.org/10.1111/j.1460-9568.2011.07613.x

Wislet-Gendebien S, Laudet E, Neirinckx V, Rogister B (2012a) Adult bone marrow: which stem cells for cellular therapy protocols in neurodegenerative disorders? J Biomed Biotechnol. https://doi.org/10.1155/2012/601560

Wislet-Gendebien S, Laudet E, Neirinckx V, Alix P et al (2012b) Mesenchymal stem cells and neural crest stem cells from adult bone marrow: characterization of their surprising similarities and differences. Cell Mol Life Sci 69(15):2593–2608. https://doi.org/10.1007/s00018-012-0937-1

Wu CC et al (2017) G-CSF-mobilized bone marrow mesenchymal stem cells replenish neural lineages in Alzheimer's disease mice via CXCR4/SDF-1 chemotaxis. Mol Neurobiol 54(8): 6198–6212. https://doi.org/10.1007/s12035-016-0122-x. Humana Press Inc.

Wu Z et al (2020) Gene therapy conversion of striatal astrocytes into GABAergic neurons in mouse models of Huntington's disease. Nat Commun 11(1):1–18. https://doi.org/10.1038/s41467-020-14855-3. Nature Research

Yuan SH, Shaner M (2013) Bioengineered stem cells in neural development and neurodegeneration research. Ageing Res Rev 12(3):739–748. https://doi.org/10.1016/j.arr.2013.04.002

Zakrzewski W et al (2019) Stem cells: past, present, and future. Stem Cell Res Ther. https://doi.org/10.1186/s13287-019-1165-5. BioMed Central Ltd

Ziemka-Nalecz M (2012) *Endogenous neurogenesis induced by ischemic brain injury or neurodegenerative diseases in adults—PubMed*. https://pubmed.ncbi.nlm.nih.gov/23377263/. Accessed 1 Apr 2021

The manufacturer's authorised representative in the EU is Springer Nature Customer Service Centre GmbH, Europaplatz 3, 69115 Heidelberg, Germany. If you have any concerns regarding our products, please contact ProductSafety@springernature.com

Printed and bound by CPI Group (UK) Ltd, Croydon, CR0 4YY
06/10/2026
02225252-0011